Business Ma

Also by Jim Dewhurst

MATHEMATICS FOR ACCOUNTANTS AND MANAGERS
BUSINESS COST−BENEFIT ANALYSIS
SMALL BUSINESS: Finance and Control (*with Paul Burns*)
SMALL BUSINESS IN EUROPE (*editor with Paul Burns*)

Business Mathematics

Jim Dewhurst

MACMILLAN
EDUCATION

© Jim Dewhurst 1988

All rights reserved. No reproduction, copy or transmission
of this publication may be made without written permission.

No paragraph of this publication may be reproduced, copied
or transmitted save with written permission or in accordance
with the provisions of the Copyright Act 1956 (as amended),
or under the terms of any licence permitting limited copying
issued by the Copyright Licensing Agency, 33-4 Alfred Place,
London WC1E 7DP.

Any person who does any unauthorised act in relation to
this publication may be liable to criminal prosecution and
civil claims for damages.

First published 1988

Published by
MACMILLAN EDUCATION LTD
Houndmills, Basingstoke, Hampshire RG21 2XS
and London
Companies and representatives
throughout the world

Printed in Hong Kong

ISBN 0-333-38409-1 (hardcover)
ISBN 0-333-38410-5 (paperback)

Contents

Preface xi
Acknowledgements xiii

1 Elementary Mathematics 1
   1.1 The concept of number 1
   1.2 Rounding and truncating 4
   1.3 Elementary algebra 5
   1.4 Operator precedence 6
   1.5 Percentages, discounts and mark-ups 7
   1.6 Fees and commissions 9
   1.7 Simple interest 10
   1.8 Ratios 11
   1.9 Exercises 12

2 Computer Number Systems 14
   2.1 Introduction 14
   2.2 Binary digits 14
   2.3 Binary calculations 16
   2.4 Computer codes 17
   2.5 Octal system 17
   2.6 Quintal system 20
   2.7 Hexadecimal system 20
   2.8 Binary coded decimals 22
   2.9 6, 8 and 16 bits 23
   2.10 Decimal-to-binary conversion 23
   2.11 Exercises 24

3 Exponents, Progressions, Present value 27
   3.1 Introduction 27
   3.2 Exponential form 28
   3.3 Computer arithmetic 29
   3.4 Logarithms 30
   3.5 Arithmetic progressions 33
   3.6 Geometric progressions 34
   3.7 Infinite geometric progressions 35
   3.8 Compound interest 35

3.9	Frequency of conversion	36
3.10	Present value	37
3.11	Interpolation	38
3.12	Annuities	40
3.13	Annuities for perpetuity	41
3.14	Deferred annuities	42
3.15	Capital investment appraisal	43
3.16	Investment appraisal example	46
3.17	Exponential smoothing	48
3.18	Exercises	50

4 Sets and Relations — 58
- 4.1 Defining a set — 58
- 4.2 Empty set, universal set, subsets — 59
- 4.3 Venn diagrams — 60
- 4.4 Union, intersection, complement, difference — 60
- 4.5 Algebra of sets and duality — 62
- 4.6 Ordered pairs, products — 64
- 4.7 Relations — 65
- 4.8 Exercises — 68

5 Matrices, Vectors and Determinants — 72
- 5.1 Matrices — 72
- 5.2 Vectors — 73
- 5.3 Matrix addition and multiplication — 74
- 5.4 Square and identity matrices — 77
- 5.5 Inversion of matrices — 78
- 5.6 Determinants — 79
- 5.7 Inversion of matrices by determinants — 80
- 5.8 Transformation matrices and their use in matrix inversion — 82
- 5.9 Multidimensional arrays — 85
- 5.10 Input–output analysis — 86
- 5.11 Exercises — 90

6 Algorithms — 93
- 6.1 Flowchart preparation — 93
- 6.2 Flowchart symbols — 93
- 6.3 Loops — 97
- 6.4 Pseudocode programs — 97
- 6.5 Exercises — 101

7 Linear and Higher-power Equations — 105
- 7.1 Use of equations — 105
- 7.2 Linear equations — 105
- 7.3 Gaussian elimination — 108
- 7.4 Matrix inversion method — 111
- 7.5 Iterative method solution of equations — 112
- 7.6 Cubic equations — 116
- 7.7 Power of four — 117
- 7.8 Other iterative methods — 118
- 7.9 Exercises — 119

8 Graphs — 122
- 8.1 Introduction — 122
- 8.2 Axes and co-ordinates — 122
- 8.3 Graph of a linear equation — 123
- 8.4 Points lying on either side of a line — 125
- 8.5 Non-linear graphs — 127
- 8.6 The circle — 128
- 8.7 The hyperbola — 130
- 8.8 The exponential function — 132
- 8.9 Graphs and business problems — 133
- 8.10 Economic approach to investment, financing and dividend decision making — 137
- 8.11 Exercises — 143

9 Calculus — 147
- 9.1 Differentiation — 147
- 9.2 The product rule — 150
- 9.3 The quotient rule — 152
- 9.4 The function of a function (or chain) rule — 152
- 9.5 The exponential and logarithmic functions — 153
- 9.6 Maxima and minima — 153
- 9.7 Differentials — 159
- 9.8 The integral calculus — 159
- 9.9 Simple integration techniques — 160
- 9.10 Calculation of the area under a curve — 162
- 9.11 Partial differentiation — 164
- 9.12 Total differentials — 167
- 9.13 Three-dimensional maxima and minima — 168
- 9.14 Maxima and minima subject to constraint — 170
- 9.15 Multivariate functions and multiple constraints — 171
- 9.16 Elasticity of demand — 172
- 9.17 Marginal analysis — 173
- 9.18 The economic order quantity model — 173
- 9.19 Exercises — 178

10	**Probability**	**186**
10.1	Probability theory	186
10.2	Finite probability spaces	187
10.3	Equiprobable spaces	188
10.4	Conditional probability	188
10.5	Independence	190
10.6	Probability and combinatorial analysis	191
10.7	Inventory stock out costs	191
10.8	Risk and uncertainty in business forecasting and decision making	192
10.9	Markov chains	203
10.10	Absorbing states	206
10.11	Combinatorial analysis	208
10.12	Permutations	209
10.13	Combinations	211
10.14	Binomial theorem	212
10.15	Exercises	212
11	**Regression Analysis**	**220**
11.1	Introduction	220
11.2	Simple regression	221
11.3	Formulae for the line of best fit or regression line	223
11.4	Application of the formulae for the line of best fit	224
11.5	Prediction and confidence limits	227
11.6	Multiple regression	228
11.7	Interpretation of the statistical data	232
11.8	Application of the multiple regression formulae	236
11.9	Extensions of the regression model	237
11.10	Dummy variables	238
11.11	Business uses of regression analysis	239
11.12	Discriminant analysis	240
11.13	Z score analysis	242
11.14	Exercises	243
12	**Linear Programming** (*Robert Ashford*)	**246**
12.1	Introduction	246
12.2	A simple problem	246
12.3	Some basic concepts	248
12.4	An algebraic approach	251
12.5	The simplex method	253
12.6	The simplex tableau	259
12.7	Exercises	264

13	**Sensitivity Analysis, Duality and the Transportation Algorithm** (*Robert Ashford*)	**270**
	13.1 Sensitivity analysis or ranging	270
	13.2 Duality	275
	13.3 The transportation algorithm	279
	13.4 Integer programming	290
	13.5 Exercises	290
Appendix A	*Tables*	299
	A.1 Logarithms	299
	A.2 Present value tables	301
	A.3 Annuity tables	302
	A.4 Probability tables	303
Appendix B	*Solutions to Exercises*	304
Appendix C	*Worked Answers*	310
Further Reading		328
Index		329

Preface

This book is intended for those students and managers whose basic knowledge of mathematics can best be described as being broadly equivalent to the (old!) 'O' level.

The approach used in this book is based partly on 'traditional' maths and partly on modern maths. This reflects the view of the author that both have advantages, and that acting together they can provide a very powerful base. This approach is also desirable for other, more pragmatic reasons. First, readers will have widely differing mathematical backgrounds; second, many professional, managerial, accountancy and banking examinations require an understanding of both these approaches. Further, most techniques these days will be implemented with the aid of a suitable computer program. This book may therefore be seen as being based on the integration of *traditional maths with modern maths* through the technique of *computers*.

The material has been divided into thirteen chapters and they have been written so that they are to an extent independent of each other.

Each chapter starts with the relevant mathematics and at the end covers appropriate *applications* to the business situation. There are exceptions to this, since sometimes the business application is more suitably dealt with in the body of the text. An example of this is in Chapter 13, where the transportation algorithm is an integral part of the mathematical text and is treated as such.

Answers are given for most of the quantitative questions (Appendix B); in addition fully-worked answers (the author's responsibility) have been provided for some questions (Appendix C). The intention here has been to provide one of these worked answers for a typical examination question in each of the main areas of the subject.

Except where necessary for a proper appreciation of the business techniques *this book does not concern itself with statistics*. Plenty of books have been written solely on this subject. There have, however, been some 'demarcation' problems! These have been particularly difficult in the general area of what might be described as 'mathematically-based operational research methods'. In order to help bridge this gap a short selected bibliography of books in this area has been provided. Because of their practical and examination importance regression analysis and linear programming have been included.

Chapters 12 and 13 have been written by Robert Ashford (of the School of Industrial and Business Studies, University of Warwick) who is a leading authority in the field of linear programming.

Many readers will certainly be working for university, professional, and other examinations. This book covers the mathematical requirements for the examinations of the Chartered Institute of Cost and Management, the Chartered Institute of Public Finance and Accounting, the Institute of Banking, the Chartered Insurance Institute, and the British Institute of Management.

Sometimes, in this book, sexist words and phrases may be used, though hopefully only where general usage sanctions it.

Finally, my thanks to Louise Harris, Shirley Clarke, Robert Ashford, Phillip Moon and Graham Jones.

J.D.

Acknowledgements

The author and publishers acknowledge with thanks permission from professional bodies including the Institute of Chartered Accountants in England and Wales, the Chartered Institute of Cost and Management Accountants, the Chartered Insurance Institute and the Chartered Institute of Public Finance and Accounting to reproduce questions from their past examination papers.

Chapter 1
Elementary Mathematics

1.1 The concept of number

Mathematics is the abstract science of quantity, and quantity is measured in *numbers*. So, since mathematics is really about numbers the first thing we have to ask ourselves is: what is a number? What, for instance, does the number 5,334 really *mean*?

This number certainly consists of *four separate digits*. Each digit is a *symbol*. In our system, normally called the decimal system (though strictly the term denary is more correct) we know that there are ten distinct digits: 0, 1, 2, 3, 4, 5, 6, 7, 8, 9. All numbers in the decimal system are expressed in terms of these ten digits, and no others.

Other systems of numbering use other digits. The *octal system* uses eight digits (those from 0 to 7 inclusive). The *hexadecimal system* uses sixteen symbols. These are the ten decimal digits mentioned above together with the first six letters of the alphabet. Hexadecimal systems and octal systems are used in computer coding. Computers, however, basically operate in the binary system. In this system, instead of there being those ten digits to which you and I are accustomed, there are a great many less – in fact only *two digits*: 0 and 1. In the binary system all numbers are expressed in terms of these two digits only. Because of the importance of computer arithmetic these days a separate chapter (Chapter 2) deals entirely with binary, hexadecimal and other systems used in electronic data processing.

So far we have mentioned systems with the standard ten digits (decimal) and other systems with sixteen digits (hexadecimal) eight digits (octal) and two digits (binary). The number of separate digits in a system is referred to as the *base*. The decimal system has a base of 10 and the bases of the hexadecimal, octal and binary systems are 16, 8 and 2 respectively.

However it is not just the advent of modern computer systems which has made us think of – and use – systems other than the decimal one to which we are accustomed. The Mayas of Central America used a base of twenty, the Sumerians and Babylonians used sixty. Some traces of this still remain with us. We still reckon time by seconds and minutes, and we still use 6×60 degrees in a circle. Quite early in history the base 10 achieved some sort of general acceptance. We do not know for sure, but it seems almost certain that this arose from the practice, natural to us, of counting on the ten fingers of our hands. Spiders are the most widespread and one of the most effective survivors in the evolutionary world. Had these eight-legged creatures (instead of us) made it to the top, they would no doubt have counted in the octal system, and found computer arithmetic natural and understandable!

When there is any doubt in which base a number is written, it is usual to show the base as a *subscript*. Strictly, we should have written all the numbers in the preceding paragraphs in this way (e.g. $6_{10} \times 60_{10}$ degrees in a circle) – the 10 is the subscript – but when the base is clear (as in most normal decimal arithmetic) it is always left out.

We revert now to our normal decimal arithmetic. The importance of the *place* (or position) of a digit is in fact obvious to anyone brought up on 'traditional' arithmetic. The answer to the question which we asked right at the very beginning – what does the number 5,334 really mean – is:

$$5 \times 1,000 + 3 \times 100 + 3 \times 10 + 4 \times 1$$

In the form of a simple table:

	Thousands	Hundreds	Tens	Units
5,334	5	3	3	4

This is called *expanded notation*, which implies that we start with the number 5,334 and then write it out to express it in this expanded form. More accurately we should see 5,334 as a *condensed* way of expressing a number consisting of 5 thousands, 3 hundreds, 3 tens and 4 units.

In every case, moving the position of the digit *one* place to the *left multiplies its value by 10*. Moving it *two* places to the *left multiplies its value by 10×10* (*i.e., 100*). The standard abbreviation or notation for 10×10 is 10^2, for $10 \times 10 \times 10$ it is 10^3, and so on. Moving the position *one* space to the *right* obviously *divides its value by 10*. Thus in 5,334 the left-hand 3 stands for '300' and the next 3 stands for '30'.

What happens when we continue to move the position of the digit to the right? The hundreds become tens

$$300 \times \frac{1}{10} = 30$$

the tens will become units

$$30 \times \frac{1}{10} = 3$$

One more step in the same direction and the units become tenths

$$3 \times \frac{1}{10} = \frac{3}{10}$$

Another move to the right and the tenths become hundredths, and so on. In

Elementary Mathematics

decimal notation:

300	Hundreds
30	Tens
3	Units
0.3	Tenths
0.03	Hundredths

It is important to note that we write 300 with two noughts at the end in order to fix the position of the 3 in the 'hundreds' column. We have to put in a decimal point and extra noughts for tenths and hundredths in a similar way – i.e., to *fix the place position of the digit relative to the decimal point*.

In the example above, all the numbers are lined up with entries in each column one beneath another. It is usual when writing down rows of numbers to do so in this way. In particular the extreme right-hand side digits (*units*) are all lined up. (This is called 'right justified'. Typed words and sentences, by contrast, are usually left justified. The writing on this page of this book (as on all book pages) is both left and right justified.) When calling out numbers we use a *left-justified approach* – i.e. we start with the left-hand side digit. (This sometimes causes problems in class when the lecturer is writing down a series of numbers, called out by a student, which he is proposing to add up. This, however, will be of little interest to the reader!)

Adding up, however, is of interest. Addition (and indeed subtraction, multiplication and division) of decimals is easy once the point has been grasped that places to the right of the decimal point are just extensions of the general rule that a move of one place to the left is equivalent to multiplication by 10, a move of one place to the right equivalent to a division by 10 (i.e., a multiplication by $\frac{1}{10}$). The rule is general. It applies to all places both after the decimal point as well as before it.

The important thing to remember in addition or subtraction is that the *decimal points must be lined up*. The addition of 26 + 3.7, for example, must be set out:

$$\begin{array}{r} 26.0 \\ 3.7 \\ \hline 29.7 \end{array}$$

In multiplication, the rule is to multiply as for whole numbers regardless of the decimal point, then add together the decimal places in each of the two numbers. Measure off from the right this number of decimal places in the product, and put the decimal point in there. Thus to multiply 1.2 × 1.1:

$$\begin{array}{r} 12 \\ \times 11 \\ \hline 132 \end{array} \quad \text{and so,} \quad \begin{array}{r} 1.2 \\ \times 1.1 \\ \hline 1.32 \end{array} \begin{array}{l} \text{(1 decimal place)} \\ \text{(1 decimal place)} \\ (1 + 1 = 2 \text{ decimal places}) \end{array}$$

In division, the divisor can be changed into a whole number first and the decimal points lined up, as in our example below. (Often this is not necessary, because the position of the decimal point in the answer can be determined easily *by inspection*.)

Divide

$$\frac{14.857}{1.79}$$

$$\frac{14.857}{1.79} = \frac{14.857 \times 100}{1.79 \times 100} = \frac{1485.7}{179}$$

$$\begin{array}{r} 8.3 \\ 179\overline{)1485.7} \\ 1432 \\ \hline 53.7 \\ 53.7 \\ \hline \end{array}$$

1.2 Rounding and truncating

For much business work the degree of precision that is required is limited, and measurements involving only two or three places of decimals are required. If the original measurement contained more than the two (say) places of decimals required, the process of dropping the extra decimals is called 'rounding off'. Sometimes the phrase 'correcting to two (or whatever it may be) places of decimals' is used.

Rounding off is not confined to decimal places. Consider the decimal 36.1. The nearest whole number to this decimal is certainly 36. The same applies to 36.2, 36.3 and 36.4. The exact position of 36.5 is ambiguous – it lies half way exactly between the two numbers 36 and 37. Conventionally such numbers are rounded to the nearest *even number* – that is, in our case to 36. (If there are additional places of decimals – e.g., 36.54 – the uncertainty is resolved, in any case, in favour of the larger whole number.) For the decimals above 36.5 – i.e., 36.6, 36.7, 36.8 and 36.9 – the process of rounding off or correcting to the nearest whole number requires that they shall be replaced by 37, *the whole number above*.

Accountants will be used in their work to rounding figures for their *final published accounts* purposes to thousands. £8,147,846 rounded to thousands will be £'s(000)8,148.

Calculations in a computer may result in more digits than can conveniently be stored in memory. Rather than rounding such numbers computers may just drop the least 'significant' digits. The most significant digit in a number is the *first (from the left) non-zero digit*. The least significant digit is the *last (again from the left) non-zero digit*. The number of significant digits is the total number of places between, and including, these two. The number of significant digits in 4,706.038 is 7 and the number of significant digits in 0.03013 is 4. The process of dropping the least significant digits (rather than rounding) is called

Elementary Mathematics

'truncating' or 'chopping'. If we truncate the first of these two numbers to 6 significant places we get 4,706.03; if we truncate the second to 3 significant places we get 0.0301. *Positive numbers* always *decrease in value* when truncated. The disadvantage of truncating (as opposed to rounding) is that when manipulating large quantities of numbers the errors can build up. Both 12.001 and 12.999 truncated to two significant figures are 12. Their (truncated) addition is 24 as opposed to a correct 25. The average error of a large quantity of truncated numbers will be half the value of the last retained digit. By contrast, average round-off error will tend to zero. Banks (or indeed any organisation using computers), and pocket calculators, are examples of 'truncators'!

1.3 Elementary algebra

So far we have discussed numbers and their elementary manipulation – that is addition, multiplication and so on. In each case we have used a particular number by way of example. We shall want to go further and use more advanced manipulation of numbers, and it will be easier if we can use symbols for numbers and so deal with the problem in a more general sense, rather than being confined to a particular figure.

It is customary to use the letters of the alphabet $(a, b, c \ldots)$ to *denote* (or 'stand for') numbers. Sometimes when we run out of these letters, or we wish to have a separate clearly distinguishable group, the Greek alphabet letters $(\alpha, \beta,$ etc.) are used in the same way. The mathematical manipulation of such general symbols is referred to as *algebra*. (The word 'algebra' comes from the Arabic: it meant – according to the OED – 'the science of restoration of normal and equating like with like'. The latter part of this is more immediately relevant to modern algebra – much of algebra does indeed consist of comparing like with like, or more shortly, writing *equations*.)

Let us start off with a few algebraic 'laws' expressed in the form of equations:

$$a \times 0 = 0$$
$$a \times 1 = a$$
$$\left.\begin{array}{r} a + b = b + a \\ a \times b = b \times a \end{array}\right\} \text{ the 'commutative' laws}$$
$$\left.\begin{array}{r} a + (b + c) = (a + b) + c \\ a \times (b \times c) = (a \times b) \times c) \end{array}\right\} \text{ the 'associative' laws}$$

Parentheses (i.e., brackets) indicate that the operations *within the brackets* are to be performed *first*. An example of the associative law would be that the equation:

$$3 \times (4 \times 2) = (3 \times 4) \times 2;$$

means that

$$3 \times 8 \quad\quad = 12 \times 2$$

Hopefully all this will be self-evident to the reader; if not, we ask him or her to 'accept' them on commonsense grounds. A number of excellent mathematical treatises have been written 'proving' these and other similar basic laws and theorems. Some proofs are more rigorous than others, the degree of rigour depending on the level of mathematical expertise. We are not proposing to prove these (or other similar laws) since this is a practical book, and not in any sense at all a theoretical text. More sensibly, these are usually stated as *axioms*, rather than laws (which neatly gets round the problem!).

It is, however, necessary to *state* these axioms or laws. Partly this is because when we discuss set theory we wish to compare and contrast these normal algebraic laws with the algebra of sets. Mainly, however, it is because we propose to build on them so that we can manipulate symbols in a more sophisticated way. One example will show this.

Consider

$$a \times b = c.$$

Conventionally we shorten this to:

$$ab = c \qquad (1.1)$$

This is an *equation* – that is both sides are, we say, *equal*. Presumably both sides will still be equal if we multiply or divide both of them by the same thing (say 2, 3 or indeed any number). If we divide both by b we have

$$\frac{ab}{b} = \frac{c}{b}$$

Since the b's cancel out on the left-hand side (LHS) (or alternatively we can just say that $b/b = 1$), we have

$$a = \frac{c}{b} \qquad (1.2)$$

Comparing (1.1) and (1.2), we see that the b at the top LHS can be transferred to the bottom RHS. This is called 'cross-multiplication'. More generally – and proved by the same approach! –, if

$$pq = rs,$$

then

$$\frac{p}{r} = \frac{s}{q}$$

1.4 Operator precedence

What is the value of $6 + 5 \times 8$? The answer is 46, because the accepted convention is that the *operation of multiplication has precedence over that of addition*. Had we wished the operation of $6 + 5$ to have had precedence, we have already seen one way by which this could have been achieved. If we put

Elementary Mathematics

brackets round the $6+5$, we have $(6+5) \times 8$; because the brackets have precedence, the answer is now 88.

These are examples of 'operator precedence'. Most operators are 'binary' operators – i.e. they require two things to operate on. Addition, subtraction, multiplication, division and 'raise to the power of' all need *two numbers* ($6+5$ or 5×8 in our example). There are some *unary operators* – the unary minus acts on a single number to change the value of (for example) 5 to -5.

In simple practical terms, computers have overall orders of precedence which are arranged to tie in with the existing algebraic conventions. Typical groupings are:

Group
1. Brackets, unary plus, unary minus, NOT, functions
2. Raise to the power (otherwise called 'exponents')
3. Multiplication, division
4. Addition and subtraction
5. Equal to ($=$)
   Less than ($<$), greater than ($>$)
   Not equal to ($<>$ or $\neq$).

Group 1 has the highest priority or precedence. All operators within each group have equal precedence; computers are programmed in this case to deal with them on a left to right basis.

1.5 Percentages, discounts and mark-ups

We have already seen that a decimal such as 0.03 stands for $\frac{3}{100}$. The fraction $\frac{3}{100}$ is commonly called 3 per cent (literally, per hundred) and is written 3% for short. Put in more general terms, x% is defined as the fraction $x/100$, i.e.:

$$x\% \text{ is } \frac{x}{100}$$

Frequently we want to find *what the per cent of a number* is; what, for example, is 50% of 18? Commonsense tells us that since 50% is $\frac{1}{2}$ ($=\frac{50}{100}$) and half of 18 is 9, the answer is 9. In the form of an equation:

$$\frac{50}{100} \times 18 = 9$$

The answer (9 in this case) is referred to as the 'amount', so that the equation above can be written:

$$\text{per cent} \times \text{base} = \text{amount}$$

To find the per cent of a sum of money, we use the same technique

$$15\% \text{ of } £42.50 \text{ is } \frac{15}{100} \times £42.50 = £6.375$$

$$106\% \text{ of } \$394 \text{ is } \frac{106}{100} \times 394 = 1.06 \times 394$$
$$= \underline{\$417.64}$$

There are a number of short cuts in finding percentages. To find 1% of a number, all that is necessary is to move the decimal point *two places to the left*: 1% of 87.2 is 0.872, and 3% of £15.50 is £0.465.

The formula

$$\text{per cent} \times \text{base} = \text{amount}$$

can be used to determine any one of the three terms when the other two are given. For instance, in the problem 'what per cent of 18 is 3', we are given the base (18) and the amount (3), and we are asked to find the percentage. The formula can be rewritten:

$$\text{per cent} = \frac{\text{amount}}{\text{base}} \left(= \frac{3}{18} \times 100, \text{ in our problem} \right)$$

Alternatively, we may wish to find the base, given the other two. For example, we might be asked: 16 is 14% of what number? Using cross-multiplication, we can rewrite the formula as

$$\text{base} = \frac{\text{amount}}{\text{per cent}} = \text{(in our case)} \frac{16}{14/100}$$
$$= \frac{16 \times 100}{14} = \underline{114.3} \text{ (correct to one decimal place)}$$

Discounts given in business are of two types:

1. *A cash discount*: this is a reduction in the amount that has to be paid, if the payment is made within a certain number of days. Typically standard terms of trade are net monthly, but if payment is made within 10 days a cash discount of 2% may be allowed. This would be written '2 net 10'.
2. *A trade discount*: this is a reduction in the invoiced price of the goods. It is usually made when goods are being sold in bulk, or under the provisions of a long-term agreement. It may, however, be for special reasons – to sell damaged or obsolete stock, to meet competition, or to attract new customers.

In both cases the discount is expressed as a percentage of the original price. All discount calculations are therefore percentage calculations.

The extra percentage that a retailer adds on to the purchase price of goods to provide for profit is called the 'mark-up' (sometimes, 'premium'). It is easy to

Elementary Mathematics

calculate this mark-up if the purchase price and the sales value are known. If a retailer sells a tennis racket for £24 which cost him £18, his profit is £6. His mark-up is

$$\frac{6}{18} \times 100 = 33\tfrac{1}{3}\%$$

Expressed as a per cent of his selling price, however, the rate will be

$$\frac{6}{24} \times 100 = 25\%$$

Mark-ups for a particular trade are fairly constant, and VAT and tax inspectors use this to check on the performance of businesses. Any substantial variations in the mark-ups as shown in the accounts submitted by a trader (either by comparison with typical trade figures or by comparison with his past performance) will be viewed with suspicion and will invite further investigation!

Until the Companies Act 1981, the accounts of a company did not have to show the *gross profit*. Format I (the most widely used of the alternative formats available) of the 1985 Act lays down that sales turnover, cost of sales and gross profit (i.e., sales turnover *less* cost of sales) *must be shown*. Cost of sales will include purchases and all factory expenses (but not Distribution and Administrative expenses which are deducted from turnover later). Cost of sales and gross profit were rarely disclosed in pre-1981 accounts on the grounds that to do so gave competitors too good an idea of the operating efficiency of the business. This hardly seems a justifiable criticism now since *all* companies above a certain size have to comply with the Act; it is therefore the same for all, and more information about a competitor should offset any disbenefit from his increased knowledge of one's own accounts.

1.6 Fees and commissions

Frequently in commerce, *fees* are expressed as percentages. Salesmen's commissions are one good example. Banks will express reservation or arrangement fees as percentages (though it must be said that the rates vary so widely that for practical purposes the charge made is really just a sum decided on by the bank manager). However in theory the fee will be expressed as a percentage of the sum to be borrowed. If, for example, the budget of a business shows that the cash deficit will reach a maximum of £150,000 in the middle of the coming year, arrangements may be made with the bank to have this sum available, perhaps by way of overdraft. An arrangement fee of (say) 1% of the sum involved (i.e., £1,500) will be charged at the time when the facility is made available.

Perhaps we should add that this fee is *additional to the interest charge* on the sum actually borrowed. That charge will be at a rate of around 2–4 per cent above the bank's own base rate. Each bank has its own base rate (as indeed do all the major financial institutions such as finance houses). The bank's base rate is the key reference point in determining overdraft, loan and deposit rates. Each bank in theory fixes its own base rate, though in practice banks tend to move their rates together. Banks' base rates are changed in response to money market conditions and also in response to informal pressure from the government through the Bank of England (sometimes referred to as 'their' City branch!).

1.7 Simple interest

One of the most important applications of percentages is in interest calculations. *Interest is a payment for the use of money for a period.* The interest rate (expressed as a percentage) is usually denoted by (r), the base sum of money is called the principal (P) and the amount of interest is (I). Replacing the words in the formula 'per cent × base = amount' by these words we have:

$$\text{interest rate} \times \text{principal} = \text{interest},$$

or in symbols

$$rP = I$$

thus the interest on £3,000 at 9% for 1 year is

$$\frac{9}{100} \times 3{,}000 = £270$$

Had the money been on loan for two years instead of one, then the two years' interest would have been

$$£270 + £270 = £540$$

For five years, the answer would obviously have been

$$£270 \times 5 = £1{,}350$$

In general, if the interest is for t years (say), the formula for the interest becomes:

$$I = rPt$$

Thus the interest on £3,000 for 18 months at 9% is

$$0.09 \times £3{,}000 \times 1.5 = £405$$

At the end of the period of 18 months, the total of the principal and the interest together is £3,000 + £405 = £3,405. This amount is called the sum (S).

Clearly

$$S = P + I,$$

and   since $I = rPt$

$$S = P + rPt = P(1 + rt)$$

Elementary Mathematics

This is the standard formula for determining the accrued value or sum (S) to which a principal (P) will accumulate at simple interest for (t) years at a rate of interest (r).

We can now rewrite this formula as:

$$P = \frac{S}{1+rt}$$

and then rephrase it to say that it now gives the 'present value' (P) of a sum (S) at simple interest rate (r), t years from now.

In the above calculations we have used time periods of years and the interest rate has been so much per cent per annum (i.e., per year). We could have chosen any period (one month, a quarter, six months) and the calculations would have been just the same, provided we had expressed the interest rate as so much per cent *for that period*. In all interest and present value calculations, the more appropriate general term is a 'period', though in practice the periods often are years.

EXAMPLE 1

Find the present value of £1,500 in 9 months' time at a simple interest rate of 12% p.a.

$$S = 1,500, \quad r = 0.12, \quad t = 0.75$$

$$P = \frac{S}{1+rt} = \frac{£1,500}{1+0.12 \times 0.75} = \frac{£1,500}{1.09} = £1,376$$

EXAMPLE 2

What is the maturity value of a promissory note of face value £6,000 due in 8 months' time, at 11% p.a. simple interest?

Here we are given

$$P = 6,000, \quad r = \frac{11}{100}, \quad t = \frac{2}{3}, \quad \text{and we need to find } S.$$

$$S = P(1+rt)$$

$$= £6,000\left(1 + \frac{11}{100} \times \frac{2}{3}\right) = 6,000\left(1 + \frac{22}{300}\right)$$

$$= 6,000\frac{(300+22)}{300} = 6,000 \times \frac{322}{300} = £6,440$$

1.8 Ratios

Earlier the point was made that 'per cent' means 'per hundred', i.e. 42% is $\frac{42}{100}$. This fraction is, of course, one number (42) divided by another (100). This phrase can be shortened to 'the ratio of 42 to 100'; in general terms, therefore, the ratio of a to b is the fraction a/b.

The ratio of one number to another is not altered if we multiply or divide both numbers by the same figure. Thus:

$$80\% = \frac{80}{100} = \frac{4}{5}$$

We can look at this conversion the other way round, from a ratio to a percentage, i.e.:

$$\frac{4}{5} = \frac{80}{100} = 80\%$$

Seen this way, a percentage is just that ratio (out of the infinite number of possible ones obtained by multiplying top and bottom of a fraction by any number) which has, as its bottom, 100.

1.9 Exercises

1. Round off to the nearest whole number: 13.125 36.49 136.5 0.44.
2. Round off to tenths: 19.92 42.08.
3. Round of to hundredths: 8.115 5.779.
4. Express as decimals: $29\frac{6}{10}\%$ $2\frac{1}{2}\%$.
5. What is: 150% of 4,000 0.5% of 7,000 3.5% of 1,200?
6. A sales manager earns £160 per week, and 2% commisission on all sales he makes. Last week his sales were £2,496. What total amount did he earn, to the nearest pound?
7. I buy some shares for £12,000 and sell them for £14,400. What is my profit expressed as a percentage?
8. I buy some goods for £200 and sell them for £325. Express my profit as percentages of both cost and selling price.
9. If I take advantage of both cash and trade discounts, what amount will I pay, if: the list price is £2,500, cash discount is 3% for payment within 10 days, and, trade discount is 25%?
10. Find the face value of the digit 6 in (a) 7,625, (b) 64,723, (c) 305.56, (d) 0.012365.
11. Rewrite in expanded notation (a) 2,468, (b) 54.322.
12. Evaluate 7.32 + 33.3 + 34.678.
13. At 4% simple interest, what is the present value today of
    (a) £2,000 due in 6 months with interest at 5%?
    (b) £3,000 due in 1 year with interest at 6%?
14. In September 1967, GEC announced its classic bid for AEI shares. Based on market values before the news was announced, GEC were offering £2.60, in cash and shares, for AEI shares at £2.175. What is the percentage increase? What is the customary margin for a bid over and above the market price?

Elementary Mathematics 13

15. When arranging the purchase/sale of a medium/small ongoing business, I know that bargaining usually begins in practice around a figure some five times the net profit before tax. I advise my accountant that I want him to make an offer of 10% under this customary figure. If average profits of the business before tax are some £42,500 a year, what will he offer on my behalf? What will he offer if I ask him to round it off to the nearest £5,000?
16. Convert the number 364 (base 10) into its equivalent binary (base 2) number.
17. Convert the binary number 101101101 into its equivalent number in base 10.
18. Remembering that the operator precedence rule applies to fractions (but that first it is necessary to change to proper fractions) calculate:

$$\frac{1\frac{1}{4} \times \frac{1}{2} + \frac{1}{3}}{\frac{1}{3} \text{ of } 5 - \frac{1}{5} \text{ of } 3}$$

19. The Systéme Internationale (SI) standardises units of length (a metre), mass (a kilogram) and so on. It also lays down decimal multiples and submultiples. A very shortened summary of these units and their multiples and symbols is:

Prefix	kilo	hecto	deca		deci	centi	milli	micro
Symbol	k	h	da		d	c	m	μ
Multiple	10^3	10^2	10	1	10^{-1}	10^{-2}	10^{-3}	10^{-6}
Length	km			m		cm	mm	μm
Mass	kg			g			mg	μg

(a) A model of a vehicle moves 50 centimetres in 1 second; what is the speed expressed in kilometres per hour?
(b) The mass of 1 mm of a computer connecting cable is 1.5 mg. If you need 800 metres of this uniform cable what would its mass be in kilograms?
(c) The plans for your new 'Post Experience' lecturing complex have been drawn to a scale of 1:750. What would be the actual length of a corridor which was 54 mm long on the plan? You are considering adding another 'Harvard Business School' style room for teaching cases. This room would be 30 m by 18 m. On the plan there is a spare space 35 mm by 25 mm. Will the room you want fit into this space?

Chapter 2
Computer Number Systems

2.1 Introduction

Our dependence on computers as tools for analysis and calculation hardly needs comment. Soon no area of life will be free from their influence. It might seem to follow from this that we would all seek to understand and come to terms with them.

It is true that a new race of computer experts has risen. Yet for many the reaction to computers is that they are too complicated and too awesome in their powers even to try to understand. Computers have almost divided us into a two-nation country. Yet the way computers *work* is basically easy to understand. It is only the sophisticated complexities of their language which may at first frighten.

2.2 Binary digits

Most of the hardware components of a computer are *two-state* in nature. By this we mean that they can either be on or off (as in a switch), or they can be clockwise or counter-clockwise magnetised. If we denote these states by the *binary digits* (i.e. 0 and 1) we see immediately that binary arithmetic is ideal for use by computers. An individual unit of information (letter, number, operator, sign, etc.) is represented in the computer by a series of these binary digits (called *bits* for short).

Any number can be easily and directly expressed in binary form as a sequence of binary digits. Large numbers – which will require only a comparatively few digits in the decimal numbering system – will, it is true, require a great deal more in binary. However once numbers have been converted into binary form a computer can begin to operate with them; when it does so, it operates like lightning – i.e., at the speed of light!

Just as the place or position values in our decimal system are in powers of 10 (i.e., 10, 10^2, 10^3), and for places on the right-hand side of the decimal point in negative powers of 10 (i.e., 10^{-1}, 10^{-2}, 10^{-3}), so in the binary system place values are the *powers of the base 2*. And just as a number in the decimal system can be expressed in expanded notation by multiplying the digits by the appropriate power of 10, so any binary number can be expressed in expanded form by multiplying the digits by the appropriate power of 2. Strictly, as with numbers to a decimal base (or indeed to any base) the reverse is true, and

Computer Number Systems

normal notation is just an abbreviated way of writing numbers in expanded notation.

Thus

$$10111 \text{ (in binary)} = 1 \times 2^4 + 0 \times 2^3 + 1 \times 2^2 + 1 \times 2 + 1$$
$$= 1 \times 16 + 0 \times 8 + 1 \times 4 + 1 \times 2 + 1 = 23 \text{ (in the decimal system)}$$

and

$$101.11 \text{ (in binary)} = 1 \times 2^2 + 0 \times 2 + 1 + 1 \times 2^{-1} + 1 \times 2^{-2}$$
$$= 1 \times 4 + 0 \times 2 + 1 + 1 \times \tfrac{1}{2} + 1 \times \tfrac{1}{4} = 5.75 \text{ (in the decimal system)}$$

Tables 2.1 and 2.2 show respectively the decimal/binary equivalents from 0 to 20 and also the values of some of the powers of 2. Using Table 2.2 and expressing binary in expanded notation makes binary to decimal conversion easy.

TABLE 2.1

Decimal number	Binary number equivalent				
	16s	8s	4s	2s	1s
0					0
1					1
2				1	0
3				1	1
4			1	0	0
5			1	0	1
6			1	1	0
7			1	1	1
8		1	0	0	0
9		1	0	0	1
10		1	0	1	0
11		1	0	1	1
12		1	1	0	0
13		1	1	0	1
14		1	1	1	0
15		1	1	1	1
16	1	0	0	0	0
17	1	0	0	0	1
18	1	0	0	1	0
19	1	0	0	1	1
20	1	0	1	0	0

TABLE 2.2

Powers of two	Decimal value
2^{10}	1024
2^9	512
2^8	256
2^7	128
2^6	64
2^5	32
2^4	16
2^3	8
2^2	4
2^1	2
2^0	1
2^{-1}	1/2 = 0.5
2^{-2}	1/4 = 0.25
2^{-3}	1/8 = 0.125
2^{-4}	1/16 = 0.0625

2.3 Binary calculations

Addition, subtraction, multiplication and division in binary are simple, too. They follow along exactly the same lines as when operating in the decimal system and they need to be set out in just the same way; in particular decimal points needs to be vertically lined up – or, in the jargon, they need to be *right justified*. There are only the two digits 0 and 1, so when we wish to go above 1, we have to go into the next place position. Just as $9+1 = 10$ (in decimal) so $1+1 = 10$ (in binary). The basic binary additions are $0+0 = 0$, $0+1$ (or $1+0$) = 1, $1+1 = 10$ (and $1+1+1 = 11$, etc.).

In the binary system, there are only two 'multiplication tables', i.e.:

The 'nought' times   *The 'one' times*
$0 \times 0 = 0$ $1 \times 0 = 0$
$0 \times 1 = 0$ $1 \times 1 = 1$

EXAMPLE 1
Add 111 to 101

$$\begin{array}{r} 111 \\ +\,101 \\ \hline -1\,100 \end{array}$$

EXAMPLE 2
111×101

$$\begin{array}{r} 111 \\ \times\,101 \\ \hline 111 \\ 000 \\ 111 \\ \hline 100011 \end{array}$$

The steps in binary *addition* (Example 1) are as follows. First, line up the sums to be added; second, add the *right-hand column* (i.e., $1+1 = 10$), put down 0 and carry 1 into the next (middle) column. For that column, too, we have $1+1 = 10$ (i.e., put down 0 again, and carry 1 into the next (left) column). There we have $1+1+1$ (brought forward) = 11, put down 1 and carry 1 into the next column, giving the addition of 1100. A check by converting into the decimal shows that $7+5 = 12$.

The binary *multiplication* (Example 2) can easily be checked. The answer, in expanded form, is $1 \times 2^5 + 1 \times 2 + 1$ (omitting the multiplications by nought which clearly are zero) $= 32 + 2 + 1 = 35$. In decimal notation, the calculation is $7 \times 5 = 35$.

Binary *subtraction* and *division* follow along the same lines. For decimal

Computer Number Systems

calculations we may have to 'borrow' 10 from the next column on the left; for binary calculations we may have to borrow, too.

EXAMPLE 3
Calculate

$$\begin{array}{r} 1110 \\ -101 \\ \hline 1001 \end{array}$$

The steps in this binary subtraction are: first, line up; second, starting from the right (since 1 cannot be taken from 0), 'borrow' from the next column. This reduces the 1 in that column to 0, and the rest of the subtractions are easy.

Frequently in binary subtractions it is not possible to borrow from the next column on the left because the digit there is zero. There is an analogous problem in decimal, but this is less frequent as there are nine non-zero digits as opposed to only one in binary. As in decimal the rule is to borrow from the *first non-zero digit to the left*. Two examples (the one on the left in decimal, the one on the right in binary) show how this is done:

$$\begin{array}{cc} \textit{Decimal} & \textit{Binary} \\ 1{,}007 & 100 \\ -54 & -1 \\ \hline 953 & 11 \end{array}$$

2.4 Computer codes

We noted earlier that binary notation has a basic disadvantage in that it is somewhat unwieldy – the number 694 in decimal notation, is 10 1011 0110 in binary notation. Even if we leave gaps after every 4 digits (a fairly common practice, analogous to our habit of putting commas after every thousand in large decimal numbers) it is still confusing and difficult to handle.

To bridge the gap between the more compact decimal notation and the binary notation which is so suitable for computers either the octal (base 8), or now more frequently the hexadecimal (base 16) system is used. Because 8 and 16 are 2^3 and 2^4 respectively there is an easy conversion to binary. Additionally both octal and hexadecimal are compact.

2.5 Octal system

INTRODUCTION

In the octal system we have *eight digits* – 0, 1, 2, 3, 4, 5, 6 and 7. Tables 2.3 and 2.4 show respectively some decimal/binary/octal equivalents, and also the values of some of the powers of 8.

TABLE 2.3

Decimal values	Binary number equivalent	Octal number equivalent
0	000	0
1	001	1
2	010	2
3	011	3
4	100	4
5	101	5
6	110	6
7	111	7
8	1 000	10
9	1 001	11
10	1 010	12
11	1 011, etc.	13
16	10 000	20
17	10 001	21
18	10 010, etc.	22
23	10 111	27
24	11 000, etc.	30
31	11 111	37
32	100 000	40

TABLE 2.4

Powers of eight	Decimal values
8^5	32,768
8^4	4,096
8^3	512
8^2	64
8^1	8
8^0	1
8^{-1}	0.125

OCTAL/BINARY CONVERSION

As can be seen from Table 2.3, octal/binary conversion is very easy. We can view each octal digit simply as a short way of writing its equivalent 3-bit (binary) value:

27 (in octal) is

10 111 (in binary)

with the octal 2 and 7 each being converted directly in its binary equivalent. Large numbers convert equally easily:

4105 (in octal) is

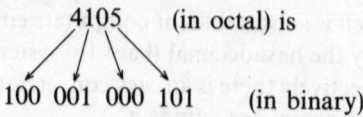

100 001 000 101 (in binary)

To reverse the process – i.e., to convert from binary into octal – we have (starting from the right) to *partition the binary number* into 3-bit groups or blocks. We can then convert each group into its equivalent octal digit:

11 111 001 010 (in binary) is

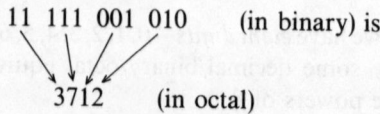

3712 (in octal)

Computer Number Systems

OCTAL DECIMAL CONVERSION

Octal arithmetic is often most easily done by expressing in expanded form, converting to decimal, and then reconverting.

To add $146_8 + 35_8$, we have:

$$146_8 = 1 \times 8^2 + 4 \times 8 + 6 = 64 + 32 + 6 = 102$$
$$35_8 = \phantom{1 \times 8^2 + {}} 3 \times 8 + 5 = \phantom{64 + {}} 24 + 5 = 29$$
$$\phantom{35_8 = 1 \times 8^2 + 4 \times 8 + 6 = 64 + 32 + {}} 131 \text{ (in decimal)}$$

To reconvert (i.e., the reverse process) we need to divide by 8, showing in each case the remainder separately:

```
                          Remainder
        8) 131
          8) 16              3
             2               0
```

The sequence of remainders in reverse order gives this octal number as 203 (i.e., $131 = 203_8$). A quick check using expanded notation shows that $203_8 = 2 \times 8^2 + 0 \times 8^1 + 3 = 2 \times 64 + 0 + 3 = 131$ (in decimal).

The method of repeated division (and using the sequence of remainders in reverse order to give the answer) is the standard way of converting a number expressed in a base back to decimal notation. The division is repeatedly made by the base number. It is, we emphasise again, the reverse of expressing a number in expanded notation in powers of the base.

For octal/decimal conversion we must remember that there are only 8 digits in octal, and beyond that we have to use *place values*. The octal addition we have just shown can in fact be done directly:

$$146_8$$
$$+35_8$$
$$\overline{203_8}$$

Starting with the right column, and referring to Table 2.3, we have $6 + 5 = 11$ (in decimal) = 13 (in octal). Carry 1 into the next column when we have $4 + 3 + 1$ (carried forward) = 8 (in decimal) = 10 (in octal). For the extreme left column $1 + 1$ (carried forward) = 2.

Although this octal addition (using comparatively simple figures) can be done directly, more complicated calculations – and particularly those involving multiplication and division – are best done by first converting to decimal via expanded notation, then making the calculation in decimal and finally reconverting, using repeated division.

2.6 Quintal system

We have concerned ourselves particularly with the octal system because of its use in computer arithmetic, but these conversion methods apply equally to any base system. Thus in the quintal (base 5) system we have:

$$4323_5 = 4 \times 5^3 + 3 \times 5^2 + 2 \times 5 + 3 = 588 \text{ (in decimal)},$$

and to convert 588 into its quintal form we divide by 5:

```
                    Remainder
        5)588
        5)117        3
        5) 23        2
            4        3
```

giving the number as 4323_5.

2.7 Hexadecimal system

In the hexadecimal (hex) system we have sixteen digits. These are the normal ten decimal digits followed by the letters of the alphabet A to F inclusive. Tables 2.5 and 2.6 show respectively some decimal/binary/hexadecimal equivalents and also some powers of 16.

'Hexadecimal' is usually shortened to hex, and from now onwards we shall use this abbreviated form. Just as octal has a unique 3-bit representation in binary, hex has a unique 4-bit representation. Thus

1A   (in hex) is

0001 1010   (in binary)

A quick check with Table 2.5 confirms this (both are equal to 26 in decimal). To reverse the process (i.e., to convert from binary to hex) we have (starting from the right) to *partition the binary number* into 4-bit groups or blocks and then convert each 4-bit group directly into hex. The resulting sequence is the hex number.

Arithmetic in hex, as in octal, is most easily done by expressing in expanded form and then reconverting. To add $13E_{16} + D8_{16}$ we have:

$$13E_{16} = 1 \times 16^2 + 3 \times 16 + 14 = 318$$

$$D8_{16} = \phantom{1 \times 16^2 + {}} 13 \times 16 + 8 = \underline{216}$$

$$\phantom{D8_{16} = 1 \times 16^2 + 13 \times 16 + 8 = {}} 534 \text{ (in decimal)}$$

To reconvert, divide by 16, showing the remainder separately after each

Computer Number Systems

TABLE 2.5		
Decimal values	Binary equivalent	Hexadecimal equivalent
0	0000	0
1	0001	1
2	0010	2
3	0011	3
4	0100	4
5	0101	5
6	0110	6
7	0111	7
8	1000	8
9	1001	9
10	1010	A
11	1011	B
12	1100	C
13	1101	D
14	1110	E
15	1111	F
16	1 0000	10
17	1 0001, etc.	11
26	1 1010	1A

TABLE 2.6	
Powers of 16	Decimal values
16^5	1,048,576
16^4	65,536
16^3	4,096
16^2	256
16^1	16
16^0	1
16^{-1}	0.0625

division:

$$\begin{array}{r} & & Remainder \\ 16\overline{)5\,34} & & \\ 16\overline{)33} & & 6 \\ 2 & & 1 \end{array}$$

The sequence of remainders in reverse order gives the hex number as 216. A quick check using expanded notation shows that:

$$216_{16} = 2 \times 16^2 + 1 \times 16 + 6 = 534 \text{ (in decimal)}$$

This hex addition can be done directly. The sum is set out below. Starting with the right column, and referring to Table 2.5, we have

$$E + 8 = 14 + 8 = 22 \text{ (in decimal)} = 16 \text{ (in hex)}.$$

Carry 1 into the next column where we have

$$3 + D + 1 \text{ (carried forward)} = 3 + 13 + 1 = 17 \text{ (in decimal)}$$
$$= 11 \text{ (in hex)}.$$

Finally carry one into the last column.

$$13E_{16}$$
$$\underline{D8_{16}}$$
$$216_{16}$$

$$C.3F_{16}$$
$$\underline{0.5D8_{16}}$$
$$C.9B8_{16}$$

The addition shown immediately above is another calculation in hex, this time with decimal numbers. The important thing in hex addition – as in *all* calculations, whatever the base – is to *line up the decimal points*.

2.8 Binary coded decimals

There are ways of expressing decimal data in binary form other than by the direct conversion into binary which we have been discussing (and which is normally called 'straight binary coding'). Another way is to encode a decimal number *digit by digit*. This is called BCD (or binary coded decimal) coding. Since there are ten different decimal digits we need at least four bits (binary coded digits) for each decimal digit. The most common BCD code (called 8-4-2-1) in fact is simply that which has the binary equivalent as its representation for each decimal digit. Table 2.7 shows this:

TABLE 2.7

Decimal digit	BCD code
0	0000
1	0001
2	0010
3	0011
4	0100
5	0101
6	0110
7	0111
8	1000
9	1001

There are of course six spare BCD codes available (1010, 1011, etc.) since if we are using four places or bits and each can take the form of 0 or 1 we have a total of $2^4 = 16$ possibilities. Ten have already been used above for the ten decimal digits (0 to 9), leaving six. Typically two of these are used in 4-bit BCD codes for the + and − signs.

BCD coding has certain advantages over straight binary coding. For one thing it is easier, since each decimal digit is directly and separately converted into its BCD code. One obvious disadvantage is that it requires more digits than straight binary – the decimal number 471 has a BCD representation of 0100 0111 0001 which is three more digits than the straight binary coding 1 1101 0111.

2.9 6, 8 and 16 bits

In 6-bit BCD (i.e., allowing six places, each of which is filled with a binary digit) we would have a total of $2^6 = 64$ possibilities or characters. This enlarged representation allows not only the ten decimal digits but also the twenty-six letters of the alphabet to be expressed. After these ten numeric and 26 non-numeric characters have been encoded there are $64 - 36 = 28$ spare characters. These are usually reserved for so called 'special characters' which include not only the $+$, $-$, and $\times$ and $\div$ signs, but also brackets, £ signs and other characters shown on the alphanumeric keyboard of a computer.

Some data processing requires both upper case and lower case letters giving $2 \times 26 = 52$ characters needed for the letters only. To cope with the enlarged need for this and a variety of special characters most computers have been operating with 8-bit codes. Two 8-bit codes are used in the computer industry today. They are:

ASCII 8 (American Standard Code for Information Interchange) – strictly a 7-bit code with one spare bit and
EBCDIC (Extended Binary-Coded Decimal Interchange Code)

The latter is mainly used by IBM and IBM-compatible equipment; the former by non-IBM systems (and this includes most of the popular microcomputers such as the BBC 'micro', and IBM PCs).

It would be interesting to continue with a discussion of the use of 8-bit codes (and even 16-bit codes which IBM is using). However, our aim is not to deal with the mechanics of electronic data processing, but with the mathematical background. Number systems with varying bases are the real concern of this chapter. We have dealt principally with binary and hexadecimal because these are two typical number systems, which also have an immense practical importance. We conclude with a note on the practicalities of conversion from a decimal to a binary number.

2.10 Decimal-to-binary conversion

To find the binary representation of a decimal number N we convert its integral part, and its fractional part, in two separate parts. Consider the decimal number $N = 108.8125$.

PART 1

To convert 108 to its binary equivalent, divide it (and each successive quotient) by 2, noting the remainders:

Divisions	Quotients	Remainders
$108 \div 2$	54	0
$54 \div 2$	27	0
$27 \div 2$	13	1
$13 \div 2$	6	1
$6 \div 2$	3	0
$3 \div 2$	1	1
$1 \div 2$	0	1

The zero quotient indicates the end of the calculations. The sequence of remainders from the bottom up is the required binary equivalent. That is, $108 = 1101100$.

PART 2

To convert 0.8125 to its binary equivalent, multiply it (and each successive fractional part) by 2, noting the integral part of the product, as follows:

$$
\begin{array}{r|l}
 & \text{Integral part} \\
0.8125 \times 2 = & 1.625 \\
0.625 \times 2 = & 1.25 \\
0.25 \times 2 = & 0.5 \\
0.5 \times 2 = & 1.0 \\
\end{array}
$$

The zero fractional part indicates the end of the calculations. The sequence of integral-part digits from the top down as shown by the arrow gives the required equivalent as 0.1101.

Putting the two together we have:

$$108.8125 = 1101100.1101$$

2.11 Exercises

1. Express the following binary numbers as decimal numbers: 11, 100, 1101, 1001, 100110.
2. Express the following decimal numbers as binary numbers: 5, 17, 50, 121.
3. Turn the decimal number 422 into the corresponding number using a base of six.
4. Add the following pairs of binary numbers, and check the answers by changing each number (and the answers) into decimal numbers:

    (a) $10 + 1$
    (b) $100 + 1$
    (c) $101 + 10$

Computer Number Systems

  (d) $111 + 10$
  (e) $111 + 110$

5. Carry out the following subtractions of binary numbers. Check your answers by changing into ordinary (decimal) numbers:

  (a) $111 - 100$
  (b) $1101 - 100$
  (c) $1101 - 111$
  (d) $10000 - 111$

6. Carry out the following multiplications of binary numbers. Check your answers by changing into ordinary numbers:

  (a) 11×0
  (b) 110×1
  (c) 11×10
  (d) 1101×101

(Remember binary multiplication and division sums are set out in just the same way as multiplication and long division in ordinary arithmetic.)

7. Carry out the following divisions of binary numbers. Check your answers by changing into ordinary numbers:

  (a) $110 \div 10$
  (b) $110 \div 11$
  (c) $11001 \div 101$
  (d) $110010 \div 101$

8. Convert 4207_8 into its equivalent binary number.
9. Convert the binary number 10101011111 to its octal equivalent (start off by partitioning the binary number into 3-bit blocks commencing at the extreme right bit; then replace each block by its equivalent octal digit).
10. Convert the binary number 1011.0101 to its octal equivalent (when partitioning the binary number into 3-bit blocks commence both to the left and to the right of the binary point; then replace each block by its equivalent octal digit).
11. Work out the following octal additions:

  (a) $5_8 + 4_8$
  (b) $7_8 + 4_8$
  (c) $1_8 + 5_8 + 6_8$

Express the answers as octal numbers.

12. Work out the following octal addition:

    $7346_8 + 5265_8$

    Express the answer as an octal number (remember to align the two numbers in the same way as for ordinary numbers, that is right justified; then add the column on the right first, carry forward any number to the top of the next column, and so on).
13. Convert the hexadecimal number $39.B8_{16}$ into its equivalent decimal number by first expressing it in expanded (hex) notation.
14. Convert the decimal fraction 0.78125 into its equivalent in hex (use the integral part approach: a zero fractional part is reached).
15. Work out the additions of the following hexadecimal numbers:

    (a) A + 9
    (b) 3 + B
    (c) 1 + 5 + C

    Express the answer as a hex number.
16. Work out the sum of the two following hexadecimal numbers:

    C868 + 72D9

    Express the answer as a hex number (again remember to align the two hex numbers in the same way as for ordinary numbers; then add the column on the right first, carry forward any number to the top of the next column, and so on).
17. Convert the binary number 10110100101110 to its hexadecimal equivalent (remember to start off by partitioning the binary number into 4-bit blocks adding any 0s that may be necessary; then replace each of the 4-bit blocks by its corresponding hex digit).
18. Convert the binary number 11100.1011011011 to its hexadecimal equivalent (use the same approach as for the previous question but remember to start both to the left and to the right of the binary point).
19. Add the following hex numbers:

    (a) $82C5_{16} + 9D86_{16}$
    (b) $4C.3E_{16} + 2.5D8_{16}$

    Express as hex numbers. Remember to line up the hex points for (b).

Chapter 3
Exponents, Progressions, Present Value

3.1 Introduction

The manipulation of symbols is helped by the use of *abbreviations*. The commonly accepted abbreviation for $a \times a$ is a^2, for $a \times a \times a$ is a^3 and more generally, for $a \times a \times a \ldots \times a$ (*n* times) is a^n, and *n* is called the *exponent* or *index*.

The process of operating on *a* or changing *a* so that it becomes a^n is called 'raising to the power of *n*'. The computer symbol is $\wedge$ or ⋀ or **. Thus a^n can also be written as $a \wedge n$ or a ⋀ n or $a**n$. These are merely other shorthand ways of saying the same thing.

If we multiply a^2 by a^3, we have:

$$a^2 \times a^3 = (a \times a) \times (a \times a \times a) = a^5$$

This demonstrates the *multiplication rule for exponents* – i.e., *add* the indices (in our examples, $2+3 = 5$).

If we divide a^5 by a^2, we have

$$\frac{a^5}{a^2} = \frac{a \times a \times a \times a \times a}{a \times a} = a \times a \times a = a^3$$

This demonstrates the *division rule for exponents* – i.e., *subtract* the indices (in our example, $5-2 = 3$).

Finally consider $(a^2)^3 = a^2 \times a^2 \times a^2 = a \times a \times a \times a \times a \times a = a^6$.

This demonstrates the rule for the *powers of exponents* – i.e., *multiply* the indices together (in our example, $2 \times 3 = 6$).

In more general terms we have

$$a^m \times a^n = a^{m+n}$$
$$a^m \div a^n = a^{m-n}$$
$$(a^m)^n = a^{mn}$$

Although we have only demonstrated these rules using one example, in which *m* and *n* were positive decimal integers, the rule applies generally. The exponents m and n can be positive or negative, integers or fractions. Some examples of these are given below.

First, consider $a^2 \div a^2$. This is clearly 1, but by the subtraction rule we know that it is $a^{2-2} = a^0$. Hence $a^0 = 1$. Since a is a symbol this should apply for all values of a. However there is one exception. If $a = 0$, we would have $0/0$ (which is not defined, that is we do not know what it is!).

Second

$$\frac{1}{a^2} = \frac{a^0}{a^2} = a^{0-2} = a^{-2}, \quad \text{so that} \quad a^{-2} = \frac{1}{a^2}$$

Although we choose to use a power '2' (for simplicity) to demonstrate these two results it must be evident that we could have used a^3 or a^4 or indeed any power of a.

Finally

$$a = a^1 = a^{\frac{1}{2}+\frac{1}{2}} = a^{1/2} \times a^{1/2}$$

This, of course, is commonly called the 'square root' (as a symbol, $\sqrt{\ }$). Hence we write $a^{1/2} = \sqrt{a}$ (and call it the square root of a) and (more generally):

$$a^{1/n} = n\sqrt{a} \quad \text{(the nth root of a).}$$

3.2 Exponential form

We still live in a world which is 'decimal' (here used in the more modern sense of numbers to the base of 10, rather than the more traditional sense of digits to the right of the 'decimal' point; the fact that there are two uses of the word can create a great deal of confusion!). It is therefore only to be expected that exponents which are powers of 10 (i.e., $a = 10$), are widely employed. One use is in logarithms – we shall come to that later – but another and more modern use is in writing numbers in exponential form for *computer manipulation*.

The sun is approximately 93,000,000 miles away from the earth. A shorter way of writing this figure is 9.3×10^7. Alternatively we could write it as 0.93×10^8 (remember moving the decimal point one place multiplies or divides by 10). The latter form is called the *normalised* exponential form. The rule for this form is that the decimal point must appear *before* the first non-zero decimal digit. (There is another exponential form much used in scientific work. The rule for this *scientific* exponential form is that the decimal point must appear *after* the first non-zero digit.) It is obvious that exponential notation could be very useful in astronomy where distances between stars, for example, are immensely greater than the earth–sun distance. Scientific exponential notation is indeed used by astronomers though such great distances are also often expressed in other forms such as light years – that is the distance light travels in a year.

Reverting to normalised exponential form, this has a particular use in computing. Taking the same figure of 93,000,000, this is first shortened to

Exponents, Progressions, Present Value 29

0.93×10^8. Computer software, however, does not have indices. This input figure is therefore expressed as $0.93E+8$. The number 0.93 is called the *mantissa* and $+8$ is called the *exponent*.

The benefit from the use of exponential form becomes greater the larger the number is. 46,000,000,000 (46 billion) is 4.6×10^{10}, i.e. $4.6\ E+10$. Light travels approximately 980,000,000 feet per second. This is equal to 0.98×10^9 (or, in its normalised exponential form, ready for input into a computer, $0.98\ E+9$). The printed output from a computer is in this same form. In computer circuit planning the distance that light travels in very short periods of time is obviously important. The term 'nanosecond' for $1/10^9$ seconds is commonly used as the time period. (Light travels just about 1 foot in 1 nanosecond.)

Numbers written in bases other than the normal decimal 10 can also be written in exponential form. (Binary numbers written in exponential form, for example, will use powers of 2 rather than powers of 10.) This type of representation can sometimes be very useful with large binary numbers which would otherwise take up a great deal of space.

Two examples of binary numbers and their normalised exponential form are set out below:

Binary number	Normalised exponential form
0.00111	0.111×2^{-2} or $0.111\ E-2$
1100000	0.11×2^7 or $0.11\ E+7$

Sometimes, when the exponents are positive, the $+$ sign is omitted; in this last example we could write the exponential form as $0.11\ E\ 7$. This would be understood as meaning $0.11\ E+7$.

3.3 Computer arithmetic

Computers usually do arithmetical calculations with numbers in *exponential form*. Addition of two numbers in this form is easy if both have the same exponent; all that is necessary is that the mantissa be added and the same exponent used:

$$0.304 \times 10^5 + 0.621 \times 10^5 = 0.925 \times 10^5$$

If the exponents are different, then one exponent (usually the *smaller*) will be adjusted:

$$0.304 \times 10^4 + 0.621 \times 10^5 = 0.0304 \times 10^5 + 0.621 \times 10^5 = 0.6514 \times 10^5$$

There is one problem that arises in these adjustments. Computers have a limited space available in each memory location. The greatest number of significant digits that a computer can store in each memory space is called the *precision* of the computer. If we assume (for purposes of illustration only) that the calculation above had been performed by a computer which could deal only with three decimal digit mantissas (precision three), then the final answer would be truncated to 0.651×10^5. The error is not inconsiderable.

The most substantial errors, however, are likely to occur in *subtraction*. Suppose we wish to find the value of $33.99 - 33.12$. Expressed in exponential form and truncating to keep in precision three, this is

$$0.339 \times 10^2 - 0.331 \times 10^2 = 0.008 \times 10^2 = 0.8$$

The true value is 0.87 and the error is therefore 0.07. Expressed as a percentage of the correct figure of 0.87 this error is

$$\frac{0.07}{0.87} \times 100 = 8\% \text{ (approx.)}$$

A relative error of 8 % is high on any showing.

The rule for *computer multiplication* is 'multiply the mantissa and add the exponents':

$$(0.361 \times 10^2) \times (0.24 \times 10^3) = (0.361 \times 0.24) \times 10^5 = 0.08664 \times 10^5$$

The rule for *computer division* is 'divide the mantissa and subtract the exponents':

$$(0.08664 \times 10^5) \div (0.24 \times 10^3) = (0.08664 \div 0.24) \times 10^2 = 0.361 \times 10^2$$

As with addition and subtraction, the final results will be expressed in normalised exponential form and will have been truncated (if necessary) in the process. The answer to the multiplication above, if the computer worked in precision three (an unrealistically low precision used, we repeat, only as an example) would be 0.866×10^4. This would be printed out as 0.866 E4.

When numbers are processed in this manner in exponential form it is customary to call this process *floating point* or *real arithmetic*. Such numbers are also called floating point or real numbers. (This is a different use of the term 'real' to that used in Chapter 4. This is confusing. This book is often like that!)

3.4 Logarithms

Suppose we call the expression a^n (i.e., a raised to the power of n), L (say) – i.e., as an equation $a^n = L$.

Clearly, there is a relationship between a, n and L. Mathematically speaking, we say that each one is some *function* of the other two – i.e., it depends on them and varies with them. We already have L expressed as a function of a and n. We can express a in terms of n and L by taking the nth root of each side – i.e., $a = \sqrt[n]{L}$.

We may wish to express n in terms of a and L. We cannot do this in terms of the conventional signs and symbols used in elementary arithmetic. Certainly, however, n does depend on a and L – if they vary, so must it.

Exponents, Progressions, Present Value

We therefore define this relationship by using a new term 'log' (short for 'logarithm'), and say that if $a^n = L$, then
$$n = \log_a L$$
It is most important to realise that when we are stating this, we are merely saying it means that $a^n = L$ – this, and nothing more.

In the equation $n = \log_a L$, a is called the *base*. Only two values for this base are used much in practice:

1. $a = 10$. Such logarithms are called common logarithms.
2. $a = e$ where e is an irrational number approximately equal to 2.71828. Such logarithms are called natural or Napierian logarithms (confusingly, e is sometimes also referred to as 'the exponential').

For purposes of calculation, the base 10 is used. Ordinary or common logarithms tables always have this base, and it was usual to omit the 10 and write $\log L$ for $\log_{10} L$ (the past tense is used since calculators have superseded logarithm tables for such work; nevertheless manipulation of logarithms, as we shall see later, is important).

By definition, if $10^n = L$, then $\log L = n$, so that putting $n = 1$ we have $\log 10 = 1$. Similarly $\log 100 = 2$ (as $10^2 = 100$), and $\log 1000 = 3$, etc. Since $10^0 = 1$, we have $\log 1 = 0$ and putting $n = -1, -2$, etc. we have $\log 0.1 = -1$, $\log 0.01 = -2$ and so on.

To find the logarithm of a product we put $A = 10^a$ and $B = 10^b$. By definition, $a = \log A$ and $b = \log B$. The product $A \times B = 10^a \times 10^b = 10^{a+b}$, giving $\log (A \times B) = a + b = \log A + \log B$.

Expressed as a rule:

1. The logarithm of the product of two numbers is the sum of the logarithms of the two numbers.

By a similar approach, we can show that:

2. The logarithm of the division or quotient of two numbers is the logarithm of the numerator (top) minus the logarithm of the denominator, (bottom).
3. The logarithm of the power of a positive number is that power multiplied by the logarithm of the number.

Common logarithm tables are restricted to logarithms of numbers lying between 1 and 10. To determine the logarithm of a number lying outside this range presents no problems:

EXAMPLE 1
Find the logarithm of 6124.

$$\text{Since } 6124 = 10^3 \times 6.124,$$
$$\log 6124 = \log (10^3 \times 6.124) = \log 10^3 + \log 6.124$$
$$= 3 + \log 6.124$$

From the tables (Appendix Table A.1, p. 300) $\log 6.124 = 0.7871$. Hence $\log 6124 = 3.7871$. The entry in the logarithm tables (i.e., 0.7871) is the *mantissa* (and, strictly, the tables ought to be called mantissa tables!), and the digit 3 is called the *characteristic*. Clearly the number (6124) determines the mantissa and the characteristic varies with the position of the decimal point. Had the number been 612.4 the characteristic would have been 2: the mantissa is the same.

For numbers less than 1, the characteristic is shown rather differently. An example will make this clear.

EXAMPLE 2
Find the logarithm of 0.02222.

$$\text{Since } 0.02222 = 10^{-2} \times 2.222$$
$$\log 0.02222 = \log (10^{-2} \times 2.222) = \log 10^{-2} + \log 2.222$$
$$= -2 + \log 2.222$$

From the tables $\log 2.222 = 0.3468$. Hence $\log 0.02222 = -3 + 0.3468$, which is written $\bar{3}.3468$, emphasising that this means $-3 + 0.3468$ and not -3.3468.

Antilogarithm tables reverse the process – that is they convert the logarithm back into the original number.

EXAMPLE 3
Find $N = (1.022)^{10}$

$$\log N = \log (1.022)^{10}$$
$$= 10 \log (1.022)$$
$$= 10 \times 0.0094$$
$$= 0.094$$
$$\text{Hence } N = 1.242$$

(Note that 1.242 is the antilogarithm of 0.094; this can be read off directly from antilogarithm tables, if they are available.)

EXAMPLE 4
Find $N = \dfrac{329.7}{9.639}$

$$\log N = \log \frac{329.7}{9.639} = \log 329.7 - \log 9.639$$
$$\log 329.7 = 2.5181$$
$$\log 9.639 = 0.9840$$

Exponents, Progressions, Present Value 33

subtracting,

$$\log N = \underline{1.5341}$$

giving $N = 34.21$.

Frequently, as here, antilogarithm tables are not available. In this case it is necessary to locate the log figure in the body of the tables and then read off the number from this. It is important in this process *not to include the characteristic*. In Example 4, we locate 0.5341 in the body of the log tables and read off the number as 3.421. We can then use the characteristic of 1 to establish the decimal place in the number (two places), though generally commonsense will indicate it equally easily.

3.5 Arithmetic progressions

Consider the sequence of numbers 5, 9, 13, 17, 21, 25, 29. If we subtract from each number (or 'term', as it is usually called) the preceding one we obtain a sequence of 4s. Each term in the original sequence must therefore be the previous term together with a fixed number added on. In this case the fixed number is 4.

More generally, if we start with a number a (say) and continue to add a fixed number d (say), for n terms we have the sequence:

$$a, a+d, a+2d, a+3d \ldots a+(n-2)d + a+(n-1)d$$

A sequence such as this, in which each term is made by adding a fixed amount to the preceding term is called an *arithmetic* progression (AP).

If we call the last term l, the term before must be $l-d$, two before $l-2d$, and so on. Suppose we now wish to find the sum of the n terms of this AP. Call the sum S. We have:

$$S = a + (a+d) + (a+2d) + \ldots + (l-2d) + (l-d) + l, \text{ and}$$
$$S = l + (l-d) + (l-2d) + \ldots + (a+2d) + (a+d) + a$$

(if we rewrite the sequence, this time starting with the last term first). Adding these two, we have:

$$2S = (a+l) + (a+l) + \ldots + (a+l) + (a+l) + (a+l)$$

i.e., $\quad 2S = n(a+l)$

and since $l = a + (n-1)d$

this gives

$$S = \frac{n}{2}(2a + (n-1)d).$$

This is the formula for the sum to n terms of an AP.

EXAMPLE 5
Find the sum of the first 35 terms of the AP 50, 46, 42, 38, ...

Here $a = 50$, $d = -4$ and $n = 35$

The formula is
$$S = \frac{n}{2}(2a + (n-1)d)$$

$$S = \frac{35}{2}(100 + 34 \times (-4))$$

$$S = \frac{35}{2}(100 - 136) = \frac{35}{2}(-36) = -630$$

We have chosen to sum up so large a number of terms that the later negative ones outweigh the initial positive ones in the total sum.

3.6 Geometric progressions

Consider the sequence of numbers 486, −324, 216, −144, 96, −64. Each term in this sequence can be obtained by multiplying the preceding term by a fixed amount, in this case −2/3, called the *common ratio*.

More generally, if the first term is a and the common ratio is r and there are n terms, we have the sequence:

$$a, ar, ar^2, ar^3 \ldots ar^{n-1}$$

A sequence such as this in which each term consists of the preceding term multiplied by a common ratio, is called a *geometric* progression (GP). If we call the sum to n terms S, we have:

$$S = a + ar + ar^2 + ar^3 + \ldots ar^{n-2} + ar^{n-1}$$

If we multiply by r

$$rS = ar + ar^2 + ar^3 + \ldots + ar^{n-2} + ar^{n-1} + ar^n$$

Subtracting:

$(S - rS) = a - ar^n$ (as all the other terms cancel out), i.e.,

$S(1 - r) = a(1 - r^n)$, giving,

$$S = \frac{a(1-r^n)}{(1-r)}$$

This is the formula for the sum to n terms of a GP.

EXAMPLE 6
A business buys a machine costing £50,000 and estimated to have an economic life of 5 years and a scrap value of £750. The business prefers the fixed percentage method of depreciation which it believes to be more realistic than the straight line method. Approximately what annual rate of depreciation should be used?

Exponents, Progressions, Present Value 35

Let r be the required rate per year. The depreciation for the first year will be $50{,}000 \times r$ and the written down value (wdv) at the end of the first year will be $50{,}000 - 50{,}000 \times r = 50{,}000\,(1-r)$. After two years the wdv will be $50{,}000\,(1-r)^2$, and so on. After 5 years the wdv will be $50{,}000\,(1-r)^5$. We require this to be 750, so:

$$50{,}000\,(1-r)^5 = 750, \text{ giving}$$

$$(1-r)^5 = \frac{750}{50{,}000} = 0.015$$

$$(1-r) = \sqrt[5]{0.015} = 0.4317 \text{ (approx.)}$$

$$r = 0.5683, \text{ or (as a percentage) } 56.83\%$$

3.7 Infinite geometric progressions

Consider the GP $1, \frac{1}{5}, \frac{1}{25}, \frac{1}{125}, \frac{1}{525}, \ldots$

Here, $a = 1$, $r = \frac{1}{5}$ and the nth term is $1 \times \{\frac{1}{5}\}^{n-1} = \{\frac{1}{5}\}^{n-1}$

Obviously, as n increases, this will rapidly get less and less. When n is sufficiently large, it will become infinitesimally small.

More generally, provided that r is less than 1 in absolute value, and we take n sufficiently large, r^n can be neglected, and we can say that in the limit r^n will be 0.

Now the sum to n terms is

$$a\,\frac{(1-r^n)}{1-r}$$

It follows that the sum to infinity (usually written S_∞) is

$$S_\infty = \frac{a}{1-r}$$

EXAMPLE 7

Find the sum to infinity of the GP:

$$\frac{1}{1.1} + \frac{1}{(1.1)^2} + \frac{1}{(1.1)^3} + \ldots$$

Here $a = \dfrac{1}{1.1}$ and $r = \dfrac{1}{1.1}$

$$S_\infty = \frac{a}{1-r} = \frac{1}{1.1} \times \frac{1}{1-\frac{1}{1.1}} = \frac{1}{1.1} \times \frac{1}{\frac{1.1-1}{1.1}} = \frac{1}{1.1} \times \frac{1}{\frac{0.1}{1.1}} = \frac{1}{0.1} = 10$$

3.8 Compound interest

In transactions in which interest is being earned on a sum of money over a period of years the interest may:

1. Either be paid over separately at the end of each year or period (this is how interest is paid on government securities such as bonds), leaving the original sum intact. This is called *simple* interest.
2. Or be allowed to remain (be 'compounded') with the principal. At the end of each year or period the principal is increased by the amount of interest earned in the year. This is called *compound* interest.

Since interest, if it is paid over separately, is always available for investment, and can earn interest, it follows that in practice compound interest is more realistic and (if the phrase be allowed) of much more interest to us!

In Chapter 1 we saw that the amount to which a principal (P) will accumulate at a simple interest rate (r) for (t) years (or periods) is S where

$$S = P(1+rt)$$

If we put $t = 1$ in this formula, the sum after one year will be

$$S_1 \text{ (say)} = P(1+r)$$

In compound interest, we assume that the interest is allowed to remain with the principal, so that this sum S_1 is the new principal. Applying the same formula, but replacing P by S_1, the sum after two years will be:

$$S_2 = S_1(1+r) = P(1+r)(1+r) = P(1+r)^2$$

Repeating the process the sum after three years will be

$$S_3 = P(1+r)^3$$

and in general the sum S_n, to which the original principal will have grown after n years, is given by

$$S_n = P(1+r)^n$$

EXAMPLE 8

£7,500 is invested at a compound interest rate of 11 %. What is the sum after 3 years?

Here $P = 7,500$, $r = 0.11$ and $n = 3$

The formula

$S = P(1+r)^n$ gives $S = 7,500(1+0.11)^3$, so that
$S = 7,500(1.11)^3 = £10,257$

3.9 Frequency of conversion

In Chapter 1 we noted that all the simple interest calculations could have used time periods other than a year, provided the interest rate was expressed as so much per cent for that period. The same applies to compound interest – indeed, we used the term 'year or period' in the preceding section.

Exponents, Progressions, Present Value

Some confusion may occur when the conversion is made other than annually. The rate of interest may be expressed as so much per period – e.g., 8 % for each half-year. Since the principal sum for the second half-year will be the initial sum plus the first half-year's interest, the effective annual rate is higher. If we take the interest on an original sum of £100 at this 8 % per half-year rate, the sum at the end of the half year is 108, and at the end of the year is 108 (1 + 0.08) = 116.64.

Those institutions (finance houses, banks, credit card agencies, etc.) who lend money are required by law to show the *effective or equivalent annual rate*. In this example, it is 16.64 % and not the nominal 16 %. This effective annual rate is called the Annual Percentage Rate (APR).

EXAMPLE 9

What sum will an initial investment of £100,000 amount to at the end of a year if it is invested at 18 % p.a. compounded quarterly?

$$P = 100,000 \quad r = 0.045 \quad n = 4$$
$$S = P(1+r)^n = 100,000 (1 + 0.045)^4 = £119,252$$

Here we have a frequency of conversion, or rate of compounding of 4. The nominal rate of 18 % is equivalent to an annual rate of (approx.) 19.25 %. Banks calculate interest on overdraft amounts on a daily basis. When the frequency of conversion and the interest rates are high, the difference between nominal and effective rates can be considerable.

3.10 Present value

We arrived at the formula $S_n = P(1+r)^n$ for determining the amount (S_n) to which a principal (P) will accumulate at a compound interest rate (r) after n years (or, more generally, is compounded n times).

We can rewrite this as

$$P = \frac{S_n}{(1+r)^n} \quad \text{or} \quad P = \frac{1}{(1+r)^n} \times S_n$$

and then rephrase it to say that it now gives the present value (P) of a sum (S_n) given (or received) n years from now, at compound interest rate (r).

EXAMPLE 10

A business proposes to purchase some machinery. Under the terms of the contract it has to pay £25,000 now and £75,000 in 2 years' time. What is the equivalent present cash value, using a 12 % interest rate compounded annually?

For the £75,000 payment we have:

$$P = \frac{1}{(1+r)^n} \times S_n = \frac{1}{(1+0.12)^2} \times 75,000 = 0.7972 \times 75,000$$
$$= £59,790$$

The present value of £75,000 is therefore £59,790, and the total equivalent present cash value is £25,000 + £59,790 = £84,790.

In the above example, in which we found the value now of £75,000 due in 2 years' time, we multiplied by the factor 0.7972. This factor converts payments or receipts made in two years' time into their present value (using an interest rate of 12%). This factor is called the *discount factor*, or *present value factor*. Clearly the factor will vary with the number of years (or periods) and the interest rate. Tables of these factors are given at the end of this book (Table A.2, p. 301).

The equation $P = \dfrac{S_n}{(1+r)^n}$ itself can be more usefully written for Capital Investment Appraisal work as

$$S_n \times \frac{1}{(1+r)^n} = P$$

Using the standard abbreviations in this type of analysis, we write this as:

$$SUM \times DF = PV$$

where *DF* is the appropriate discount factor and *PV* is the present value. This is the basic equation for this sort of work.

3.11 Interpolation

In most practical compound interest and present value work the appropriate factor is obtained from sets of tables (Table A.2, p. 301). These tables may not provide the exact rates of figures required, and occasionally a precise answer is necessary. In such a case recourse may be made either to the formulae or to the method of *interpolation*. Since interpolation has a wider application than solely for interest and present value calculations we shall give a simple example here. First we use the formula, then interpolation from tables.

EXAMPLE 11

My wife Joan wants to be worth the magic figure of £100,000 by the end of the next ten years. Her present savings are (I believe!) £50,000. What annual interest rates must she seek?

The formula is:

$$S = P(1+r)^n$$

that is, $100{,}000 = 50{,}000(1+r)^{10}$

$$(1+r)^{10} = 2$$
$$(1+r) = \sqrt[10]{2}$$
$$r \simeq 7.17\%$$

Exponents, Progressions, Present Value

Alternatively we may use tables. We know that the number of years is ten and that the 'answer' we require to find in the body of the tables is 2. In practice the most detailed tables commercially available are five-'figure' tables.

A very short extract from such tables would show:

Years/Periods	Interest rate	
n	7%	8%
1		
2		
3		
.		
.		
.		
9		
10	1.96715	2.15892

Clearly the required rate is between 7 and 8 per cent. Diagramatically, the position is as shown on the left of Figure 3.1. The relevant part of this figure is shown in expanded form in the diagram on the right. As can be seen from Figure 3.1, the graph of the rate against the sum or accumulated value is not a straight line and this is always the case when interest is compounded. However for moderately low interest rates – and certainly for short distances – a 'straight line' assumption is not unrealistic. The diagram on the right makes this assumption.

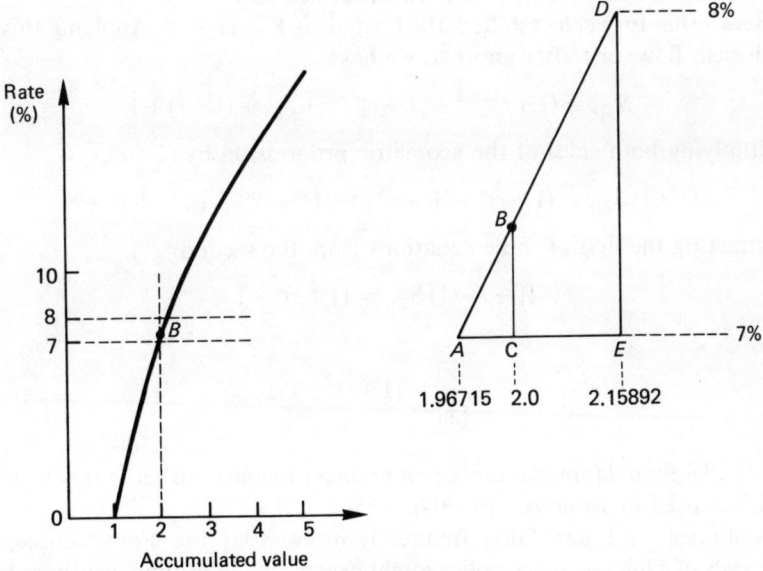

Fig. 3.1

In the diagram on the right of Figure 3.1, *ABC* and *ADE* are similar triangles (one common angle and each has a right angle). Hence, using ratios:

$$\frac{BC}{DE} = \frac{AC}{AE}$$

$$\frac{BC}{1} = \frac{2.0 - 1.96715}{2.15892 - 1.96715} = \frac{0.03285}{0.19177}$$

Giving $\qquad BC = 0.1713$

Therefore, assuming that the straight line assumption is justified, the 'exact' rate is $7 + 0.1713 = 7.1713 \%$.

3.12 Annuities

When *equal* payments (or receipts) of money are made annually (or, more generally, periodically) the term used is 'annuity.' The sum or accumulated value of an annuity for n years (or periods) at a rate of interest r is often denoted by $S_{\overline{n}|r}$. For an amount each time of (say) R the sum/accumulated value is therefore:

$$R \times S_{\overline{n}|r}$$

We now derive the general formula for an annuity. It is convenient to arrive at this formula for receipts or payments of 1 each time rather than R. This means the figures tie up with the usual tables (which are for unit cash flows) and also avoids the bother of including R in each line below!

Recall that for *each* cash flow the formula is $S = (1+r)^n$. Applying this for each cash flow, one after another, we have:

$$S_{\overline{n}|r} = (1+r)^{n-1} + (1+r)^{n-2} + \ldots + (1+r) + 1$$

Multiplying both sides of the geometric progression by $(1+r)$,

$$(1+r)S_{\overline{n}|r} = (1+r)^n + (1+r)^{n-1} + (1+r)^{n-2} + \ldots + (1+r)$$

Subtracting the first of these equations from the second:

$$(1+r-1)S_{\overline{n}|r} = (1+r)^n - 1$$

Giving

$$S_{\overline{n}|r} = \frac{(1+r)^n - 1}{r}$$

This is the formula for the sum of an ordinary annuity, and it is this formula which is used in Table A.3 (p. 302).

Annuities are found fairly frequently in everyday life. For example, the proceeds of a life insurance policy might be invested to produce yearly or half-yearly receipts of cash for life. Or to borrow money for a mortgage on a house

Exponents, Progressions, Present Value

you may agree to make monthly payments. Strictly this latter is an 'annuity certain' (that is, payments are unconditional, that is, they do not end on death).

In capital investment appraisal work, many of the expected cash receipts in the future will probably be estimated as being equal each year (at any rate after the initial build-up period). This is the 'easy' assumption to make; estimation becomes increasingly difficult the further we proceed into the future. It might therefore be argued that annuities (and the formulae for annuities) have an important place in such work. In practice this is not so. With the availability of micros and calculators repetitive calculations for each year's cash flow are easy. With this caveat, we nevertheless proceed.

For most 'annuities' in capital investment appraisal work the period is one year, and we shall generally assume this to be the case from now on. There are some special forms of annuities. The two main types are:

(a) Those for *perpetuity* – that is, receipts or payments go on forever.
(b) *Deferred* – that is, they start late.

We shall discuss these after arriving at the general formula for the present value of an ordinary annuity.

The present value of an annuity for n years at a rate of interest r is often denoted by $A_{\overline{n}|r}$.

Recall that for each unit cash flow the formula is

$$P = \frac{1}{(1+r)^n}$$

Applying this for each cash flow, one after another, we have:

$$A_{\overline{n}|r} = \frac{1}{(1+r)} + \frac{1}{(1+r)^2} + \frac{1}{(1+r)^3} + \ldots + \frac{1}{(1+r)^{n-1}} + \frac{1}{(1+r)^n}$$

This too, is a geometric progression; multiplying both sides of this equation by $\frac{1}{(1+r)}$ and subtracting one equation from the other we find that:

$$A_{\overline{n}|r} = \frac{[1-(1+r)^{-n}]}{r}$$

This is the *formula for the present value of an ordinary annuity*.

3.13 Annuities for perpetuity

We can now determine the value of an annuity for perpetuity – that is, one in which the equal sums of money continue forever.

In effect, this is the sum to infinity of a geometric progression with common ratios $\frac{1}{(1+r)}$. Using the formula $S_\infty = \frac{a}{1-r}$ and the 'r' and the 'a' in this formula both equal to our $\frac{1}{(1+r)}$, the present value of an annuity is $\frac{1}{r}$.

EXAMPLE 12

What is the present value of £1 received every year for ever at a 10 % interest rate?

Here $r = 0.1$. Hence the present value of an infinite stream of cash receipts of £1 is $1/0.1 = £10$.

EXAMPLE 13

The Bank of England decides to run a national lottery with two major prizes. The first prize is £1,000 per week for ever! The second is £1,000 per week for life.

Assuming a weekly interest rate of 17/52 % (not, of course, quite equal to 17 % p.a.) and an average life expectancy of 30 years, and ignoring the effects of inflation, what are the present values of the first and second prizes?

First prize

$$PV = 1{,}000 \times \frac{1}{r} = 1{,}000 \times \frac{1}{0.00327}$$

$$= £305{,}810 \text{ (approx.)}$$

Second prize

Here $r = 30 \times 52 = 1560$ which is too large a figure for most sets of annuity tables! Reverting to the formula:

$$A_{\overline{n}|r} = \frac{[1-(1+r)^{-n}]}{r}$$

and putting $r = 0.00327$ and $n = 1560$, we have as the present value:

$$\frac{1-(1.00327)^{-1560}}{0.00327}$$

We can calculate $(1.00327)^{-1560}$ by the use of either

(a) A suitable calculator.
(b) A micro.
(c) Logarithms (remembering that $\log a^b = b \log a$).

Whichever method we use gives the value of the second prize to be £303,045 (approx.), which is not all that much less than the value of perpetuity!

3.14 Deferred annuities

This is an annuity where the cash flows start after a period of years. We shall give one example of a deferred annuity calculation.

EXAMPLE 14

Building a new ski lift up Mount Cook in New Zealand will result in estimated net cash takings of $500,000 p.a. (assume all cash flows at year ends).

Exponents, Progressions, Present Value

Since the ski lift will not be operational for three years the first net cash receipt will be at the end of year 4. Take it that the ski lift will go on for another ten years and that the last net cash receipt will be at end of year 13. Ignoring tax and inflation, and assuming an interest rate of 20% per annum, what will be present value (PV) of the net cash takings from the operation of the ski lift?

PV of net cash flows years 4 to 13
$= PV$ of net cash flows years 1 to 13
Less PV of net cash flows years 1 to 3
$= 500{,}000 \, [A_{\overline{13}|0.20} - A_{\overline{3}|0.20}]$
$= 500{,}000 \, (4.533 - 2.106)$
$= \$1{,}213{,}500$

3.15 Capital investment appraisal

We now deal directly with the problems involved in capital investment. We start off with two examples of present-value calculations, given in the form of questions, and then consider the implications of our answers and their use in project appraisal work.

EXAMPLE 15
What is the present value of equal sums of £1,000 received at the end of each of three consecutive years? Money spare can command 10 per cent.
Using the basic equation (SUM × DF = PV) for each of the years we have:

Year	Sum £	×	Discount factor (10%)	=	Present value £
1	1,000		0.909		909
2	1,000		0.826		826
3	1,000		0.751		751

Total present value   £2,486

The present value is therefore £2,486.

It is easy to check that these two – i.e., £1,000 at the end of each year for three years and £2,486 now, at 10 per cent – are the same, by finding what they are each worth after three years (columns A and B in Table 3.1).

Alternatively, we can imagine that the £2,486 is lent for three years at 10 per cent, and the capital is returned, intact, at the end of the third year (column C). These are two ways of looking at it. Remembering that money spare is worth 10 per cent, the comparisons are shown in Table 3.1.

In this example, the sums of money are the same each year. When this is the case, it is possible to shorten the calculation dramatically. If we multiply the sum of £1,000 by the factor of 2.486 we arrive at the present value. This factor is the sum of the three individual years' factors and can be read off directly from the cumulative or annuity tables. Table A.3, p. 302 gives the figure as 2.487 (the difference is due to 'rounding').

TABLE 3.1

	A £1,000 for 3 years	B £2,486 now	C £2,486 lent
Sum at start	—	2,486	
Interest for first year		249	249
Sum received after one year	1,000	—	—
Total at end of one year	1,000	2,735	249
Interest for second year	100	274	{ 25 249
Sum received after two years	1,000	—	
Total at end of two years	2,100	3,009	523
Interest for third year	210	301	{ 52 249
Sum received after three years	1,000	—	2,486
Total at end of three years	£3,310	£3,310	£3,310

EXAMPLE 16

At what interest rates will it be worth while to invest £41,500 in a business project which is likely to bring in the following estimated sums of cash:

Year	Sum (£)
1	15,000
2	20,000
3	30,000

The interest rate is not known, so we have to proceed by the method of trial and error. If we try 22 and 24 per cent we have (using the factors given in Table A.2):

Year	Sum (£)	DF (22%)	Present value (£)	DF (24%)	Present value (£)
1	15,000	0.820	12,300	0.806	12,090
2	20,000	0.672	13,440	0.650	13,000
3	30,000	0.551	16,530	0.524	15,720
		Total present values	£42,270		£40,810

Exponents, Progressions, Present Value

What do these figures mean? The effect of multiplying a cash sum by a discount factor is to bring it back to its *present-day value*. A direct comparison can be made with the sum paid out on the project 'now'. (The advantage of choosing now as the time when the comparisons are to be made is fairly obvious – we are making the decision now!).

To summarise:

		At 22%	At 24%
	Cash in: PV of cash flows	42,270	40,810
Less	Cash out on investment	41,500	41,500
	Net present value (NPV)[a]	£ 770	£ (690)

[a] That is, PV of cash flows *less* investment cost.

Discount factors always fall as interest rates rise, so for rates lower than 22 per cent the NPV will always be positive. For rates higher than 24 per cent it will always be negative: the higher the rate, the greater will be this gap.

Around 23 per cent the NPV will be nil – i.e., the present value of the cash flows will be equal to the project investment sum. The 'exact' rate could be determined by interpolation (and this calculation gives the rate to be 23.07 per cent).

Example 15 showed that when the present value is precisely equal to the investment amount, the discount rate was that rate of return that the estimated cash flows were actually earning on the project. This rate of return is usually called the *internal rate of return* (IRR), or sometimes the DCF rate, though the term DCF is also used to describe any type of capital investment appraisal analysis that properly takes into account the time value of money. To make matters worse, the term used in the USA for internal rate of return is yield!

Example 15 also brings out an important point concerning the IRR calculations. In these comparisons, money spare was always reinvested at the 10 per cent rate – i.e., at the internal rate. This is the 'reinvestment assumption' and is important when comparing the validity of IRR against the present-value approach for making capital investment appraisals.

In real life, estimates of future cash flows cannot be determined with any accuracy and we need techniques to deal with risk and uncertainty. Cash flows, too, will typically go on for a much greater number of years than are shown in these examples. However, the principle is still the same. Tax and inflation will also both affect the values of the net cash flows.

IRR analysis has one apparent advantage over present-value analysis. Once the rate of return on a project has been established, a quick and accurate method for comparing projects is available, for clearly the higher the rate, the better the project. Present-value analysis takes no account of the *size of the investment in the project*. A large project will tie up much more money than a smaller project and therefore may not be preferable even though it shows a higher absolute NPV figure. The solution to this problem of how to rank projects under PV analysis is to use the profitability index (PI). This index is defined as:

$$PI = \frac{PV \text{ of cash benefits}}{Investment \text{ cost}}$$

The Profitability index really gives the benefit per £ (or other unit of money) invested. It is conceptually a more satisfactory method of ranking projects than the internal rate of return approach. This is because IRR analysis assumes that spare funds (if any) during the life of the project are reinvested at the project's *own internal rate of return*. This reinvestment assumption is not likely to be very realistic. The project's own rate of return should be fairly high for investment to be seriously considered (rates of around 20 to 25 per cent or higher are typical these days), and it is not likely that spare funds will be able to earn this amount.

In Example 15 the yearly cash inflows were equal. When this is the case it is sometimes more convenient to 'annualise' the initial investment cost and compare this annual cash cost with the yearly cash benefits, rather than the other way round (as in conventional IRR or PV analysis). If we assume, now, that the investment amount is £2,100 and that a 10 per cent criterion rate is appropriate, then the annualised cost of £2,100 over a 3-year period is simply

$$\frac{2,100}{2.487} = £844.39 \text{ p.a.}$$

(again using the same factor given in Table A.3).

Each year is now strictly comparable. Each year the benefit is £1,000 and the equivalent cost is just over £844. *Annualised cost* or *annualised equivalent*, gives the same go-ahead signal as conventional NPV analysis would (since NPV = £ (2,486 − 2,100), which is positive).

A number of corporations, including British Telecom, use annualised cost, on the grounds that the concept is simpler to apply than the usual DCF techniques.

3.16 Investment appraisal example

An engineering workshop is proposing to invest in a new robot costing £38,000. Purchase of this robot will enable the engineering company to dispose of its existing machine for an estimated sum of £12,000. The life of the robot is expected to be some 6 years. Savings will be mainly labour savings from using one man less and will amount, in all, to approximately £9,150 per year. No redundancy payments will be incurred. Taxable profits are being made. No regional development grants are available. At a 20 per cent cut-off rate, is investment in the new robot worthwhile?

Capital allowances consist of a writing down allowance of 25% on the reducing balance.

Exponents, Progressions, Present Value

Corporation tax is at 35%.

Capital allowances* are therefore:

Year	25% of	Capital allowance	Reducing balance
1	26,000	6,500	19,500
2	19,500	4,875	14,625
3	14,625	3,656	10,969
4	10,969	2,742	8,227
5	8,227	2,057	6,170
6	6,170	1,543	4,627

*Rounding up for intermediate values: a common practice for tax calculations

At the end of year 6 it is assumed that the machine will be sold for an estimated salvage value of £4,000 so that a balancing allowance of (4,627 − 4,000 =) 627 will be given in year 7. A suitable layout for the calculation is:

Year	Cash flow from savings (a)	Tax on savings (b)	Capital allowances (CAs) (c)	Tax on CAs (d)	Net cash flows (e)	DF (20%)	PV of cash flows
1	9,150		6,500		9,150	0.833	7,622
2	9,150	3,203	4,875	2,275	8,222	0.694	5,706
3	9,150	3,203	3,656	1,706	7,653	0.579	4,431
4	9,150	3,203	2,742	1,280	7,227	0.482	3,483
5	9,150	3,203	2,057	960	6,907	0.402	2,777
6	9,150	3,203	1,543	720	6,667	0.335	2,233
7		3,203	627	540	(2,663)	0.279	(743)
8				219	219	0.233	51

Total PV of cash flows   25,560

Note that col. (e) = col. (a) *less* col. (b) *plus* col. (d).

Net investment in robot is:

	Robot	38,000
Less	sale old machine	12,000
		26,000
Less	scrap value robot (est.) after 6 years 4,000 × 0.335	1,340
	Net investment	24,660

Since the present value of the cash flows is greater than the net investment, replacing the old machine by a robot is indicated.

3.17 Exponential smoothing

In Chapter 11 we discuss the use of regression analysis as a general tool for forecasting sales. Here we consider a technique of proven use for forecasting which is specifically short term in nature. Stock levels and production and consumption figures are typical applications.

Suppose we have a relatively stable business with – we assume initially – no obvious long-term trends or seasonal or other cyclical factors. The best guide to future levels of some figure such as inventory may well be the immediate past figures. By way of example let us say that the figures for the past three months have been:

$$\begin{array}{ll} \text{January} & 48{,}000 \\ \text{February} & 49{,}000 \\ \text{March} & 53{,}000 \end{array}$$

If we attach equal importance to all three, our best overall estimate for April will be:

$$\tfrac{1}{3} \times (48{,}000 + 49{,}000 + 53{,}000) = 50{,}000$$

However, we might well attach more weight to recent figures. In that case we might give as the best estimate some figure such as:

$$\tfrac{1}{6} \times (1 \times 48{,}000 + 2 \times 49{,}000 + 3 \times 53{,}000) = 50{,}833$$

The distinguishing feature of exponential smoothing is that the ratio which the weight for any one month bears to the weight for the previous month is a *constant factor*, and this relationship is true for all the weights, regardless of which month is chosen. So that had we chosen weights of 1, 2, and 4 (say) instead of 1, 2, 3 we would have had a primitive form of exponential weighting, the constant ratio in this case being $\tfrac{1}{2}$. More than just three sets of past figures will be used in practice: indeed in theory the figures go back infinitely far.

However long the sequences of past figures may be, the estimation of the next figure by exponential smoothing is remarkably easy. As our starting point, assume that our estimate for (say) this month's stock of a product is 5,900 units. Later, the actual figure becomes available. If this is (say) 6,100, there is an error in our estimate of 200. The exponentially weighted estimate for the *following* month's stock is simply:

$$5{,}900 + a(200)$$

where a is a constant, which typically lies between 0.1 and 0.3. If we take it to be 0.2, the estimate for next month's stock is:

$$5{,}900 + 0.2(200) = 5{,}940$$

The next month's actual figure is unlikely to be exactly this. If it turns out to be

Exponents, Progressions, Present Value

5,980, there is an error of 40, and the calculations of the estimate for the month after that will be:

$$5{,}940 + 0.2(40) = 5{,}948, \text{ and so on}$$

To determine the value of the constant which should be used for any particular case, we try a series of values between 0.1 and 0.3, and apply these to past figures to see which gives estimates that agree best with actual past results. Whatever value of a is finally chosen, it will remain constant for immediate future calculations. The interesting thing in these calculations of the next months' estimates is that the figures for the earlier months enter only indirectly into the computation – through the estimate for the month.

The basic rule therefore is this. The estimate for the next month is the current month's estimate, plus a constant multiplied by the error in the estimate for the current month.

Further modifications to the basic formula take account of:

1. A known overall trend (e.g., the general inflationary rise in prices or sales over the years).
2. Known seasonal or cyclical fluctuations (e.g., seasonal variations in the sales of ice cream, weed killers, camping equipment, etc.; weekly variations in airline ticket sales, due to high weekend bookings, etc.).

The proof of the basic rule is simple. Suppose the actual figures for last month were x_t. Using the same notation the actual figures for the month before (that is $(t-1)$) will be x_{t-1}, and so on. If the weight for the last month is a_1, for the previous month is a_2, and so on, the estimate m_t for the current month $(t+1)$ will be

$$m_t = a_1 x_t + a_2 x_{t-1} + a_3 x_{t-2} + \ldots$$

The estimate for the last month, m_{t-1} (using the same notation) will have been arrived at by a similar computation.

$$m_{t-1} = a_1 x_{t-1} + a_2 x_{t-2} + a_3 x_{t-3} + \ldots$$

Since the weighting is exponential,

$$\frac{a_2}{a_1} = \frac{a_3}{a_2} = \frac{a_4}{a_3} \ldots \text{ etc.} = b \text{ (say)}$$

so that, cross-multiplying

$$a_2 = a_1 b$$
$$a_3 = a_2 b = a_1 b^2, \text{ and so on}$$

we can now rewrite our estimates as

$$m_t = a_1 x_t + a_1 b x_{t-1} + a_1 b^2 x_{t-2} + \ldots$$
$$m_{t-1} = a_1 x_{t-1} + a_1 b x_{t-2} + a_1 b^2 x_{t-3} + \ldots$$

Multiplying the second estimate by b gives

$$bm_{t-1} = a_1 b x_{t-1} + a_1 b^2 x_{t-2} + a_1 b^3 x_{t-3} + \ldots$$

Subtracting this last equation from the first estimate

$$m_t - bm_{t-1} = a_1 x_t$$

Now, whatever the type of weighting may be, all the coefficients, or weights, must add up to 1, i.e.:

$$a_1 + a_2 + a_3 + a_4 + \ldots = 1$$

Substituting the value $a_1 b$ for a_2, $a_1 b^2$ for a_3, and so on, this becomes

$$a_1 + a_1 b + a_1 b^2 + a_1 b^3 + \ldots = 1, \text{ i.e.:}$$
$$a_1 (1 + b + b^2 + b^3 + \ldots) = 1$$

If we have a large number of past figures available, this series will have a large number of terms, in fact nearly an infinite number. The sum to infinity of the geometric progression in brackets is

$$\frac{1}{1-b}$$

Hence

$$a_1 \times \frac{1}{1-b} = 1$$

$$a_1 = 1 - b, \text{ or } b = 1 - a_1$$

The equation $m_t - bm_{t-1} = a_1 x_t$, becomes (after substituting in and rearranging):

$$m_t = m_{t-1} + (1-b)(x_t - m_{t-1})$$

or in words, the estimate for the *next* month is the current month's estimate plus a constant $(1-b)$ times the error in the estimate for the current month.

3.18 Exercises

Worked answeres are provided in Appendix C to question(s) marked with an asterisk.

1. Simplify:

$$a^7 \times a^5 \qquad \frac{a^2 \times a^4}{a^9} \qquad \left\{\frac{1}{a^2}\right\}^5 \qquad (1.12)^8 (1.12)^9$$

2. Simplify:

$$b^{1/3} \times b^{1/2} \qquad (b^{2/3})^{-6}$$

3. Using logarithms, find:

$$3.14 \times 0.7357 \qquad 263.5 \div 7.25 \qquad \frac{8.37 \times 0.867}{0.0911}$$

Exponents, Progressions, Present Value

4. Using logarithms, find:

$$\sqrt[4]{468} \qquad \sqrt{\frac{3.187 \times 0.1508}{215}} \qquad \sqrt{\frac{785}{4 \times 3.14}} \qquad \sqrt[3]{0.00563}$$

5. The economic order size (*n*) for a batch of manufactured units is given by the formula:

$$n = \sqrt{\frac{2ds}{rw}}$$

where *w* = factory cost per unit
d = estimated yearly demand for the product in units
r = stock carrying cost, expressed as a fraction of average stock value
s = set-up costs – i.e., the special costs involved in producing a batch.

What is the economic order size if the special cost of each batch is £50, the annual requirements are 40,000 units, the stock-carrying cost is 0.2, and the factory cost is £5 per unit?
(Note that this formula can also be used for determining the economic quantity to purchase. The formula for the economic order quantity is prooved in Chapter 9.)

6. Find the 15th term and the sum of the first 15 terms of the arithmetic progression 54, 50, 46, 42 . . .
7. Find the sum of the first 10 terms of the geometric progression 4, 8, 16, 32, 64 . . .
8. Find the sum of the first 12 terms of the geometric progression 4, −8, 16, −32, 64 . . .
9. A machine which costs £24,000 is estimated to have a scrap value of £1,800 after 6 years. Assuming that depreciation each year is to be based on a constant percentage of the written-down value at the end of the preceding year, what must this percentage be?
10. A lathe costs £20,000. Depreciation is calculated monthly at the rate of 5% of the written-down value at the beginning of the month. What is the depreciated book value after 24 months?
11. A mortgage of £8,000 is authorised by a bank on 1 January 1986. Repayment is to be made in twenty equal instalments each due at the end of a year with the first due on 31 December 1986. Interest (compound) is 12% per annum. What is the amount of each instalment?
12. Two possible alternative projects each cost £50,000. Their estimated cash inflows are as follows:

Year	Project A	Project B
1	20,000	Nil
2	20,000	10,000
3	20,000	20,000
4	20,000	60,000

No other cash flows are to be considered. Ignore tax. Compute the net present values for Project A and for Project B at rates of 10%, 20% and 30%.

Plot the points for each project on a graph with the rates along the x axis and net present value along the y axis. From the graph how can you read off the approximate internal rate of return (IRR) of each project?

13. A machine cost £31,500 and its depreciated book value at the end of 8 years is £6,500.

    If the constant percentage (on written-down book value) method is used for depreciation, find:

    (a) this percentage rate, and
    (b) the written-down book value at the end of the fifth year.

14. A surface testing machine drops a small steel ball from a standard height of 135 cm. On the test surface in question the ball rebounds each time to two-thirds of the height from which it falls. Determine:

    (a) How far it will rise on the sixth rebound.
    (b) How far, in total, it will have travelled when it hits the surface for the eighth time.
    (c) How far, in total, it will have travelled before it finally comes to rest on the surface.

15. (a) Your company has decided to set up a fund for its employees with an initial payment of £2,750 which is compounded six monthly over a four-year period at 3.5% per six months.

    Required:

    (i) Calculate the size of the fund to two decimal places at the end of the four years.
    (ii) Calculate the effective annual interest rate, to two decimal places.

    (b) The company has purchased a piece of equipment for its production department at a cost of £37,500 on 1 April 1984. It is anticipated that this piece of equipment will be replaced after five years of use on 1 April 1989. The equipment is purchased with a five-year loan, which is compounded annually at 12%.

Exponents, Progressions, Present Value   53

     Required:

       (i) Determine the size of the equal annual payments.

      (ii) Display a table which shows the amount outstanding and interest for each year of the loan.

 (c) If in (b) the £37,500 debt is compounded annually at 12% and is discharged on 1 April 1989 by using a Sinking Fund Method, under which five equal annual deposits are made starting on 1 April 1984 into the fund paying 8% annually,

      (i) determine the size of the equal annual deposits in the sinking fund, and

     (ii) display a table which demonstrates the growth of the loan and the sinking fund.

Extract from net present value tables

Period	8% Discount factor	12% Discount factor
0	1.0000	1.0000
1	0.9259	0.8929
2	0.8573	0.7972
3	0.7938	0.7118
4	0.7350	0.6355
5	0.6806	0.5674

(*The Association of Certified Accountants*, December 1984)

16. A finance director estimates that his company will have to spend £0.25 million on new machinery in two years from now. Two alternative methods of providing the money are being considered, both assuming an annual rate of interest of 10%.

 (a) A single sum of money, £A, to be set aside and invested now, with interest compounded every six months. How much should this single sum be, and what is the *effective* annual rate of interest?

 (b) £B to be put into a reserve fund every six months, starting now. If interest is compounded every six months, what should £B be, in order that the £0.25 million will be available in two years from now?

(*Institute of Cost and Management Accountants*, May 1984)

17. In the near future a company has to make a decision about its computer, C, which has a current market value of £15,000. There are three possibilities:

    (i) sell C and buy a new computer costing £75,000;
   (ii) overhaul and upgrade C;
  (iii) continue with C as at present.

Relevant data on these decisions are given below:

Decision	Initial outlay	Economic life	Re-sale value after 5 years	Annual service contract plus operating cost (*payable annually in advance*)
	£	years	£	£
(i)	75,000	5	10,000	20,000
(ii)	25,000	5	10,000	27,000
(iii)	0	5	0	32,000

Assume the appropriate rate of interest to be 12% and ignore taxation.

You are required, using the concept of net present value, to find which decision would be in the best financial interest of the company, stating why, and including any reservations or assumptions.
(*Institute of Cost and Management Accountants*, May 1986)

18. (a) Your board is considering two alternative investments. The life of both is 5 years and your company can borrow for either project at 15% per annum. Neither project will have any value at the end of the 5 years.
One investment will cost £250,000 and is expected to generate a cash flow over its life of £140,000, £120,000, £110,000, £100,000 and £90,000 in successive years. The other will cost £150,000 and will produce an annual cash flow of £70,000.
Which project would you recommend to the board?
(b) Explain in simple terms, by way of a formal report to the board members, the basis of the method of appraisal you employ in reaching your decision.
Indicate any limitations of the technique you have used in assessing the profitability of these projects which you consider may affect significantly the reliability of your recommendation to the board.
(*Institute of Chartered Accountants in England and Wales*, November 1985)

19. (a) The discounted cash flow method of investment appraisal has been defined as 'the basis of an evaluation procedure applied to a capital project to estimate its merit as an investment'.
Explain and discuss this definition.

Exponents, Progressions, Present Value 55

    (b) Use the discounted cash flow method to appraise a proposed project for which you are given the following estimated figures:

Initial capital cost at the beginning of the first year	£40,000
Cost of borrowing	15% p.a.
Cash flow in each of the years 1 to 5	£15,000
	£12,000
	£10,000
	£10,000
	£ 8,000
Value at end of year 5	£ 5,000

    (c) What are the key factors to be considered in assessing the reliability of the discounted cash flow method of appraisal?

(*Institute of Chartered Accountants in England and Wales*, May 1985)

20. A company is considering investing in a new manufacturing facility with the following characteristics:

    A initial investment £350,000 – scrap value Nil;
    B expected life 10 years;
    C sales volume 20,000 units per year;
    D selling price £20 per unit;
    E variable direct costs £15 per unit;
    F fixed costs excluding depreciation £25,000 per year.

The project shows an internal rate of return (IRR) of 17%. The managing director is concerned about the viability of the investment as the return is close to the company's hurdle rate of 15%. He has requested a sensitivity analysis.

You are required to:

    (a) recalculate the internal rate of return (IRR) assuming each of the characteristics A to F above, in isolation, varies adversely by 10%;
    (b) advise the managing director of the most vulnerable area likely to prevent the project meeting the company's hurdle rate;
    (c) explain what further work might be undertaken to improve the value of the sensitivity analysis undertaken in (a);
    (d) re-evaluate the situation if another company, already manufacturing a similar product, offered to supply the units at £18 each – this would reduce the investment required to £25,000 and the fixed costs to £10,000.

(*Institute of Cost and Management Accountants*, May 1986)

21.* A local government department is proposing to purchase a mechanical hedge-cutting machine to replace existing manual work. Manpower savings are greatest when using a machine with the largest cutting blade but the initial purchase cost increases more than proportionately to the width of the cut achieved.

The relevant cost details are:

Cutter size	12"	18"	24"	30"	36"
	£000	£000	£000	£000	£000
Purchase cost	20	30	45	70	100
Annual operating cost savings after deduction of depreciation	8	10	11	8	6

Depreciation – straight line method over ten years
Taxation – assume no tax allowances or tax payments.

The local council will not authorise any proposed investment which does not yield a discounted cash flow (DCF) return of 15% per annum. All proposals meeting this target are further subject to an overall total investment limit.

You are required to:

(a) calculate the discounted cash flow (DCF) yield for each of the five machines;
(b) state the largest size machine which could be purchased and meet the 15% investment criterion;
(c) list *five* factors that would influence the decision on this particular investment if there are more investment projects than funds available; at least *two* should be in favour of proceeding with it and at least *two* supporting deferment.

(*Institute of Cost and Management Accountants*, May 1985)

22. A company with a 12% time value rate for its capital investment appraisals, and limited investment funds is evaluating the desirability of several investment proposals:

Project	Initial investment £	Life (in years)	Constant end of year cash inflow £
A	60,000	2	37,520
B	40,000	5	13,200
C	40,000	3	20,000
D	20,000	9	4,000
E	60,000	10	13,200

Questions

(a) Rank each project according to its DCF rate (internal rate of return), net present value, and profitability index.

(b) Explain the cause of any difference in ranking in these three figures.
(c) What projects should be selected if only £100,000 is available for investment? Explain your choice of projects.
(d) Give *three* reasons (either theoretical or practical) for favouring DCF(IRR) compared with NPV, and *three* reasons for preferring NPV to DCF. *Explain* your reasons in some detail.

23. Your marketing director, a lady known for her flair in talking anybody into anything (but a person not particularly interested in finance or figures, which subjects she regards as at best boring and irrelevant) calls you into her office and tells you this:

'I want to buy my daughter a newish F type Jag as a 21st birthday present. It only costs £24,000 and Mr Brown the finance director there says we can pay for it over 24 months by a loan at a "low low" interest rate of only 10%. Payments are calculated as follows (here she pulls a dirty envelope out of her pocket and reads):

Annual interest 10% of 24,000 = £2,400 p.a.
Total due = Principal + 2 years' interest
$$= 24,000 + 2 \times 2,400 = £28,800$$
Monthly repayments $= \dfrac{28,800}{24} = £1,200$ per month

Isn't that good!'
(a) Calculate the effective rate of interest per annum on this loan.
(b) Draft a reply to your marketing director in one paragraph explaining in very simple terms why the effective rate is nearly twice the 'low low 10%' rate which she thinks it is.

Chapter 4
Sets and Relations

4.1 Defining a set

Any collection of objects is a *set*. The objects are called the *elements* or *members* of the set. It is normal to use upper case letters (i.e. capitals) to denote the sets and lower case letters for the elements of the sets. If an element p is a member of a set A, we write:

$$p \in A$$

and if p is not a member we write:

$$p \notin A$$

A set is completely determined or defined when *all its members are specified*. This is called the principle of extension. There are two ways in which a particular set can be defined. The first is by specifying its *members*:

$$B = \{2, 4, 6\}$$

Note that the members are enclosed in brackets and separated by commas. The second method is to state the *property characterising the elements*. We could therefore express this set B equivalently by writing:

$$B = \{x : x \text{ is a positive even number less than } 8\}$$

where the colon is to be read as 'such that'.

The latter method is rather cumbersome for our example and it is usual when there are only a few elements to list them. In some cases this might not be possible, particularly if the set is infinite – that is it contains an infinite number of elements. Consider the set:

$$C = \{x : x \text{ is an even number}\}$$

This contains an infinite number of elements. Sometimes although we cannot and do not list all the members we can nevertheless make clear what set we mean by putting down just a few. In our case we could write:

$$C = \{2, 4, 6, 8, \ldots\}$$

it is clear now that we mean the infinite set of even numbers.

Sets and Relations

It is not necessary for the elements of a set to be numbers, though this is easily the most common in practice. The collection of all the known planets is a set, the letters in the Cyrillic alphabet form another set, and so do all the people born in the world in 1984. The latter set would have to be specified by its property since although in theory the members of this set could be determined (and hence the set expressed by listing its members) in practice this would be so difficult as to be impossible. The *principle of abstraction* states that a set can be *precisely* determined by stating the *property common to all its elements*, as in this example.

4.2 Empty set, universal set, subsets

The set with no elements is called the *empty or null set*. It is written $\varnothing$. It follows from the principle of extension that there can be only one null or empty set.

At the other end of the spectrum we frequently wish to consider a set which contains *all possible elements or elements* similar to those under consideration. In the preceding section, the universal set corresponding to $B = \{2, 4, 6\}$ would be the universal set $C = \{2, 4, 6 \ldots\}$. The universal set is written U. Note that though there is only one empty set, there are many universal sets. The appropriate universal set is determined by the set under investigation. In plane geometry when the members are particular points, the universal set will be all the points on the plane. When dealing with individual businesses, the universal set may be all businesses (either in the world or in that particular country, as appropriate).

If all the members or elements of one set are members of another then the first set is said to be a *subset* of the second set. Still staying with the example above, $B = \{2, 4, 6\}$ is clearly a subset of $C = \{2, 4, 6, \ldots\}$, and we write:

$$B \subset C \text{ (or } C \supset B\text{)}$$

if B were not a subset of C we would write:

$$B \not\subset C$$

A given set may have a number of subsets. For instance the set $B = \{2, 4, 6\}$ has the following subsets:

$$\varnothing \ \{2\} \ \{4\} \ \{6\} \ \{2,4\} \ \{4,6\} \ \{2,6\} \ \{2,4,6\}$$

It is usual, as in this example, to regard as one of its subsets *the set itself* (here $\{2, 4, 6\}$). Strictly, we should use the symbols $\subseteq$ and $\not\subseteq$. If the set itself is excluded from its subsets then the term *proper* subset is appropriate, and of course $\subset$ and $\not\subset$ are then absolutely correct.

The set which consists of all the subsets of a set (i.e. for B all the sets immediately above) is called the *power set*. For a set, such as B, containing 3 elements, the number of subsets in its power set is $8 = 2^3$. For a set containing m members or elements, its power set will contain 2^m elements. This follows because for each member of the set there are two possibilities – either to be included or not to be included in any particular subset. The total number of possibilities (i.e., the number of subsets) is therefore 2^m.

4.3 Venn diagrams

At this stage it is helpful to introduce the concept of Venn diagrams. These are a particularly useful form of visual presentation for sets. Typically the appropriate universal set is shown as a rectangle. Within that rectangle a particular set is represented by a circle or disk. In Figure 4.1 we show the three basic alternatives when we are considering two sets D and E. In (a) the sets do not intersect – i.e., they have no members in common. Such sets are said to be *disjoint*. An example of this would be if D was the set $\{1, 2\}$ and E the set $\{10, 11\}$. In (b) one set is entirely contained within an other. Clearly E is a subset of D – i.e., $E \subset D$. An example would be if D was the set $\{6, 7, 8\}$ and E the set $\{6, 7\}$.

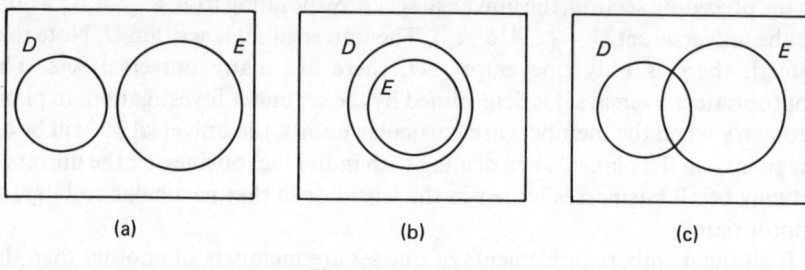

Fig. 4.1 (a), (b), (c)

In (c) the sets do intersect – i.e., they have common members – but they also have their own members or elements which are not members of the other set. An example would be if D was the set $\{2, 4, 6, 8\}$ and E the set $\{6, 8, 10, 12\}$. It is important to note that in all Venn diagram work the size of the areas in the circles or discs and the number of members or elements are not directly proportional. The areas only *represent* the sets diagrammatically.

4.4 Union, intersection, complement, difference

In Figure 4.2 we show two identical illustrations of two sets F and G which intersect.

The *union* of these sets is defined as that set whose members belong *either* to F or to G or to F and G. This is shown by the shaded area in Figure 4.2(a), and it is written $F \cup G$.

The *intersection* of these sets is defined as that set whose members belong to both F and G. It is shown by the shaded area in Figure 4.2(b) and it is written $F \cap G$.

Sets and Relations

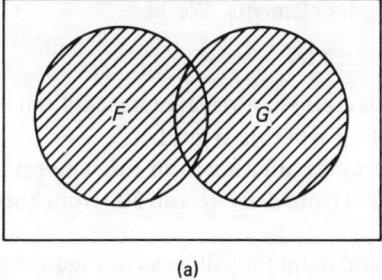 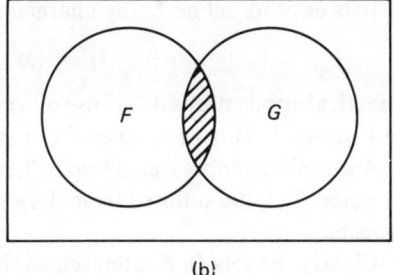

Fig. 4.2 (a), (b)

It is obvious that the layout of the sets in Figure 4.2 is similar to that in Figure 4.1(c). This layout is only one (though the most interesting one) of the three typical layouts for two sets. Both 'union' and 'intersection' are general terms, and apply to all three layouts.

In Figure 4.1(a) the union of D and E will be the areas of both discs. The intersection will be the null set. In Figure 4.1(b) the union will be the set D, and the intersection the set E.

The *difference* of F and G is defined as the set of elements whose members belong to F *but not to* G. In Figure 4.2(b) it is the area within F to the left of the shaded area (in Figure 4.1(a) the difference of D and E is the set D, and in Figure 4.1(b) it is the area of D without the disc E). The term 'relative complement of G with respect to F', means the same things as the difference of F and G. The difference of F and G is read as 'F minus G' and written as $F - G$ or $F \backslash G$.

It will be recalled that all the sets under consideration in any particular situation are all subsets of a universal set U. If we imagine that the set F is expanded to become the universal set U, we have the situation shown in Figure 4.3. The difference of U and G is the shaded area.

This shaded area is now referred to as the 'absolute complement' (or, more usually, just the *complement*) of G. It is written G^c (or, alternatively, G' or $\bar{G}$).

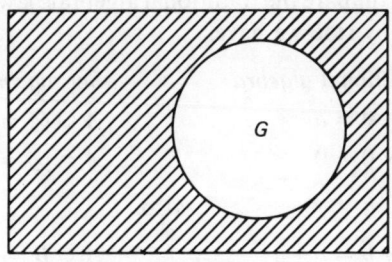

Fig. 4.3

It is easy to define G^c by characterising its elements. We have
$$G^c = \{x: x \in U, x \notin G\}$$
Practical applications of the use of Venn diagrams as a visual representation of sets abound. This one, taken from sampling work, is typical.

A sampling report stated that in 1,800 samples suffering from either defect D or defect E, 1,400 suffered from defect D, 800 from defect E and 300 from both defects.

Clearly the sets D, E intersect, so the Venn diagram will be as in Figure 4.4.

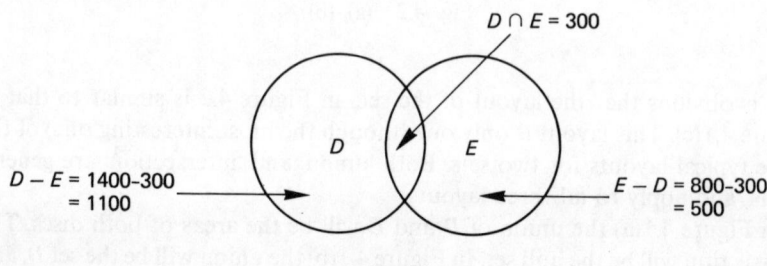

Fig. 4.4

The total, $D \cup E$, should be, we are told, 1,800, but from the diagram it is $1,100 + 300 + 500 = 1,900$. The data therefore are inconsistent. In a simple case such as this, with only two sets involved, it is fairly easy to work this out by commonsense methods. When there are three or four sets the Venn diagram approach is easy and helpful.

4.5 Algebra of sets and duality

In Chapter 1 we discussed 'operations' (such as $+$, $-$, $\times$, and their use in 'conventional' algebra). In set theory work the symbols $\cup$, $\cap$ and c (the complement), can be seen as operators acting on the sets. Thus the symbol $\cup$ acts on the sets A and B to give another set $(A \cup B)$ – the *union* of A and B. And, much as there are algebraic laws which relate to traditional operators, so in set theory we have laws for the algebra of sets. These laws are set out in Table 4.1 below.

It is interesting to compare the traditional algebraic laws with these set laws. If we put them side by side we have:

Traditional algebra	*Set algebra*
$a + 0 = a$	$A \cup \emptyset = A$
$a \times 0 = 0$	$A \cap \emptyset = \emptyset$

also,

$a + b = b + a$	$A \cup B = B \cup A$
$a \times b = b \times a$	$A \cap B = B \cap A$

Sets and Relations

TABLE 4.1

	Idempotent laws
$A \cup A = A$	$A \cap A = A$
	Identity laws
$A \cup \emptyset = A$	$A \cap U = A$
$A \cup U = U$	$A \cap \emptyset = \emptyset$
	Commutative laws
$A \cup B = B \cup A$	$A \cap B = B \cap A$
	Associative laws
$(A \cup B) \cup C = A \cup (B \cup C)$	$(A \cap B) \cap C = A \cap (B \cap C)$
	Distributive laws
$A \cup (B \cap C) = (A \cup B) \cap (A \cup C)$	$A \cap (B \cup C) = (A \cap B) \cup (A \cap C)$
	Complement laws
$A \cup A^c = U$	$A \cap A^c = \emptyset$
$U^c = \emptyset$	$\emptyset^c = U$

which seems to imply a similarity between 0 and $\emptyset$, $+$ and $\cup$, and $\times$ and $\cap$. However:

$$a + a = 2a \quad \text{whereas} \quad A \cup A = A$$
$$a \times a = a^2 \quad\quad\quad\quad\quad A \cap A = A$$

So the similarity is limited.

In fact set theory algebra has a very close similarity with the algebra of propositions (truth statements, etc.). It also forms the basis of Boolean algebra (a rather specialised algebra which has important applications in switching circuits and logic gates for computer hardware); indeed, this set theory approach may be regarded as providing the groundwork for an appreciation of computer algebra.

All the laws in Table 4.1 are arranged in two columns. The reason for this arrangement is that all the laws in each column can be seen to be identical if we make the following substitutions.

$$\cap \text{ for } \cup$$
$$\emptyset \text{ for } U$$

or vice-versa. This ability to replace or substitute in each law one operator by another is called the *principle of duality*, and the replacement operator is called the dual of the original operator. In set algebra, the principle of duality states that if E (say) is an identity then its dual E^x (say) is also an identity where E^x is obtained from E by substituting for each operator its dual. This principle of duality extends beyond set theory into, for example, Boolean algebra. It has a parallel in linear programming, discussed later in Chapter 13.

4.6 Ordered pairs, products

In general, the order in which the members of any set are specified is immaterial. Thus the sets $\{2, 4\}$ and $\{4, 2\}$ are the same. However in certain circumstances, when we have two elements (as in the sets above) we may wish to take the *order* into account. Clearly we need some way of saying this. We therefore define an ordered pair as consisting of two elements (say) a, b in *that* sequence. Such an ordered pair would be written (a, b). Another ordered pair (l, m) would be equal to it only if $a = l$ and $b = m$.

Ordered pairs have a particular use in *products*. We define the product of two sets A and B as the set of all ordered pairs (a, b) where a is a member of A and b is a member of B. We write the product as:

$$A \times B = \{(a, b): a \in A, b \in B\}$$

and the abbreviation A^2 is used for $A \times A$.

Perhaps the most important application of ordered pairs and product sets is in geometry. Consider two sets $A = \{1, 2, 3\}$ and $B = \{4, 5\}$. The product $A \times B$ is the set:

$$\{(1, 4), (1, 5), (2, 4), (2, 5), (3, 4), (3, 5)\}$$

Figure 4.5 shows these ordered pairs as points on a plane graph. For example the first ordered pair $(1, 4)$ is the point P, whose co-ordinates would be $(1, 4)$ using the normal notation for two-dimensional plane graphs.

We now deal with *real numbers*. Real numbers here are whole numbers or fractions but not complex or imaginary numbers, that is numbers containing

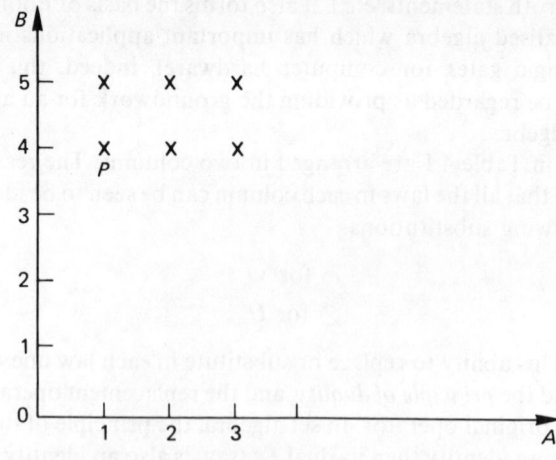

Fig. 4.5

Sets and Relations

$i = \sqrt{-1}$. Such complex or imaginary numbers are interesting and not particularly difficult to manipulate, but they have little place in the real (yet another use of this word !) world in which this book fancies it operates. So we now suppose that $\underline{R}$ represents the set of all real numbers. $\underline{R}^2 = \underline{R} \times \underline{R}$ will be the set of all ordered pairs of all real numbers. It is evident that $\underline{R}^2$ is now represented by *all* the points on the plane. $\underline{R}^2$ is called the Cartesian plane (or, more shortly, the plane).

In Chapter 1 we referred to mathematics in terms of numbers, and though this book does not try for any theoretical rigour, it might be thought that there (right at the start!) we went a little off the rails, since geometry apparently was excluded from our 'definition'. In the set theory approach, geometry is seen as a natural visual representation of the sets of all real numbers. Using the same notation, $\underline{R}^3$ will denote the usual three-dimensional space, and there is a natural extension, too, to $\underline{R}^n$ for n-dimensional space. Hence 2, 3 or even n-dimensional geometry can be seen as arising from the manipulation of a particular set (i.e., the set of all real numbers).

4.7 Relations

Consider again the two sets A and B and their product $A \times B$. This is the set of all ordered pairs (a, b), where a is a member of A – i.e. $a \in A$ – and similarly $b \in B$. A *binary relation* (usually shortened to 'relation') R will exist if for each ordered pair (a, b), one of the following is true:

a is related to b; this is written as $a\,R\,b$, or
a is not related to b; this is written as $a\,\cancel{R}\,b$

A simple example will best explain this somewhat abstract concept. Consider the set of all men in my road and the set of all women. The product of this set is the set of all ordered pairs (i.e., one from each set). A relationship R (marriage) exists between all these ordered pairs since in every case

a particular man is married to a particular women, or
a particular man is not married to a particular woman.

Let us restrict the sets even further for simplicity, so that the set A is just myself Jim and (confusingly!) Jimmy next door, and B is my wife, Joan and Jimmy's wife Yvonne. The set of all ordered pairs is

(Jim, Joan)	$a\,R\,b$
(Jim, Yvonne)	$a\,\cancel{R}\,b$
(Jimmy, Joan)	$a\,\cancel{R}\,b$
(Jimmy, Yvonne)	$a\,R\,b$

where R is the same relationship throughout – that is, marriage.

Consider the term $<$ (i.e., 'less than'). When applied to two numbers it indicates a relationship – e.g., $3 < 4$ and $2 < 5$. Let us now apply this to two sets which consist respectively of the numbers of employees in firms in Birmingham and Coventry. Then, for any ordered pair (the first element being the number of employees in that firm in Birmingham and the second element being the number of employees in a firm in Coventry) either $a\,R\,b$ or $a\,R\!\!\!/\,b$. This relationship symbolised by '<' is really one of 'order of size' or more shortly 'order' (where the word 'order' is used here in its normal sense). It is quite general in its application, since for any ordered pair (a, b) of real numbers, either $a < b$ or $a \not< b$. From a mathematical point of view it is helpful to see relations as special types of sets – namely sets containing all ordered pairs of objects which satisfy the relationship. In this example, the relation 'less than' is the set L (say) of pairs (a, b) of real numbers for which $a < b$. In set notation

$$L = \{(a, b): a \text{ and } b \text{ are real numbers}: a < b\}.$$

Finally, suppose the set $A\{y, n\}$ covers the two possibilities that a firm is taken over (y) or not taken over (n) in the next year, and the set $B\{i, s, r\}$ the three likely economic conditions (inflation, stability and recession) to which the firm may be exposed. The product $A \times B$ consists of all the ordered pairs of A and B. It can conveniently be illustrated by a tree diagram, as in Figure 4.6.

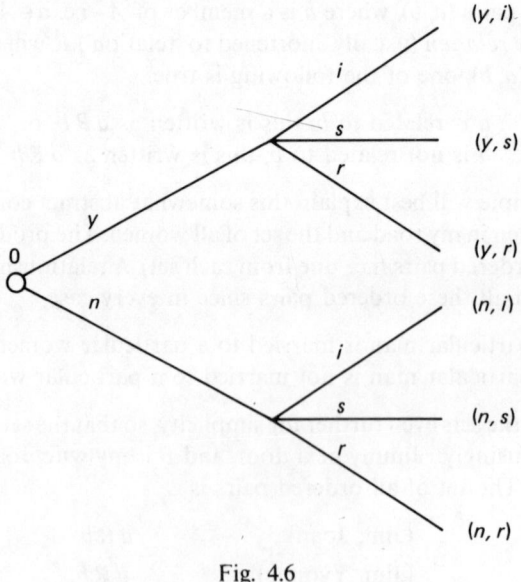

Fig. 4.6

It is important to realise that in this tree diagram we are merely showing an effective *visual representation of certain facts*. It is, however, a very helpful presentation and later (in Chapter 10) we use it for further manipulation and analysis.

Sets and Relations

The pictorial representation of relations by *graphs*, however, is outstandingly the most important representation of all from the mathematical point of view. We have already noted that a relation can be seen as a set, and also that $\underline{R}^2 = \underline{R} \times \underline{R}$ is the plane of a two-dimensional graph.

The relation is itself a subset of $\underline{R}^2$. We can picture this relationship set by looking at those points which represent it. This pictorial representation is called the *graph of the relation*.

Typically this relation will consist of all ordered pairs of real numbers which satisfy some equation. In general terms we write $F(x, y) = 0$ and read this as 'function of x, y equals 0'. Looked at from the point of view of the x and the y there is a relationship between these two, and it is expressed by this function.

A good example of such a function is the equation

$$x^2 + y^2 = 25.$$

This implies a relation between any pair of ordered points which satisfy this equation. Since there is a relationship it is also possible to see it as the *dependence* of one variable on another – that the value of y, for instance, depends on the value of x. For example, if x is 3, y will be ± 4. The graph of this relation is a circle whose radius is 5, and whose centre is at the origin (i.e., at the point (0, 0) where the x and y axes *intersect*).

In this relationship there are an infinite number of ordered points which satisfy this relationship (i.e., the function $x^2 + y^2 = 25$). In Chapter 8 we consider further the circle and other curves.

We revert now to the earlier situation where there are a limited number of ordered points and consider the various ways in which we can picture such a relation. If we take the same sets A and B where $A = \{1, 2, 3\}$ and B is $\{4, 5\}$ we have already shown, in Figure 4.5 one way in which we can go about this. This diagram showing the product $A \times B$ as co-ordinates on a graph is usually called a *co-ordinate diagram*. However, there are other ways. We could, for instance, form a rectangular array with rows for the elements of A and columns for the elements of B. Any ordered pair which has an element from A, and another from B will now be represented by the entry 1 at the intersection of the appropriate row and column. In our case we would have:

	4	5
1	1	1
2	1	1
3	1	1

It may be that the relationship between A and B does not extend to all ordered pairs. If for instance we have $R = \{(1, 4), (1, 5), (2, 4)\}$ only, then that implies $\not R$ for the remaining three ordered pairs. In this case we would have:

	4	5
1	1	1
2	1	0
3	0	0

where the fact that there is no relationship for the pairs (2, 5) (3, 4) (3, 5) is indicated by putting in a 0 at the appropriate intersection. This form of setting out the relationship is called a *matrix*.

There is yet another convenient way. If we represent the sets A and B by two disjoint discs and put in the elements 1, 2, 3 on A and 4, 5 in B, we can indicate the fact that there is a relationship R between an element in A and another element in B by drawing an arrow from one to another. This is called an *arrow diagram*. In our last example, where there is a relationship only between (1, 4), (1, 5) and (2, 4) we can show this as in Figure 4.7.

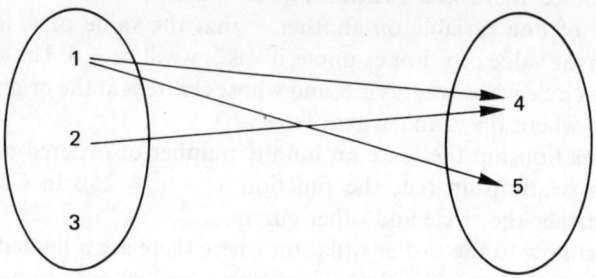

Fig. 4.7

There are no arrows between 2 and 5, and 3 and 4 or 5 because the relationship R does not exist between these elements. More accurately, R is the relationship.

To summarise, we have shown three ways of giving a pictorial representation of a relationship R between the members of two finite sets:

1. Construct the *co-ordinate diagram*.
2. Draw up a rectangular array (or, as it is often called a *matrix*).
3. Construct an *arrow diagram*.

4.8 Exercises

1. A market survey of 500 men and 500 women who watched one, and only one, breakfast TV showed that the three programmes (A, B, C) were customarily viewed as follows:

	A	B	C
Men	150	60	210
Women	70	140	120

Sets and Relations

    (a) How many men do not watch any breakfast TV?
    (b) How many women do not watch any breakfast TV?
    (c) How many people prefer programme C?

2. In question 1 above, find the numbers of elements in:

    (a) The set of men who prefer B or C.
    (b) The set of men who prefer A or B or C.
    (c) The set of men who prefer A or B.

3. In question 1 above, if:

S is the set of men who prefer A,
T is the set of men who prefer B, and
R is the set of men who prefer C,

and if the set in 2(a) above is denoted by K, and the set in 2(c) is denoted by L, show that:

    (a) $S \cup K = L \cup R$,
    (b) $S \cup K = S \cup (T \cup R)$, and
    (c) $L \cup R = (S \cup T) \cup R$.

Hence verify the associative law, that is:

$$S \cup (T \cup R) = (S \cup T) \cup R$$

4. A survey reported that:

12% of those interviewed were married women with no children,
30% were married women with children,
44% were married men, and
80% were married.

Is this possible?

5. If we use A' for the complement of A in S, and B' for the complement of B in S, show that (with a similar notation),

    (a) $(A \cup B)' = A' \cap B'$
    (b) $(A \cap B)' = A' \cup B'$.

(These laws are referred to as *De Morgan's Laws*.)

6. If N is the set of all positive numbers, that is:

$$N = \{1, 2, 3, 4 \ldots\}$$

list the elements of the following sets:

    (a) $A = \{x : x \in N, 3 < x < 12\}$
    (b) $B = \{x : x \in N, x \text{ is even}, x < 15\}$
    (c) $C = \{x : x \in N, 4 + x = 3\}$

7. For the following sets:

$$\emptyset, \quad A = \{1\}, \quad B = \{1,3\}, \quad C = \{1,5,9\}, \quad D = \{1,2,3,4,5\},$$
$$E = \{1,3,5,7,9\}, \quad U = \{1,2,\ldots,8,9\}$$

Give the correct symbol $\subset$ or $\not\subset$ to insert between each pair of these sets:

(a) $\emptyset, A$ (c) B, C (e) C, D (g) D, E
(b) A, B (d) B, E (f) C, E (h) D, U

8. If

$$U = \{1, 2, \ldots, 8, 9\} \text{ and}$$
$$A = \{1, 2, 3, 4, 5\}$$
$$B = \{4, 5, 6, 7\}$$
$$C = \{5, 6, 7, 8, 9\}$$
$$D = \{1, 3, 5, 7, 9\}$$
$$E = \{2, 4, 6, 8\}$$
$$F = \{1, 5, 9\}$$

What is:

(a) $A \cup B$ and $A \cap B$,   (b) $B \cup D$ and $B \cap D$,
(c) $A \cup C$ and $A \cap C$,   (d) A^c,   (e) B^c,
(f) $A \cap (B \cup E)$,   (g) $(B \cap F) \cup (C \cap E)$.

9. Write the dual of each of these:

(a) $(A \cup B) \cap (A \cup B^c) = A \cup \emptyset$
(b) $(A \cap U) \cup (B \cap A) = A$

10. Prove the identities $(U \cap A) \cup (B \cap A) = A$ (by the use of Table 4.1), and $(\emptyset \cup A) \cap (B \cup A) = A$.

11. In Stoneleigh Close there are 48 houses. Each householder bought at least one of three daily papers (F, G, D). You are given the following information:

30 householders bought F
25 householders bought G
15 householders bought D
7 householders had both F and G; 6 had both G and D
16 householders bought G only.

Determine how many in the Close bought:

(a) F and D but not G
(b) All three daily papers.

12. In a survey of 120 local farmers it was found that 50 read *New Forest* magazine (N), 52 read *The Farmer* (T) and 52 read *Farming Weekly* (F). Also 18 read both N and F, 22 read both N and T, 16 read both T and F and 16 read none of these three magazines.

   (a) Determine the number of farmers who read all three magazines
   (b) Draw a suitable Venn diagram and fill in the correct number of farmers in each of its eight regions
   (c) Find the number of farmers who read precisely one magazine.

Chapter 5
Matrices, Vectors and Determinants

5.1 Matrices

What is a matrix? Almost every text dealing with this subject approaches it in a different way. In Chapter 4 we saw the matrix format as simply a way of representing a relationship. Let us follow up this approach by way of an example.

Suppose a company has three operating departments, and that quarterly budgets for the sales of each department are produced. Set out with quarterly periods reading from left to right and the departments 1, 2, 3 put down vertically we might have figures such as those shown in Table 5.1. The estimated sales (S, say) of department 2 in quarterly period 3 is 605. Using subscript notation (important for much mathematical work, but essential in this area) we would write this as S_{23}, and say

$$S_{23} = 605$$

Similarly $S_{24} = 609$, $S_{31} = 1070$, $S_{32} = 1110$, and so on.

TABLE 5.1

		Quarterly periods			
		1	2	3	4
Departments	1	807	819	827	855
	2	601	603	605	609
	3	1070	1110	1190	1300

In this area of quarterly budgets the connection or relationship between department 2 and quarterly period 3 is – we are saying – represented by the entry of 605 in the appropriate row and column of the matrix. If we replace the sales figures by symbols we have the matrix:

Matrices, Vectors and Determinants

$$\begin{array}{cccc} S_{11} & S_{12} & S_{13} & S_{14} \\ S_{21} & S_{22} & S_{23} & S_{24} \\ S_{31} & S_{32} & S_{33} & S_{34} \end{array}$$

Note that in this subscript notation the *row* is put first *before the column*. This is purely conventional. Conventions, however, do tend to avoid confusion. It is conventional for, instance, to refer to entries on the left-hand side of accounts as debits and those on the right as credits. This convention is worldwide, though many conventions in accountancy (such as which side or in which way to present assets in a Balance Sheet) are not. When conventions are not standardised confusion can arise as Swedish motorists used to experience (until Sweden changed to driving on the right) when they crossed the borders into Norway and Finland, and some British motorists going to the continent still do!

The entries in a matrix are usually enclosed in two curved lines (rather as in a convex lens). This, too, is purely conventional. The general matrix with m rows and n columns can therefore be written as:

$$A \text{ (say)} = \begin{pmatrix} a_{11} & a_{12} & \ldots & a_{1n} \\ a_{21} & a_{22} & \ldots & a_{2n} \\ \ldots & \ldots & \ldots & \ldots \\ \ldots & \ldots & \ldots & \ldots \\ a_{m1} & a_{m2} & \ldots & a_{mn} \end{pmatrix}$$

The element in the ith row and the jth column is a_{ij}. If we call the matrix above A, we sometimes can avoid writing it out in full as above and write more shortly $A = (a_{ij})$, provided the circumstances are such that this abbreviation is clearly understood. A matrix, such as the one above with m rows and n columns, is called an 'm by n' matrix and this is written as $m \times n$.

5.2 Vectors

If in the example at the beginning of this chapter the company had had only the first operating department, its sales for the four quarterly periods could have been expressed in matrix notation as:

$$(807, 819, 827, 855)$$

A matrix with only one row such as this is called a *row vector*. If the company had had, after all, three operating departments, and we wished to express only the first quarter's sales in matrix notation, we would write this as:

$$\begin{pmatrix} 807 \\ 601 \\ 1070 \end{pmatrix}$$

This is called a *column vector*. Row and column vectors can best be seen as special cases of matrices which respectively have only one row and one column.

The elements or entries in these vectors and matrices are *numbers* (using the term to include all numbers, not just digits). When we are dealing with vectors and matrices we often wish to deal, as well, with numbers which are not entries in a vector or matrix. These 'normal' numbers are then referred to as *scalar numbers* to differentiate them from those numbers which are part of a matrix or vector. Scalars and vectors or matrices can be combined; for instance the product of the row vector (6, 7) and the scalar 2 is the row vector (12, 14). More generally, the product of a scalar k and a vector $u = (u_1, u_2, \ldots u_n)$ is:

$$ku = (ku_1, ku_2, \ldots, ku_n)$$

In the more general case of multiplication by a scalar k and a matrix A we follow along the same lines. The product of a scalar k and the above matrix A is:

$$kA = \begin{pmatrix} ka_{11} & ka_{12} & \ldots & ka_{1n} \\ ka_{21} & ka_{22} & \ldots & ka_{2n} \\ \ldots & \ldots & \ldots & \ldots \\ \ldots & \ldots & \ldots & \ldots \\ ka_{m1} & ka_{m2} & \ldots & ka_{mn} \end{pmatrix}$$

A vector whose entries are all zero is called a *zero vector*, and a matrix whose entries are all zero is called a *zero matrix*. A zero matrix is usually denoted by 0. Thus the 2×3 zero matrix is:

$$0 = \begin{pmatrix} 0 & 0 & 0 \\ 0 & 0 & 0 \end{pmatrix}$$

5.3 Matrix addition and multiplication

Two matrices A and B can be added if they are of the same size (i.e., the same number of rows and columns). The sum, written as $A + B$ is a matrix obtained by adding the corresponding elements in A and B. If we take the following two 2×3 matrices we would have:

$$\begin{pmatrix} 1 & 2 & 3 \\ 4 & 5 & 6 \end{pmatrix} + \begin{pmatrix} 2 & 4 & 6 \\ 8 & 10 & 12 \end{pmatrix} = \begin{pmatrix} 3 & 6 & 9 \\ 12 & 15 & 18 \end{pmatrix}$$

Matrix subtraction is similar; the matrix $-A$ is defined as $(-1)A$ – i.e., the matrix A with the signs of all its elements reversed.

Matrix multiplication is less simple. The product of two matrices A and B is defined *only* if the number of columns in A is the same as the number of rows in B. Thus if A (say) is an $m \times p$ matrix and B (say) a $p \times n$ matrix the product $A \times B$ is a matrix C. It is an $m \times n$ matrix. We write:

Matrices, Vectors and Determinants

$$\begin{pmatrix} a_{11} & \cdots & a_{1p} \\ \cdot & \cdots & \cdot \\ a_{i1} & \cdots & a_{ip} \\ \cdot & \cdots & \cdot \\ a_{m1} & \cdots & a_{mp} \end{pmatrix} \begin{pmatrix} b_{11} & \cdots & b_{1j} & \cdots & b_{1n} \\ \cdot & \cdots & \cdot & \cdots & \cdot \\ \cdot & \cdots & \cdot & \cdots & \cdot \\ \cdot & \cdots & \cdot & \cdots & \cdot \\ b_{p1} & \cdots & b_{pj} & \cdots & b_{pn} \end{pmatrix} = \begin{pmatrix} c_{11} & \cdots & c_{1n} \\ \cdot & \cdots & \cdot \\ \cdot & c_{ij} & \cdot \\ \cdot & \cdots & \cdot \\ c_{m1} & \cdots & c_{mn} \end{pmatrix}$$

where

$$c_{ij} = a_{i1}b_{1j} + a_{i2}b_{2j} + \ldots + a_{ip}b_{pj}$$

That is (for C), the element in the ith row and jth column is found by adding up all the products of the ith row of A and the corresponding elements in the jth column of B. This is often referred to as 'row into column multiplication'.

The product

$$\begin{pmatrix} 1 & 3 \\ 2 & 1 \end{pmatrix} \begin{pmatrix} 2 & 0 & 4 \\ 3 & 2 & -6 \end{pmatrix}$$

$$= \begin{pmatrix} 1 \times 2 + 3 \times 3 & 1 \times 0 + 3 \times 2 & 1 \times 4 + 3 \times (-6) \\ 2 \times 2 + 1 \times 3 & 2 \times 0 + 1 \times 2 & 2 \times 4 + 1 \times (-6) \end{pmatrix} = \begin{pmatrix} 11 & 6 & -14 \\ 7 & 2 & 2 \end{pmatrix}$$

To obtain the elements in the first *row* of the product matrix we multiplied the first *row* (i.e., (1, 3)) by the *columns* $\begin{matrix} 2 & 0 & 4 \\ 3 & 2 & -6 \end{matrix}$ in sequence (e.g., the element in the first row second column is obtained by multiplying the first row (1 3) by the second column $\begin{pmatrix} 0 \\ 2 \end{pmatrix}$ – in detail this is the first element (1) multiplied by first element (0) and adding the second element (3) multiplied by the second element (2)).

To obtain the elements in the second row we multiplied the second row (2 1) by the same columns.

Reverting to the general case where we are considering the product of two matrices we can use the normal multiplication sign and write

$$A \times B = C$$

or more shortly (provided its meaning is understood) $AB = C$. We noted that the element in the ith row and jth column of C is:

$$c_{ij} = a_{i1}b_{1j} + a_{i2}b_{2j} + \ldots + a_{ip}b_{pj}$$

and we can write this, too, more shortly as

$$\sum_{k=1}^{k=p} a_{ik}b_{kj}$$

The $\sum$ (Greek capital letter sigma) is a *summation symbol*. It is another arithmetical abbreviation or shorthand; it, too, is particularly useful in matrix and vector work. It should be read as 'the sum of all $a_{ik}b_{kj}$ where k is first put equal to 1, then to 2, and so on all the way up to $k = p$, or more shortly 'the sum of $a_{ik}b_{kj}$ from $k = 1$ to $k = p$.

Now that we have discussed matrix addition, subtraction and multiplication (and, by implication, the same manipulations for vectors too, since vectors are just special cases of matrices) it is interesting to compare how these operations compare with those in normal arithmetic and those for sets. In Chapter 4, the algebra of sets was compared with traditional algebra. We can now see how similar matrix algebra is too. The comparisons are

Traditional algebra	Matrix algebra
$a + 0 = a$	$A + 0 = A$
$a \times 0 = 0$	$A \times 0 = 0$ (if 0 is the suitable zero matrix)
$a + b = b + a$	$A + B = B + A$

(though $A + B$, for example, will not even be defined unless the two matrices to be added have the same number of rows and columns).

Also, though $a \times b = b \times a$, $AB \neq BA$, in general (and, indeed, BA will not even be *defined* unless the B has the same number of columns as A has rows). And since AB, too, will not be defined unless A's columns and B's rows are identical in number, even to be able to compare AB and BA we need B to be an $s \times r$ matrix if A is an $r \times s$ matrix. Even when this is the case, and the products are therefore defined, in general BA will not equal AB.

By way of example, suppose

$$A = \begin{pmatrix} 1 & 6 \\ 3 & 5 \end{pmatrix} \text{ and } B \text{ is } \begin{pmatrix} 4 & 0 \\ 2 & 1 \end{pmatrix}$$

then,

$$AB = \begin{pmatrix} 1 \times 4 + 6 \times 2 & 1 \times 0 + 6 \times 1 \\ 3 \times 4 + 5 \times 2 & 3 \times 0 + 5 \times 1 \end{pmatrix} = \begin{pmatrix} 16 & 6 \\ 22 & 5 \end{pmatrix}$$

whereas,

$$BA = \begin{pmatrix} 4 & 0 \\ 2 & 1 \end{pmatrix} \begin{pmatrix} 1 & 6 \\ 3 & 5 \end{pmatrix} = \begin{pmatrix} 4 \times 1 + 0 \times 3 & 4 \times 6 + 0 \times 5 \\ 2 \times 1 + 1 \times 3 & 2 \times 6 + 1 \times 5 \end{pmatrix} = \begin{pmatrix} 4 & 24 \\ 5 & 17 \end{pmatrix}$$

Clearly AB and BA are not equal.

Many of the other algebraic laws are followed by matrices. For example whenever the products and sum are defined, then:

$$(AB)C = A(BC), \text{ and}$$
$$A(B+C) = AB + AC$$

which perhaps helps to restore one's faith in *matrices* to some extent!

Matrices, Vectors and Determinants

5.4 Square and identity matrices

A matrix which has the same number of rows as it has columns is called a *square matrix*. A square matrix which has n rows and n columns is said to be of order n.

A square matrix differs from other matrices in that it has *diagonals*. The diagonal leading from top left to bottom right is called the *main diagonal*.

A square matrix which has all its elements zero except for those along its main diagonal and these are all 1, is called a *unit matrix* or *identity matrix*. The unit matrix of order 3 is therefore:

$$\begin{pmatrix} 1 & 0 & 0 \\ 0 & 1 & 0 \\ 0 & 0 & 1 \end{pmatrix}$$

We can multiply the unit matrix by any square matrix of the same order. A quick check with a particular example will show that this multiplication leaves the matrix unchanged, and this is true whatever order the multiplication takes place.

The unit matrix is usually denoted by I, so what we are saying is that

$$AI = A, \text{ and}$$
$$IA = A$$

Clearly the unit matrix plays much the same role in matrix multiplication as the number 1 does in traditional multiplication (a further and very important parallel, again helping to restore any shattered faith!)

Since a square matrix can be multiplied by itself (as the rows and columns are equal in number), it follows that we can form *powers*.

Thus we can define (if A is a square matrix)

$$A^2 = AA, \quad A^3 = A^2A = AAA$$

A^1 is defined as A and $A^0 = I$ (the unit matrix)

When we say 'define' we mean only that we use the term A^2 as a shorthand for AA (itself a shorthand for $A \times A$). Similarly A^3 is a shorthand for AAA. It may be felt that this is rather unnecessary since AA, for example, is almost as brief and simple to write as A^2. However when we have the multiplication $AAA \ldots A$ (n times) it is clearly shorter to write A^n. Moreover, as we shall see, we can manipulate the A^2, A^3, A^n, etc. in much the same as say x^2, x^3, x^n, etc. in traditional algebra. The visual representation of A^2, A^3, etc. was dealt with in Chapter 4.

Since we can add matrices and also perform multiplication of matrices by scalars we can define *functions* such as

$$f(A) = a_0 + a_1 A + a_2 A^2 + \ldots + a_n A^n$$

(where a_0, a_1, etc. are normal scalar numbers). That is, we repeat, $f(A)$ means (or is a shorthand for) the expression in A on the right of the 'equals' sign. *This, and nothing more.*

5.5 Inversion of matrices

In traditional arithmetic we see division as the *reciprocal* or *inversion* of multiplication. If we consider 1 divided by 3, to take a simple example, we say that

$$3 \times \tfrac{1}{3} = 1 \text{ (and, of course, } \tfrac{1}{3} \times 3 = 1)$$

In matrix algebra, we define inversion using the same approach.

We say that the inverse of a *square* matrix A is 'another' square matrix B where:

$$AB = I \text{ and } BA = I$$

where I is the identity matrix of the same order.

Such a matrix B, if it exists, is *unique*. It is called the inverse of A and it is usually written as A^{-1}. Consequently we have:

$$AA^{-1} = I \text{ and } A^{-1}A = I$$

Consider these two square matrices of order 2:

$$\begin{pmatrix} 3 & 10 \\ 2 & 7 \end{pmatrix} \text{ and } \begin{pmatrix} 7 & -10 \\ -2 & 3 \end{pmatrix}$$

Multiplying these two together, we get:

$$\begin{pmatrix} 3 & 10 \\ 2 & 7 \end{pmatrix}\begin{pmatrix} 7 & -10 \\ -2 & 3 \end{pmatrix} = \begin{pmatrix} 21-20 & -30+30 \\ 14-14 & -20+21 \end{pmatrix} = \begin{pmatrix} 1 & 0 \\ 0 & 1 \end{pmatrix}, \text{ and}$$

$$\begin{pmatrix} 7 & -10 \\ -2 & 3 \end{pmatrix}\begin{pmatrix} 3 & 10 \\ 2 & 7 \end{pmatrix} = \begin{pmatrix} 21-20 & 70-70 \\ -6+6 & -20+21 \end{pmatrix} = \begin{pmatrix} 1 & 0 \\ 0 & 1 \end{pmatrix}$$

Clearly these two matrices are *inverses one of the other*. Although we do not prove it here, it is true that if one product is equal to the identity matrix, then the other product (that is, the product of the two with the order of multiplication reversed) must be equal to the identity matrix, too. It is therefore necessary only to test one product to see if the matrices are inverses of each other. In formal terms, if $AB = I$, then BA must equal I, too.

We have now completed matrix addition, subtraction, multiplication and inversion. The benefits from the use of matrix algebra are substantial. One important example, as we shall see, is in the solution of *linear equations*. However for the moment we pass to the consideration of determinants, which are analogous (but different) to matrices; they are different for many reasons, one of which is that they are used in traditional algebra.

Matrices, Vectors and Determinants

5.6 Determinants

A square group of numbers set out in matrix format but enclosed by straight (rather than curved) lines is called a *determinant*. A determinant is *not* a matrix, but a *number*. The determinants of orders, 1, 2, 3 are defined to be

$$|a_{11}| = a_{11}$$

$$\begin{vmatrix} a_{11} & a_{12} \\ a_{21} & a_{22} \end{vmatrix} = a_{11}a_{22} - a_{12}a_{21}$$

$$\begin{vmatrix} a_{11} & a_{12} & a_{13} \\ a_{21} & a_{22} & a_{23} \\ a_{31} & a_{32} & a_{33} \end{vmatrix} = a_{11}a_{22}a_{33} + a_{12}a_{23}a_{31} + a_{13}a_{21}a_{32} \\ - a_{13}a_{22}a_{31} - a_{12}a_{21}a_{33} - a_{11}a_{23}a_{32}$$

The value of a determinant is the *number* defined as above on the right.

To remember the definition (which is important!) this little diagram will be helpful for the determinant of order 2:

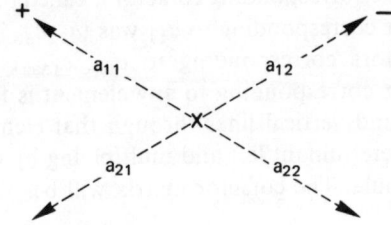

For the determinant of order 3 the simplest way is to see it as the three elements in the first row each multiplied by that determinant of the second order which is left after drawing a horizontal and a vertical line through the element. Additionally the *signs* of the elements of the first row start with + and then change as we move along. The general determinant of order 3 is therefore:

$$+ a_{11}(a_{22}a_{33} - a_{23}a_{32}) = + a_{11}a_{22}a_{33} - a_{11}a_{23}a_{32}$$
$$- a_{12}(a_{21}a_{33} - a_{23}a_{31}) = - a_{12}a_{21}a_{33} + a_{12}a_{23}a_{31}$$
$$+ a_{13}(a_{21}a_{32} - a_{22}a_{31}) = + a_{13}a_{21}a_{32} - a_{13}a_{22}a_{31}$$
  ↑   ↑
 1st row   2nd order determinants
 elements

and these add up to the definition already given.

It is not necessary to chose only the three elements in the first row (and multiply them by the appropriately-signed determinant). We can arrive at the same result by choosing any row or column, and multiplying the elements in it by the determinants formed in a similar way.

These (lower) second order determinants are called *minors* (minor determinants). We can neatly avoid the problem of the sign by defining a cofactor as the minor together with the appropriate sign. In the example, above, the minor corresponding to a_{12} is $(a_{21}a_{33} - a_{23}a_{31})$ and the cofactor is $-(a_{21}a_{33} - a_{23}a_{31})$.

Determinants of higher orders are defined in the same way – that is, by taking the elements in the top row (or any row or column) and multiplying by the determinant left over when we draw horizontal and vertical lines through that element, together with the appropriate sign – i.e., by the minor and the appropriate sign. More easily, we can say it is the product of the element and its associated cofactor. The rule for the appropriate sign is simple. Moving into more general terms, the cofactor of the element a_{ij} (that is, the element in the ith row and jth column) of a determinant is the minor of a_{ij} multiplied by $(-1)^{i+j}$. It follows that if i and j add up to an even number the cofactor and minor are the same. If i and j add up to an odd number the cofactor is the negative of the minor.

The matrix formed from an original matrix by substituting in the place of each of its elements the corresponding cofactor is called the *cofactor matrix*. In our case, the cofactor corresponding to a_{11} was $(a_{22}a_{33} - a_{23}a_{32})$. Let us call this c_{11}. The cofactors corresponding to c_{12}, c_{13}, etc. will be similarly defined. The cofactor corresponding to any element is found, we repeat, by drawing horizontal and vertical lines through that element, calculating the value of the minor determinant left and multiplying by the appropriate sign found from the formula. The cofactor matrix will be:

$$\begin{pmatrix} c_{11} & c_{12} & c_{13} \\ c_{21} & c_{22} & c_{23} \\ c_{31} & c_{32} & c_{33} \end{pmatrix}$$

Cofactor matrices will be useful to us for matrix inversion in the next section. Determinants of orders 1, 2 and 3 are much used; determinants of higher orders are less frequently met.

5.7 Inversion of matrices by determinants

In some text books a determinant is shown like a matrix but with the term 'det' preceding it. Thus the determinant $\begin{vmatrix} 9 & 3 \\ 1 & 4 \end{vmatrix}$ is written as 'det $\begin{pmatrix} 9 & 3 \\ 1 & 4 \end{pmatrix}$'.

This terminology overemphasises the closeness between matrices and determinants, nevertheless the two are related in many ways. One example of this is the proposition that a matrix is invertible only if its corresponding determinant is *non-zero*.

We do not prove this statement – we are not in the business of rigorous mathematical proofs! – but to demonstrate it we show how to go about finding

Matrices, Vectors and Determinants

the inverse of the general matrix of order 2. Suppose this matrix is:

$$A = \begin{pmatrix} a & b \\ c & d \end{pmatrix}$$

We wish to find the matrix $\begin{pmatrix} x & y \\ z & w \end{pmatrix}$ (say), such that

$$\begin{pmatrix} a & b \\ c & d \end{pmatrix}\begin{pmatrix} x & y \\ z & w \end{pmatrix} = \begin{pmatrix} 1 & 0 \\ 0 & 1 \end{pmatrix}$$

Multiplying out the product of the two matrices we have:

$$ax + bz = 1 \qquad ay + bw = 0$$
$$cx + dz = 0 \qquad cy + dw = 1$$

If we multiply both equations in the upper row by d and both equations in the second row by b, we have,

$$adx + bdz = d \qquad ady + bdw = 0$$
$$cbx + dbz = 0 \qquad cby + dbw = b$$

Subtracting the lower rows from the higher, we get

$$adx - cbx = d \qquad \text{and} \qquad ady - cby = -b$$

giving:

$$x = \frac{d}{ab - bc} \qquad \text{and} \qquad y = \frac{-b}{ad - bc}$$

Since $|A| = \begin{vmatrix} a & b \\ c & d \end{vmatrix} = ad - bc$, by the definition of the determinant, we have

$$x = \frac{d}{|A|} \qquad \text{and} \qquad y = \frac{-b}{|A|}$$

Substituting these values into the original equations, we get in a similar fashion

$$z = \frac{-c}{|A|} \qquad \text{and} \qquad w = \frac{a}{|A|}$$

The inverse of A, that is the matrix $\begin{pmatrix} x & y \\ z & w \end{pmatrix}$ is therefore:

$$A^{-1} = \frac{1}{|A|}\begin{pmatrix} d & -b \\ -c & a \end{pmatrix}$$

Hence to obtain the inverse of a square matrix of order 2, it is necessary to:
1. interchange the elements on the main diagonal, and
2. change the sign of the other two elements, and
3. divide by the determinant corresponding to the original matrix.

Since we cannot divide by 0, it follows that the inverse of this general 2 × 2 matrix exists only if the determinant is non-zero. Though once again we will not prove it, the same rule in fact applies for the inversion of square matrices of all orders – that is, inversion is possible only if the *corresponding determinant is non-zero*.

Inversion of matrices of higher orders than 2 is helped by two more definitions. The new matrix formed by interchanging the row and columns of an original matrix is called its *transpose*.

Earlier in this chapter we defined the cofactor matrix, of a matrix of order 3, as:

$$\begin{pmatrix} c_{11} & c_{12} & c_{13} \\ c_{21} & c_{22} & c_{23} \\ c_{31} & c_{32} & c_{33} \end{pmatrix}$$

The transpose of this matrix will be:

$$\begin{pmatrix} c_{11} & c_{21} & c_{31} \\ c_{12} & c_{22} & c_{32} \\ c_{13} & c_{23} & c_{33} \end{pmatrix}$$

The term used for the transpose of a cofactor matrix is *adjoint*.

We can now say (but again not prove) that the inverse of a matrix A is given by the formula:

$$A^{-1} = \frac{1}{|A|} \text{ adjoint (A)}$$

or, in words, the inverse of a matrix A is the reciprocal of the determinant of its coefficients multiplied by the transpose of its cofactor matrix.

The use of the inverse of a matrix in the solution of linear equations is given in Chapter 7.

5.8 Transformation matrices and their use in matrix inversion

If we multiply a square matrix A by the identity matrix I of the same order, the result, as we know, is A. That is

$$IA = A$$

If two of the columns of the identity matrix are interchanged, and we multiply A by this matrix on the left, the result is the interchanging of two rows of A. On the other hand if we multiply A on the right two columns are interchanged. Thus:

$$\begin{pmatrix} 1 & 0 & 0 \\ 0 & 1 & 0 \\ 0 & 0 & 1 \end{pmatrix} \begin{pmatrix} 3 & -6 & 3 \\ 9 & -4 & \frac{1}{2} \\ 2 & 1 & 2 \end{pmatrix} = \begin{pmatrix} 3 & -6 & 3 \\ 9 & -4 & \frac{1}{2} \\ 2 & 1 & 2 \end{pmatrix} \quad \text{– the matrix is unchanged}$$

$$\begin{pmatrix} 0 & 1 & 0 \\ 1 & 0 & 0 \\ 0 & 0 & 1 \end{pmatrix} \begin{pmatrix} 3 & -6 & 3 \\ 9 & -4 & \frac{1}{2} \\ 2 & 1 & 2 \end{pmatrix} = \begin{pmatrix} 9 & -4 & \frac{1}{2} \\ 3 & -6 & 3 \\ 2 & 1 & 2 \end{pmatrix} \quad \text{– the first two rows are interchanged}$$

$$\begin{pmatrix} 3 & -6 & 3 \\ 9 & -4 & \frac{1}{2} \\ 2 & 1 & 2 \end{pmatrix} \begin{pmatrix} 0 & 1 & 0 \\ 1 & 0 & 0 \\ 0 & 0 & 1 \end{pmatrix} = \begin{pmatrix} -6 & 3 & 3 \\ -4 & 9 & \frac{1}{2} \\ 1 & 2 & 2 \end{pmatrix} \quad \text{– the first two columns are interchanged}$$

If we wish to multiply a whole row of a matrix by a number this can be done by multiplying on the left by a matrix which is the same as the identity matrix except that the element on its principal diagonal corresponding to that row is altered to the number required. Similarly postmultiplying by the same matrix will multiply the whole column. Thus:

$$\begin{pmatrix} 1 & 0 & 0 \\ 0 & 4 & 0 \\ 0 & 0 & 1 \end{pmatrix} \begin{pmatrix} 3 & 2 & 4 \\ 5 & 6 & 1 \\ 2 & 8 & -1 \end{pmatrix} = \begin{pmatrix} 3 & 2 & 4 \\ 20 & 24 & 4 \\ 2 & 8 & -1 \end{pmatrix} \quad \text{– the second row is multiplied by 4}$$

$$\begin{pmatrix} 3 & 2 & 4 \\ 5 & 6 & 1 \\ 2 & 8 & -1 \end{pmatrix} \begin{pmatrix} 1 & 0 & 0 \\ 0 & 4 & 0 \\ 0 & 0 & 1 \end{pmatrix} = \begin{pmatrix} 3 & 8 & 4 \\ 5 & 24 & 1 \\ 2 & 32 & -1 \end{pmatrix} \quad \text{– the second column is multiplied by 4}$$

A multiple of a row can be added or subtracted to another row by multiplying on the left by a suitable transformation matrix. This matrix is arrived at by replacing the appropriate zero element in the identity matrix by that number. Postmultiplying by this transformation matrix gives a column addition or subtraction. Thus:

$$\begin{pmatrix} 1 & 0 & 0 \\ 0 & 1 & 0 \\ 4 & 0 & 1 \end{pmatrix} \begin{pmatrix} 4 & 5 & 2 \\ 3 & 4 & 1 \\ 2 & 6 & -1 \end{pmatrix} = \begin{pmatrix} 4 & 5 & 2 \\ 3 & 4 & 1 \\ 18 & 26 & 7 \end{pmatrix} \quad \text{– to the third row has been added four times the first row}$$

$$\begin{pmatrix} 4 & 5 & 2 \\ 3 & 4 & 1 \\ 2 & 6 & -1 \end{pmatrix} \begin{pmatrix} 1 & 0 & 0 \\ 0 & 1 & 0 \\ 4 & 0 & 1 \end{pmatrix} = \begin{pmatrix} 12 & 5 & 2 \\ 7 & 4 & 1 \\ -2 & 6 & -1 \end{pmatrix} \quad \text{– to the first column has been added four times the third column}$$

In all the examples so far we have restricted ourselves to multiplication by one transformation matrix only. There is no reason why this process cannot be repeated. For example if we call the successive transformation matrices $E_1 E_2 E_3$ and the original matrix A, then the final product will be B, given by:

$$B = E_3 E_2 E_1 A, \quad \text{or symbolically} \quad B = EA,$$

where E stands for the successive transformations $E_1 E_2 E_3$.

Clearly by the choice of suitable transformation matrices, a matrix can be transformed almost at will. Transformation matrices are frequently referred to, loosely as transition matrices.

We can compute the inverse of a matrix A, using transition matrices. Consider the matrix:

$$A \mid I$$

i.e., the matrix obtained by writing the identity matrix I on the right of A (but leaving a dividing line between the two for demarcation purposes). Such a matrix is called an augmented matrix.

We now apply a series of *row* transformations to this matrix – i.e., we multiply it by a series of suitable transformation matrices $E_1 E_2 E_3$, etc. all of which together we write symbolically as E. These transformation matrices must be chosen so that $EA = I$, or, in other words, the A part of the augmented matrix is transformed to I, the identity matrix.

If we do this we have:

$$E(A \mid I) = EA \mid EI = I \mid E$$

Since
$$EA = I \text{ (as we have arranged } E \text{ to do this)}$$
and
$$EI = E \text{ (since pre- or postmultiplying by an identity matrix does, by definition, leave the matrix unchanged)}$$

But if $EA = I$, then E is the inverse of A by the definition of an inverse. Hence the steps for finding an inverse are:

1. Add an identity matrix to the right of the original matrix
2. Apply appropriate row transformations to the combined (i.e., augmented) matrix so that the original matrix is transformed into I (the identity matrix)
3. The matrix on the right will then be the inverse.

Thus, to invert:

$$A = \begin{pmatrix} 2 & 1 \\ 1 & 3 \end{pmatrix}$$

Add an identity matrix to make the augmented matrix:

$$A \mid I = \begin{pmatrix} 2 & 1 & | & 1 & 0 \\ 1 & 3 & | & 0 & 1 \end{pmatrix}$$

and for simplicity let us call the top row right across of this augmented matrix R_1 and the bottom row R_2. Successive transformations of each row will be denoted by dashes (thus, first $R_1 R_2$, then $R'_1 R'_2$ then $R''_1 R''_2$, etc.).

Matrices, Vectors and Determinants

FIRST TRANSFORMATION: Multiply R_1 by $\frac{1}{2}$

$$E_1(A|I) = \begin{pmatrix} 1 & \frac{1}{2} & \frac{1}{2} & 0 \\ 1 & 3 & 0 & 1 \end{pmatrix} \begin{array}{l} (R'_1 = \frac{1}{2}R_1) \\ (R_2 \text{ is unchanged}) \end{array}$$

SECOND TRANSFORMATION: Subtract R'_1 from R_2

$$E_2E_1(A|I) = \begin{pmatrix} 1 & \frac{1}{2} & \frac{1}{2} & 0 \\ 0 & \frac{5}{2} & -\frac{1}{2} & 1 \end{pmatrix} \begin{array}{l} (R'_1 \text{ is unchanged}) \\ (R'_2 = R_2 - R'_1) \end{array}$$

THIRD TRANSFORMATION: Multiply R'_2 by $\frac{2}{5}$

$$E_3E_2E_1(A|I) = \begin{pmatrix} 1 & \frac{1}{2} & \frac{1}{2} & 0 \\ 0 & 1 & -\frac{1}{5} & \frac{2}{5} \end{pmatrix} \begin{array}{l} (R'_1 \text{ is unchanged}) \\ (R''_2 = \frac{2}{5}R'_2) \end{array}$$

FOURTH TRANSFORMATION: Subtract $\frac{1}{2}R''_2$ from R'_1

$$E_4E_3E_2E_1(A|I) = \begin{pmatrix} 1 & 0 & \frac{3}{5} & -\frac{1}{5} \\ 0 & 1 & -\frac{1}{5} & \frac{2}{5} \end{pmatrix} \begin{array}{l} (R''_1 = R'_1 - \frac{1}{2}R''_2) \\ (R''_2 \text{ is unchanged}) \end{array}$$

The augmented part of this is:

$$\begin{pmatrix} \frac{3}{5} & -\frac{1}{5} \\ -\frac{1}{5} & \frac{2}{5} \end{pmatrix}$$

and this is the inverse of A.

A quick check shows that:

$$\begin{pmatrix} 2 & 1 \\ 1 & 3 \end{pmatrix} \begin{pmatrix} \frac{3}{5} & -\frac{1}{5} \\ -\frac{1}{5} & \frac{2}{5} \end{pmatrix}$$

does equal

$$\begin{pmatrix} 1 & 0 \\ 0 & 1 \end{pmatrix}$$

5.9 Multidimensional arrays

Before discussing multidimensional arrays let us first summarise some of the work. We have already noted the apparent similarity between a matrix and a determinant. The similarity however is *in appearance only*. The determinant $\begin{vmatrix} 9 & 3 \\ 1 & 4 \end{vmatrix}$ *only* means $36 - 3 = 33$. It is the same as $\begin{vmatrix} 7 & 2 \\ 1 & 5 \end{vmatrix}$ or, for that matter $100 - 67$. All these equal 33, too. The matrices $\begin{pmatrix} 9 & 3 \\ 1 & 4 \end{pmatrix}$ and $\begin{pmatrix} 7 & 2 \\ 1 & 5 \end{pmatrix}$ are not equal, however. In matrix multiplication (or addition, subtraction or inversion) multiplying any other matrix by these two matrices would, in general, give entirely different results. This is because in a matrix the *position and value of each and every element* is important.

If we revert to the example given in 5.1 at the start of this chapter, where we had three departments and the sales for each department for four quarterly periods were shown, each figure mattered in its own right. At the start of Section 5 we restricted the operating departments to only one. This gave a matrix with only one row and this, we noted, was conventionally called a vector. Our original matrix was a 3×4 matrix; it is said to have *two dimensions* (that is both rows and columns). A vector has *one dimension only* (either one row or one column).

Instead of reducing the number of dimensions (as we do when we go from a matrix to a vector), we could go in the opposite direction and increase them to three. It will be difficult to *present* such a situation in table or array form (or, indeed, in any format on this page); that is because this page is two-dimensional. Nevertheless we can easily imagine that corresponding to (say) the second row and the third column of our original matrix instead of the one entry (it happens to be 605) we might wish to have two entries depending on different factors. This might be because the original entry was the historic sales figure and the other entry was inflation-adjusted, or perhaps the original was one person's estimate, and the second someone else's, or maybe the original was the forecast and the second the actual. It does not matter. All we are saying is that corresponding to each of the 12 (3×4) entries in the original matrix format we might have a second group of figures. We now have a three-dimensional format. The original 3×4 two-dimensional matrix or array has been expanded into a $3 \times 4 \times 2$ three-dimensional array. It is obvious that arrays of more than three dimensions can exist, though beyond three dimensions a pictorial visualisation is not possible.

Subscript notation is particularly helpful when dealing with 2, 3 or higher dimension arrays. We used the subscript notation for our original 3×4 two-dimensional array. The subscripts went from s_{11} to s_{34}. In our $3 \times 4 \times 2$ three-dimensional arrays the subscripts would go from s_{111} to s_{342}. The second or alternative figure, for example, in the second row and third column, would be, in subscript notation S_{232}. Three-dimensional arrays are sometimes called *tensors*.

Data is frequently stored and dealt with by computers in array form. Computer programs, however, must be typed (or punched) on one line so, for input purposes, the subscript notation has to be amended. Typically the subscripts are put in parentheses, so that these would now be typed as:

$$S(1, 1, 1) \ldots S(2, 3, 2) \ldots S(3, 4, 2)$$

Computers can handle arrays with high numbers of dimensions (in practice, only by transforming them into vectors first); the only disadvantage is the practical one that multi-dimension arrays are voracious memory eaters!

5.10 Input–output analysis

In classical economic analysis it is assumed that a country's economy can be divided into a number of industries which produce goods and services. In

Matrices, Vectors and Determinants

practice it is often not easy to make this classification. Many businesses, too, are multi-product producers, and problems of defining products in any case can be difficult. These practical problems are not considered here.

The end product of all these industries' work is to final consumer demand. Many industries, of course, supply other industries – that is, they produce *intermediate products*. After further processing, however, these products too reach the consumer.

In *Leontieff input–output analysis* it is further assumed that changes in demand for any of the final outputs from any industry affect the outputs that it requires from other industries according to a particular pattern. This assumption can be written as:

$$x_{ij} = a_{ij}x_j \tag{5.1}$$

where, x_j is the total output of industry j
x_{ij} is the output of industry i to industry j
a_{ij} is constant for a given i and j

The constant a_{ij} is called the *Leontieff input–output constant*. The implication of using such a constant in equation (5.1) above is that the demand that industry j makes of industry i depends on (that is, it is 'directly variable with') the output of industry j. An example may make the significance of this assumption clear. Suppose we are concerned with two industries, the brick industry and the estate house construction industry. Most estate houses these days are built to a fairly standard pattern. Even if this is not so, and there is some substantial variability, let us assume that the average number of bricks required is constant at (say) 3,500 per house. The Leontieff approach merely assumes that this number remains the same, i.e., 'constant'. If 10% more estate houses are made, then 10% more bricks will be required. However, here again we are not concerned with any arguments as to the accuracy of this assumption. We take it to be correct and proceed with the mathematics.

Reverting to our definition, if there are n industries in our classification, then:

$$x_i = \sum_{j=1}^{n} x_{ij} + y_i \tag{5.2}$$

for all values of i from 1 to n (since total output of an industry must either go as inputs to other industries or to final demand)
where y_i is the output of industry i to final demand.

Putting in the value for x_{ij} from equation (5.1), we have:

$$x_i = \sum a_{ij}x_j + y_i$$

where j varies from 1 to n)
that is

$$x_i - \sum a_{ij}x_j = y_i$$

Now this is true for all values of i from 1 to n, that is, in detail:

$$x_1 - a_{11}x_1 - a_{12}x_2 - a_{13}x_3 \ldots - a_{1n}x_n = y_1$$
$$x_2 - a_{21}x_1 - a_{22}x_2 - a_{23}x_3 \ldots - a_{2n}x_n = y_2$$
$$\vdots$$
$$x_n - a_{n1}x_1 - a_{n2}x_2 - a_{n3}x_3 \ldots - a_{nn}x_n = y_n$$

Grouping up the x_1s in the first equation, the x_2s in the second and so on, these equations in matrix form are:

$$\begin{pmatrix} (1-a_{11}) & -a_{12} & -a_{13} & \ldots & -a_{1n} \\ -a_{21} & (1-a_{22}) & -a_{23} & \ldots & -a_{2n} \\ - & - & - & & - \\ - & - & - & & - \\ -a_{n1} & -a_{n2} & -a_{n3} & \ldots & (1-a_{nn}) \end{pmatrix} \begin{pmatrix} x_1 \\ x_2 \\ - \\ - \\ x_n \end{pmatrix} = \begin{pmatrix} y_1 \\ y_2 \\ - \\ - \\ y_n \end{pmatrix}$$

If we call the matrix of input–output coefficients A, so that:

$$A = \begin{pmatrix} a_{11} & a_{12} & a_{13} & \ldots & a_{1n} \\ a_{21} & a_{22} & a_{23} & \ldots & a_{2n} \\ - & - & - & & - \\ - & - & - & & - \\ a_{n1} & a_{n2} & a_{n3} & \ldots & a_{nn} \end{pmatrix}$$

then the matrix on the left is:

$$I - A$$

where I is the indentity matrix of order n.

If we call the column vector of total outputs X, and the column vector of outputs to final demand (i.e., 'sales') Y, then:

$$(I - A)X = Y$$

giving

$$X = (I - A)^{-1}Y$$

assuming the inverse of $(I - A)$ exists.

This inverse, $(I - A)^{-1}$, will be some matrix B (say) where:

$$B = \begin{pmatrix} b_{11} & b_{12} & b_{13} & \ldots & b_{1n} \\ b_{21} & b_{22} & b_{23} & \ldots & b_{2n} \\ - & - & - & & - \\ - & - & - & & - \\ b_{n1} & b_{n2} & b_{n3} & \ldots & b_{nn} \end{pmatrix}$$

Matrices, Vectors and Determinants 89

and we have

$$X = BY$$

which gives the values for $x_1, x_2, x_3 \ldots x_n$ – i.e. each industry's output, as being equal to the sum of the elements of final demand sales of industries premultiplied by the coefficients of the matrix B (the inverse matrix of $I - A$). For instance, the output of the first industry is:

$$x_1 = b_{11}y_1 + b_{12}y_2 + b_{13}y_3 + \ldots + b_{1n}y_n$$

and similarly for other industries. This is an important result since for given or required sales to final demand of each industry we can find, (provided the Leontieff input–output constants are known), the *required total outputs* for each industry.

National input–output models will deal with many industries. We can show the principle, however, by taking a simple three-industry example.

$$\text{Take } A = \begin{pmatrix} 0.125 & 0.333 & 0.250 \\ 0.500 & 0.167 & 0.250 \\ 0.250 & 0.167 & 0.250 \end{pmatrix}$$

where A is the given matrix of input–output coefficients

$$I - A = \begin{pmatrix} 0.875 & -0.333 & -0.250 \\ -0.500 & 0.833 & -0.250 \\ -0.250 & -0.167 & 0.750 \end{pmatrix}$$

We can then calculate, by one of the methods suitable for matrix inversion, that:

$$(I - A)^{-1} = \begin{pmatrix} 2.000 & 1.000 & 1.000 \\ 1.500 & 2.036 & 1.179 \\ 1.000 & 0.786 & 1.929 \end{pmatrix}$$

If we take it that final demand is given by the column vector:

$$Y = \begin{pmatrix} 1{,}001 \\ 3{,}407 \\ 1{,}505 \end{pmatrix}$$

We have:

$$X = \begin{pmatrix} 2.000 & 1.000 & 1.000 \\ 1.500 & 2.036 & 1.179 \\ 1.000 & 0.786 & 1.929 \end{pmatrix} \begin{pmatrix} 1{,}001 \\ 3{,}407 \\ 1{,}505 \end{pmatrix}, \text{ i.e.}$$

$$X = \begin{pmatrix} 2002 + 3407 + 1505 \\ 1502 + 6937 + 1774 \\ 1001 + 2678 + 2903 \end{pmatrix} = \begin{pmatrix} 6914 \\ 10213 \\ 6582 \end{pmatrix}$$

which gives directly the outputs given (or required) from each industry as being 6914, 10213, 6582 respectively.

There are two further steps in input–output analysis which are of interest.

First, the vector $X - Y$ gives directly that output of each industry which goes to *other industries* as part of the production process, rather than to final demand.

Second, we can easily arrive at the marginal effect of having to produce one extra unit of final demand. Reverting to the general formulation (with n industries), we have:

$$X = (I - A)^{-1} Y$$

then:

$$\Delta X = (I - A)^{-1} \Delta Y$$

where ΔX and ΔY are the respective small increases.

Hence, corresponding to the need to produce (say) one extra unit of industry n's for sale, the extra outputs required by the individual industries will be given by putting $\Delta Y = Y_n = 1$ (and $Y_1 = Y_2 = \ldots Y_{n-1} = 0$), in the general matrix above.

5.11 Exercises

1. If $A = \begin{pmatrix} 3 & 6 \\ 1 & 4 \end{pmatrix}$ and $B = \begin{pmatrix} 3 & 2 \\ 2 & 4 \end{pmatrix}$

    Calculate:

    (a) $A \times B$
    (b) $B \times A$

2. If $A = (1, 2, 3)$

    $B = \begin{pmatrix} 4 \\ 5 \\ 6 \end{pmatrix}$

    Calculate:

    (a) $A \times B$
    (b) $B \times A$

3. If $A = \begin{pmatrix} 1 & 2 \\ 0 & 1 \end{pmatrix}$

    Calculate:

    (a) A^2
    (b) A^3
    (c) A^5

Matrices, Vectors and Determinants

4. If $A = \begin{pmatrix} 1 & 0 \\ 1 & 0 \\ -1 & 1 \end{pmatrix}$ and $B = \begin{pmatrix} 0 & 1 & 0 \\ 1 & 0 & 1 \end{pmatrix}$

   Calculate:
   (a) $B \times A$
   (b) $A \times B$

5. For question 4 above, answer these questions
   (a) Is $B \times A = A \times B$?
   (b) Is $B \times A = I$?

6. Is it possible to multiply two rectangular matrices together to yield an identity matrix?
7. Is our definition of an inverse restricted to square matrices?
8. Find the values of the following:

   (a) $\begin{pmatrix} 1 & 6 \\ -3 & 5 \end{pmatrix} \begin{pmatrix} 4 & 0 \\ 2 & -1 \end{pmatrix}$

   (b) $\begin{pmatrix} 1 & 6 \\ -3 & 5 \end{pmatrix} \begin{pmatrix} 2 \\ -7 \end{pmatrix}$

   (c) $\begin{pmatrix} 2 & -1 \\ 1 & 0 \\ -3 & 4 \end{pmatrix} \begin{pmatrix} 1 & -2 & -5 \\ 3 & 4 & 0 \end{pmatrix}$

9. Conventionally, the transpose of a matrix X is denoted by X^T. Using this same notation, find the values of:

   (a) Y^T (b) YY^T (c) Y^TY, where $Y = \begin{pmatrix} 1 & 2 & 0 \\ 3 & -1 & 4 \end{pmatrix}$

10. If $X = \begin{pmatrix} 1 & 3 \\ 2 & -1 \end{pmatrix}$ and $Y = \begin{pmatrix} 2 & 0 & -4 \\ 3 & -2 & 6 \end{pmatrix}$

    (a) state if the product matrix XY exists
    (b) state if the product matrix YX exists
    (c) find the value of any of these matrices which do exist.

11. By taking very simple examples for the matrices A, B and C, demonstrate that for your cases:

    (a) $(AB)C = A(BC)$
    (b) $A(B+C) = AB + AC$
    (c) $(B+C)A = BA + CA$
    (d) $k(AB) = (kA)B = A(kB)$, where k is a scalar.

    Take care in choosing the matrices so that their sums and products are defined.

12. Find the value of the following determinants of order 2:

    (a) $\begin{vmatrix} 3 & -2 \\ 4 & 5 \end{vmatrix}$ (b) $\begin{vmatrix} -1 & 6 \\ 0 & 4 \end{vmatrix}$ (c) $\begin{vmatrix} a-b & b \\ b & a+b \end{vmatrix}$

13. Find the value of each of the following determinants of order 3:

    (a) $\begin{vmatrix} 1 & 2 & 3 \\ 4 & -2 & 3 \\ 0 & 5 & -1 \end{vmatrix}$ (b) $\begin{vmatrix} 4 & -1 & -2 \\ 0 & 2 & -3 \\ 5 & 2 & 1 \end{vmatrix}$

14. Find the inverse of the following matrix:

    $$\begin{pmatrix} 3 & 5 \\ 2 & 3 \end{pmatrix}$$

Chapter 6
Algorithms

Frequently in life we meet problems which, because of their very size and complexity, are difficult for us to solve. One way to cope with these sorts of problems is to break them down into smaller problems each of which can be solved by the application of logic. Breaking down the problem and using a step-by-step approach results in a series of logical instructions one following from the other. We call this an *algorithm*.

The *OED* defines an algorithm as 'a procedure or set of rules for calculation or problem-solving' and adds 'now especially with a computer'. This addition is important. Though it may be easier for us to solve a difficult problem by this logical step-by-step approach, a computer *always* needs to be programmed by means of a series of instructions – i.e., by an algorithm.

6.1   Flowchart preparation

Algorithms can be presented as a straight list of instructions (usually numbered from 1 onwards for ease of reference). However a much more helpful approach is to present these steps by a chart in which each of these instructions or steps is represented by a box, circle, or other geometrical shape, and the logical sequence from the first step to the second and so on is indicated by arrows or flowlines. Such a presentation is called a *flowchart*. Flowcharts, as a means of visual presentation, have a wider application than we have implied so far. They are much used in production, planning and control, office procedure planning, systems network analysis, and so on. However we shall confine ourselves to their applications in a step-by-step approach to problem solving.

6.2   Flowchart symbols

Because the steps in an algorithm do not all perform the same function, it is customary to use different geometrical shapes to indicate what *action* or *function* is taking place. For computer programming, these shapes or symbols are fairly standardised: (see Figure 6.1).

Unfortunately in other flowchart work different symbols are often used for performing much the same functions. For instance, decision trees often use the rectangle for a decision choice and a circle for a chance event. However,

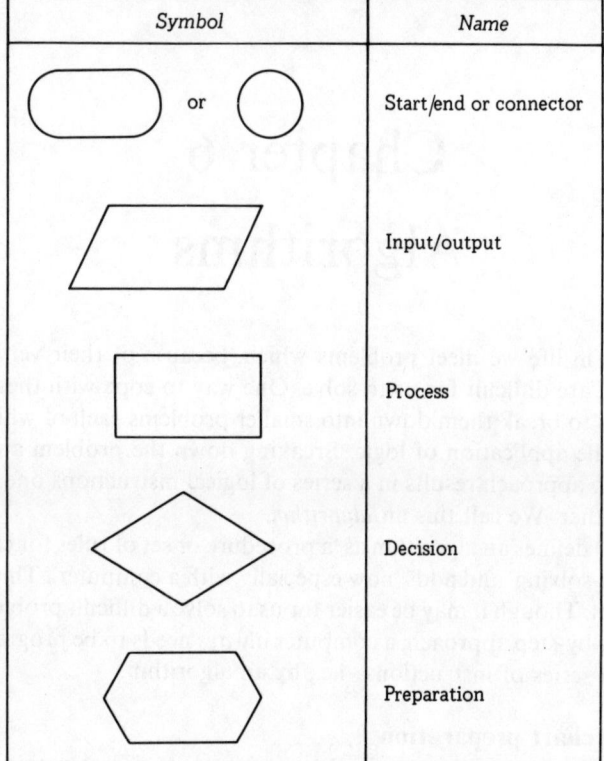

Fig. 6.1

mindful of (and agreeing with) the *OED*'s view we shall concentrate only on the computer programming presentation of an algorithm by a flowchart.

Before considering the layout of a computer program flowchart we need to discuss briefly the meaning of each symbol.

The *start/end or connector symbol* hardly needs explanation. A flowchart conventionally starts at the top and finishes at the bottom. This is so obvious and understandable that both the start and the end symbols are often omitted. The *connector* – usually a small circle – is used either when flowlines are complicated and cross each other, or when the flowchart moves from one page to another and the connecting link needs to be indicated.

The parallelogram is used for *input/output* purposes. Thus we write:

INPUT A,B

Algorithms

to show that two items of data are to be input to memory. Later we could write:

$$\boxed{\text{PRINT A,B}}$$

and the computer (after it had requested, and we had given, the data) would solemnly print out these values for us. The terms INPUT and PRINT are the ones used in BASIC. There are many other program languages than BASIC and often they use different terms for the same thing. (READ and WRITE are typical replacements for INPUT and PRINT.)

The rectangle is used to indicate a *processing operation*. The typical processing operation is an *assignment operation*. In this operation, a value is assigned to a variable:

$$\boxed{\text{LET Z} = 5}$$

In BASIC the word LET can be omitted so we are left with Z = 5. We could now produce an output (say):

$$\boxed{\text{PRINT Z} + 10}$$

in which case the computer would print out 15. Far more complicated arithmetic expressions can of course be used than in this simple example. The variable itself (which must be put down first followed by the 'equals' sign) can be manipulated. Thus we can write:

$$\boxed{Z = Z + 1}$$

This is a nonsense in algebraic terms. However the computer will understand by this assignment statement that it is to increase the value of Z by 1. The new value for Z will be 6. Clearly the effect of reiterative use of this will be to increase the value of Z until told to stop. This is easy for computers; their chief characteristic is that they are good (and immensely quick) at *iterative procedures*.

In computer flowchart work, however, there is another use for the process symbol and this is in providing a *macro instruction*. This is a shorthand for a set of programmed instructions. An example of this would be:

$$\boxed{\begin{array}{l}\text{FIND MEAN AND STANDARD DEVIATION}\\ \text{OF FIGURES INPUTTED}\end{array}}$$

One of the useful properties of a computer is that it is able to compare two figures and *decide* whether they are equal or which of the two is greater. In practice, this comparison and consequential decision is made by putting a question to the computer.

An example of a flowchart including this decision, and showing how the computer might then be programmed to operate is shown in Figure 6.2.

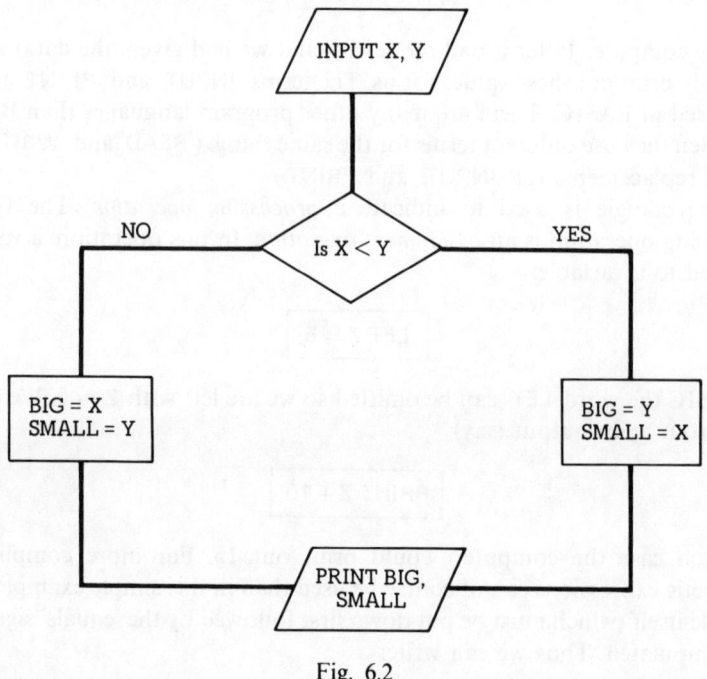

Fig. 6.2

A decision symbol is the only symbol with more than one exit. In this case, depending on which exit is taken, a *different process* will take place. In each case it is an assignment statement where a value is assigned to the variable. Finally we have an *output*. It is evident that the effect of all this is that the computer will determine which is the larger of the two numbers inputted, and print them out with the larger first.

The *preparation symbol* is not always used by programmers. Some prefer to use the process symbol. In a sense this is understandable since the preparation symbol is used for a special type of process. Frequently in programming we need to assign an initial value to a variable before subjecting it to some procedure or process. In the example above in which we increased the value of the variable (Z) iteratively by 1, we had to start off with an initial value. In our case it was 5 and we could have used a preparation symbol and written

$$Z = 5$$

The preparation symbol is widely used in counting or summing up a large

Algorithms

number of variables. If we wish to find the sum of the salaries of all employees in a business we could write initially:

$$\langle \text{SUM} = 0 \rangle$$

and then

$$\boxed{\text{SUM} = \text{SUM} + \text{SALARY}}$$

as we input from records one by one the salaries of employees.

6.3 Loops

Frequently in computer work a particular procedure is required to be repeated many times. Broadly there are two ways in which the limit to the repetition can be defined. In the first the procedure is repeated until some value (called the terminating condition) is reached. In BASIC this is called REPEAT ... UNTIL. An alternative way is to make the computer repeat the procedure a fixed number of times. In BASIC this is called FOR ... NEXT. In practice, there is not much difference between the two methods.

One example of the first type would be if we wished to list the positive odd integers which, along with their squares, do not exceed a certain number N (say). This figure, N, will be inputted into the program at the start. We use a variable K (say) whose initial value will be 1 and which will be increased by 2 at a time (to keep it as an odd number). In each case we shall evaluate the square of K, i.e. $K^2 = L$ (say). The terminating condition will be when L exceeds N. A flowchart, with the reiterative procedure (called a DO loop) and conventionally indicated as a dotted line is shown in Figure 6.3.

An alternate way is to test for $L > N$ at the *beginning* of the loop. This avoids the last line of a print out showing the first value for L which in fact has failed.

6.4 Pseudocode programs

Computers cannot handle flowcharts. They can act on a series of instructions which they can understand and which are presented in sequence – i.e., inputted line by line. Pseudocodes are a transformation of a flowchart into a suitable sequential list. However pseudocodes still need some further translation into the appropriate high-level (BASIC, FORTRAN, COBOL, etc.) language which will be used in the program before the computer can handle them. Pseudocodes are, broadly, *language independent*. They can either be seen as an intermediate step between a flowchart and a particular computer program ready for input, or as an alternate way of presenting algorithms.

Pseudocode instructions are to be followed in order – that is, they start at the first line and end with the last. As with flowcharts the start and end instructions

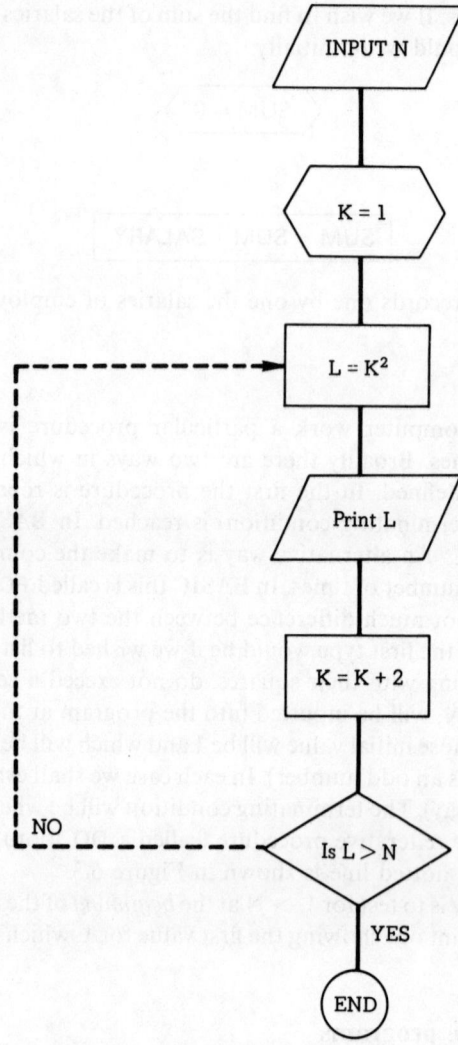

Fig. 6.3

are usually omitted. Pseudocodes use two special types of operation:

(a) Selection logic.
(b) Iteration logic.

Selection logic is the structure which is used to deal with a *choice*. It replaces the decision symbol in a flowchart. In the most usual situation there are only two exits from the decision point, so that there are just the two alternatives. Generally if a certain condition applies, then one procedure is followed,

Algorithms

otherwise another procedure is indicated. The pseudocode for this is:

>IF condition
>>(Procedure A)
>
>ELSE
>>(Procedure B)
>
>ENDIF

Conventionally the procedures are inset. This helps in the identification of this structure, often referred to by the term IFTHENELSE.

Earlier we gave a flowchart for comparing two numbers, then deciding which is the larger, and finally printing them out with the larger number first. The pseudocode program for this would be as in Figure 6.4.

>INPUT X, Y
>IF X < Y
>>BIG = Y
>>SMALL = X
>
>ELSE
>>BIG = X
>>SMALL = Y
>
>ENDIF
>
>PRINT BIG, SMALL

>Fig. 6.4

Pseudocode structures for handling the situation where there is more than one exit from a decision do exist. The structure for three exits follows along much the same lines (see Figure 6.5).

Once again the procedures are recessed for easy identification of the structure.

Iteration logic is the way a pseudocode problem handles loops in flowcharts. A loop is a procedure followed a number of times. We noted that broadly there are two ways in which a limit to the repetition can be defined. In the first the procedure is repeated until the termination condition is reached or (which is much the same thing) while the prior condition still exists. In pseudocode language these are written as:

>DOUNTIL condition is reached
>
>DOWHILE the earlier condition exists

In Figure 6.3 we showed the flowchart for listing the positive odd integers whose squares do not exceed a certain number, N (say). We can use either of

IF condition
   (Procedure A)
ELSEIF condition
   (Procedure B)
ELSE
   (Procedure C)
ENDIF

Fig. 6.5

two pseudocode programs to show this depending on whether we prefer DOUNTIL or DOWHILE. They are as in Figure 6.6.

```
INPUT N                INPUT N
K = 1                  K = 1
DOUNTIL L > N          DOWHILE L < N
   L = K²                 L = K²
   PRINT L                PRINT L
   K = K + 2              K = K + 2
ENDDO                  ENDDO
```

Fig. 6.6

The second way to limit the number of times the loop procedure is done is simply to instruct it to do it for a *fixed number of times*. Thus the flowchart for Figure 6.3 could also be presented as a DO loop in the pseudocode set out in Figure 6.7.

```
INPUT N
DO K = 1 to N by 2
   L = K²
   PRINT L
ENDDO
```

Fig. 6.7

Though in this book we are not directly concerned with computer work as such, it may be of interest to show one example of the final step in the sequence 'flowchart → pseudocode → program'. The (BBC) BASIC program corresponding to this last pseudocode is as in Figure 6.8.

Algorithms

```
INPUT N
K = 1
REPEAT
K = K + 2
L = K ↑ 2
UNTIL L > N
PRINT K, L
```

Fig. 6.8

6.5 Exercises

1. The formulae for the perimeter P, and the area A of a triangle whose sides are of lengths a, b, c are:

    (i) $P = a + b + c$
    (ii) $A = \sqrt{s(s-a)(s-b)(s-c)}$

    where $s = \dfrac{a+b+c}{2}$

    Construct a pseudocode program with inputs a, b, c and outputs the perimeter and the area.

2. A local car dealer has just installed a microcomputer. He used CODE = 1 for new cars and CODE = 2 for secondhand cars and CODE = 3 for all car accessories. His salesmen's commissions are as follows:

    (i) On a new car – 3% of the sales price with a maximum of £300.
    (ii) On a secondhand car – 5% of the sales price with a minimum of £75.
    (iii) On all car accessories – a straight 6% of the price.

    Construct a pseudocode program with inputs CODE and PRICE and output COMMISSION.

3. The monthly charge for local telephone calls in Atlanta City is:

    (i) For up to 100 calls – $9.00.
    (ii) Plus 7¢ per call for each of the next 100 calls.
    (iii) Plus 5¢ per call for each call beyond 200.

    Construct a pseudocode program with input the number of LOCAL calls and output the CHARGE.

4. The condition that three numbers a, b, c could represent the lengths of the sides of a triangle is that the following shall apply:

    (i) $a < (b + c)$
    (ii) $b < (a + c)$
    (iii) $c < (a + b)$

Construct a pseudocode program with inputs a, b, c, and output YES or NO according to whether a triangle can or cannot be formed.

5. For each of the following inputs state:

   (i) if an array is defined, and
   (ii) if so what are the dimensions and numbers of elements.

   (a) INPUT TEST (J)
   $J = 1$ to 82

   (c) INPUT NAME, RATE
   HOURS

   (b) INPUT B(J,K,L)
   $J = 1$ to 7
   $K = 1$ to 9
   $L = 1$ to 5

   (d) INPUT N $A_{J,K}$
   $J = 1$ to N
   $K = 1$ to $N+1$

6. A saleswoman's commission is 2.5% of her monthly sales with an extra bonus of £100 if her sales exceed £20,000.
   Write a pseudocode program which gives the names of the saleswoman and calculates her COMMISSION.

7. Members of the Institute of Accountants in Rutland have to pay an annual subscription of £R 85.00 together with an additional levy which rises with the size of their firm.
   For 'small' firms (that is those with 6 partners or less) these additional levies are:

   (i) for a sole practitioner – nil
   (ii) for a 2/3 partner firm – £R 5.00
   (iii) for a 4/6 partner firm – £R 10.00

   (a) Draw a flowchart with INPUT number of partners and OUTPUT total annual dues (that is annual subscription plus levy, if any).
   (b) Construct a pseudocode program for the above.

8. The following computer program is written in the language BASIC and has been constructed to establish the present value of a project to the nearest £ over a number of years.

```
10   REM PRESENT VALUE PROGRAM
20   PRINT "ENTER NUMBER OF ANNUAL CASH FLOWS"
30   PRINT "AFTER INITIAL PAYMENT"
40   INPUT N
50   PRINT "ENTER RATE OF INTEREST AS A PERCENTAGE"
60   INPUT R
70   LET A = 0
80   PRINT "ENTER INITIAL OUTFLOW"
90   INPUT P
100  LET A = −P
110  LET D = 1
```

Algorithms 103

```
120  FOR K = 1 TO N
130  PRINT "ENTER 1 FOR INFLOW OR -1 FOR OUTFLOW"
140  INPUT F
150  PRINT "ENTER CASHFLOW"
160  INPUT C
170  LET D = D*(1 + R/100)
180  LET A = A + (F*C)/D
190  NEXT K
200  LET V = INT (A)
210  LET C = A - V
220  IF C > = 0.5 THEN LET V = V + 1
230  PRINT "PRESENT VALUE = £"; V
240  STOP
250  END
```

Given that the numbers for input to the program are 3, 10, 12000, 1,6600, 1,6000, 1,4500

Required:

(a) What is the action of the program at

   (i) line 40
   (ii) line 60
   (iii) line 70
   (iv) line 90
   (v) line 100
   (vi) line 110

(b) What is the effect of the instruction at line 120 taken with the instruction at line 190?

(c) What is the action of the program at line 170?

(d) The content of the address A is 2339.5943 when the program reaches line 200.

   Trace the action of the program from line 200 to line 230

(*The Association of Certified Accountants*, December 1984)

9. Write programs for the following in BASIC (or any other suitable high-level language, if you prefer).

   (a) *A VAT calculator* – VAT is currently 15%. Your program, however, should include the option of changing the current rate. Ensure that the results are printed rounded to the nearest penny.

   (b) *A loan repayment schedule for fixed monthly repayments* – assume initially that the interest rate is *fixed*. If the interest rate changes the program can still be used by re-running with a 'new' loan equal to the reduced amount borrowed and of course, the new interest rate.

(c) *Fixed asset, written down values and depreciation amounts* – in the form of a table for both (i) the straight-line method and (ii) the reducing balance method.
(d) *A four-period moving average* – this program is particularly useful for providing updated averages over the last four periods (typically weeks or months) for sales.
(e) *Factorisation of a number* – use the method of repeated division. First try to divide by 2, initially on the original number, then (if appropriate) on the number remaining and so on until the remaining number is not divisible by 2. Then repeat the whole process with 3, then with 5 etc.
(f) *Matrix multiplication* – enter dimensions of matrices in the form of arrays and remember that two-dimensional arrays are voracious memory eaters! Include in the program a check to ensure first that the multiplication is possible.
(g) *Calculation of internal rate of return on an investment project* – base your program on that shown in question 8.

Chapter 7
Linear and Higher-power Equations

7.1 Use of equations

We have already used equations on a number of occasions, and throughout this book we shall continue to use them. Equations frequently summarise important facts. Thus the fact that the simple interest (I) at an interest rate (r) on a principal (P) is the rate multiplied by the principal can be summarized as $I = rP$. Equations can also be descriptive and useful for management decisions. In Chapter 3 we saw how the $I = rP$ equation can be helpful in capital investment appraisal decisions. Equations can provide models either at macroeconomic level – for instance, the input output model for the economy (Chapter 5) – or at the microeconomic level. One example of the latter is that for a product produced by a firm, total costs are $a + bx$, where a are the fixed costs, b the variable costs per unit and x the number of units produced (provided that certain basic assumptions and conditions are met). If we call total costs y, these costs are then given by the equation $y = a + bx$.

7.2 Linear equations

In much work in the management field the type of equations used are *linear equations*. The term 'linear' means 'appertaining to a line' and the derivation arises from the fact that the graph of a linear equation in two variables is a *straight line*. Here we are concerned with the more formal definition that a linear equation is an equation in a number of variables with no higher powers of any variables than the first. This means that x and y can be included (if x and y are the variables) but not x^2, xy, y^2, x^3, y^3, etc. (all of which have powers higher than the first). Frequently in this chapter, we shall require more variables than two. We shall still keep to the convention (particularly useful for trying to understand a strange equation!) that the earlier letters in the alphabet (e.g. a, b, c) are used for *constants* (i.e., they do not change), and the later letters (e.g., x, y, z) for *variables* (that is, their value may vary). In order to allow for large numbers of different constants and variables we can use subscripts. Constants will be, for example, $a_1 b_1$ (or even a_{11}, a_{12}, etc. with two subscripts) and variables might be x_1, y_1 (or x_{11}, x_{12}, etc.).

Typically the values of the constants in an equation are known; the aim is to find the values of the *variables* (otherwise called 'unknowns') in terms of the constants. We start with the simplest case of a linear equation in one unknown, then deal with linear equations in two and three unknowns, and finally consider the general case where there are n unknowns.

Any linear equation in one unknown (say) x must be of the form – or must easily be reduced to – $ax = b$, where a and b are constants. Cross-multiplying, we have:

$$x = b/a$$

This is the solution to the equation – that is, we have found the value of x which satisfies it. Provided $a \neq 0$ we see that there is one single value for the unknown.

For linear equations in two unknowns (say x_1, x_2) we need two equations in order to be able to solve them. Later, in Chapter 8, we shall see (as we noted in our definition of the term 'linear') that a single linear equation in two unknowns is graphically represented by a *straight line*. Graphically, too, the solution of two linear equations is represented by the intersection of the lines representing them. However, we are solely concerned now with the algebraic solution.

Any two linear equations in two unknowns are of the form – or can easily be reduced to the form:

$$a_{11}x_1 + a_{12}x_2 = b_1$$
$$a_{21}x_1 + a_{22}x_2 = b_2$$

We need to find x_1 and x_2 in terms of the constants (as and bs).

To obtain these values the best approach in general is to multiply the first equation by a_{22} throughout, and the second by a_{12}. We have:

$$a_{11}a_{22}x_1 + a_{12}a_{22}x_2 = a_{22}b_1$$
$$a_{12}a_{21}x_1 + a_{12}a_{22}x_2 = a_{12}b_2$$

Subtraction now gives:

$$(a_{11}a_{22} - a_{12}a_{21})x_1 = a_{22}b_1 - a_{12}b_2$$

Hence, provided

$$a_{11}a_{22} - a_{12}a_{21} \neq 0$$

$$x_1 = \frac{a_{22}b_1 - a_{12}b_2}{a_{11}a_{22} - a_{12}a_{21}}$$

and similarly

$$x_2 = \frac{a_{11}b_2 - a_{21}b_1}{a_{11}a_{22} - a_{12}a_{21}}$$

Linear and Higher-power Equations

It is easy to see that the denominator (bottom line) of these solutions is that determinant which is formed from the coefficients of x_1 and x_2. Let us call this determinant $|A|$ for short. So that:

$$\begin{vmatrix} a_{11} & a_{12} \\ a_{21} & a_{22} \end{vmatrix} = a_{11}a_{22} - a_{12}a_{21} = |A|$$

Using this determinant as a base we now consider two other determinants. The first is made by replacing the first column (which are the coefficients of x_1) by the column made up of the constants b_1 and b_2. The second is made by replacing the second column (which are the coefficients of x_2), also by the column made up of the constants b_1 and b_2. If we call these two determinants $|A_1|$ and $|A_2|$ respectively for short, what we are saying is:

$$\begin{vmatrix} b_1 & a_{12} \\ b_2 & a_{22} \end{vmatrix} = a_{22}b_1 - a_{12}b_2 = |A_1|, \text{ and}$$

$$\begin{vmatrix} a_{11} & b_1 \\ a_{21} & b_2 \end{vmatrix} = a_{11}b_2 - a_{21}b_1 = |A_2|$$

The solution to the linear equations in two unknowns x_1 and x_2 can now be neatly written as:

$$x_1 = \frac{|A_1|}{|A|} \quad \text{and} \quad x_2 = \frac{|A_2|}{|A|}$$

Any three linear equations in three unknowns are of the form – or can easily be reduced to the form:

$$a_{11}x_1 + a_{12}x_2 + a_{13}x_3 = b_1$$
$$a_{21}x_1 + a_{22}x_2 + a_{23}x_3 = b_2$$
$$a_{31}x_1 + a_{32}x_2 + a_{33}x_3 = b_3$$

and by analogy with the two-equation system we can see that the solution will be

$$x_1 = \frac{|A_1|}{|A|} \quad x_2 = \frac{|A_2|}{|A|} \quad x_3 = \frac{|A_3|}{|A|}$$

where

$$|A| = \begin{vmatrix} a_{11} & a_{12} & a_{13} \\ a_{21} & a_{22} & a_{23} \\ a_{31} & a_{32} & a_{33} \end{vmatrix} \quad |A_1| = \begin{vmatrix} b_1 & a_{12} & a_{13} \\ b_2 & a_{22} & a_{23} \\ b_3 & a_{32} & a_{33} \end{vmatrix}$$

$$|A_2| = \begin{vmatrix} a_{11} & b_1 & a_{13} \\ a_{21} & b_2 & a_{23} \\ a_{31} & b_3 & a_{33} \end{vmatrix} \quad |A_3| = \begin{vmatrix} a_{11} & a_{12} & b_1 \\ a_{21} & a_{22} & b_2 \\ a_{31} & a_{32} & b_3 \end{vmatrix}$$

Any particular determinant in the numerator (top line) of this group is formed by replacing the appropriate column of coefficients of x by the column made up of the coefficients b_1, b_2 and b_3. Thus for $|A_2|$ we replace the second column entries (which are the coefficients of x_2) by the column of the b coefficients. More generally the value of $|A_i|$ is the value of $|A|$ with the ith column replaced by the column of the b coefficients.

Linear equations in n unknowns are of the form – or can easily be reduced to the form:

$$a_{11}x_1 + a_{12}x_2 + \ldots + a_{1n}x_n = b_1$$
$$a_{21}x_1 + a_{22}x_2 + \ldots + a_{2n}x_n = b_2$$
$$\ldots\ldots\ldots\ldots\ldots\ldots\ldots\ldots\ldots\ldots\ldots\ldots\ldots\ldots\ldots$$
$$a_{n1}x_1 + a_{n2}x_2 + \ldots + a_{nn}x_n = b_n$$

By analogy, again, we can see that the solution will be:

$$x_1 = \frac{|A_1|}{|A|} \ldots x_i = \frac{|A_i|}{|A|} \ldots x_n = \frac{|A_n|}{|A|}$$

where $|A|$ is the determinant formed by all the coefficients of $x_1, x_2 \ldots x_n$ and $|A_i|$ is the same determinant but with the ith column replaced by the column of the b coefficients.

This solution of linear equations by the use of the determinants has an elegant appearance. It is known as *Cramer's rule* after its discoverer. Notwithstanding its appearance this approach is not much used for solving equations with greater than three unknowns. More efficient methods – Gaussian elimination and matrix inversion – exist, and it is to these that we now turn our attention.

7.3 Gaussian elimination

This technique which is simple and straightforward (if rather tedious) has two parts. In the first the original equations are replaced by new equations each of which (except the first) has one less unknown. The final form of this part is a set of equations in triangular form. In the second part these equations are solved, starting with the last one, by the technique of *back substitution*. We show this technique, first by way of an example with three unknowns, and then for the general case of n equations with n unknowns.

Consider these three equations with three unknowns x, y and z.

$$4x - 12y - 8z = 24 \ldots \quad (7.1)$$
$$2x - 4y + 2z = 18 \ldots \quad (7.2)$$
$$-3x + 8y + 9z = -9 \ldots \quad (7.3)$$

Linear and Higher-power Equations 109

The first stage in the first part (the 'triangularisation' part) is to replace equations (7.2) and (7.3) by two new equations in which x has been eliminated. We do this by using equation (7.1) in combination with equation (7.2), and then with equation (7.3).

To eliminate x from (7.1) and (7.2), we multiply (7.1) by the negative of the quotient of the coefficients of x in (7.1) and (7.2). These coefficients are 4 and 2. The negative of their quotient is $-\frac{2}{4}$. Hence we multiply (7.1) by $-\frac{2}{4}$ and we have:

$$[(7.1) \times (-\tfrac{2}{4})] \quad -2x + 6y + 4z = -12$$
$$[(7.2)] \quad 2x - 4y + 2z = 18$$

Adding $\quad 2y + 6z = 6,\quad$ which is a new equation: we call it (7.2').

Similarly to eliminate x from (7.1) and (7.3), we multiply (7.1) by $-(-\tfrac{3}{4}) = \tfrac{3}{4}$, and we have:

$$[(7.1) \times (\tfrac{3}{4})] \quad 3x - 9y - 6z = 18$$
$$[(7.3)] \quad -3x + 8y + 9z = -9$$

Adding $\quad -y + 3z = 9,\quad$ which is a new equation: we call it (7.3').

We now have reduced the original system to these three equations:

$$4x - 12y - 8z = 24 \ldots \tag{7.1}$$

$$2y + 6z = 6 \ldots \tag{7.2'}$$

$$-y + 3z = 9 \ldots \tag{7.3'}$$

The second and final stage in this first part is to replace (7.3') by a new equation in which y has been eliminated. We do this by using (7.3') in combination with (7.2').

To eliminate y we multiply (7.2') by the negative of the quotient of the coefficients of y in (7.2') and (7.3'). Hence we multiply (7.2') by $-(-\tfrac{1}{2})$ (i.e. $\tfrac{1}{2}$) and we have:

$$y + 3z = 3$$
$$-y + 3z = 9$$

Adding $\quad 6z = 12,\quad$ which is a new equation: we call it (7.3'').

We now have a set of equations in *triangular form*. They are:

$$4x - 12y - 8z = 24 \ldots \quad (7.1)$$
$$2y + 6z = 6 \ldots \quad (7.2')$$
$$6z = 12 \ldots \quad (7.3'')$$

This concludes the first part.

In the second part we *solve backwards* – i.e., from the last equation (for z), then from the next one (for y), and finally in the first one (for x).

The last equation gives $z = 2$ (by cross-multiplying).

Substituting this value in (7.2'), we have

$$2y + 12 = 6$$

giving

$$y = -3$$

Substituting these values for x and y in (7.1), we have:

$$4x + 36 - 16 = 24$$

giving

$$x = 1$$

Hence the method of back substitution gives the solution as $x = 1$, $y = -3$ and $z = 2$.

In the more general case, we have n unknowns and n equations:

$$a_{11}x_1 + a_{12}x_2 + \ldots + a_{1n}x_n = b_1 \quad (7.1)$$
$$a_{21}x_1 + a_{22}x_2 + \ldots + a_{2n}x_n = b_2 \quad (7.2)$$
$$\ldots\ldots\ldots\ldots\ldots\ldots\ldots\ldots\ldots\ldots\ldots\ldots$$
$$a_{n1}x_1 + a_{n2}x_2 + \ldots + a_{nn}x_n = b_n \quad (7.n)$$

The first stage of the first part (the triangularisation part) is to replace equations (7.2), (7.3) ... etc. by new equations from which x_1 has been eliminated. We do this by using equation (7.1) in combination with equation (7.2), then equation (7.3), and so on for all the remaining equations.

To eliminate x_1 from (7.1) and (7.2) we multiply (7.1) by the negative of the quotient of the coefficients of x_1 in (7.1) and (7.2). These coefficients are a_{11} and a_{21}. The negative of their quotient is $-a_{21}/a_{11}$. Hence we multiply by $-a_{21}/a_{11}$. We then add this equation and equation (7.2). The resulting new equation (7.2') will have had x_1 eliminated from it.

We repeat this process for all the remaining equations. For example to eliminate x_1 from (7.1) and the kth equation multiply (7.1) by $-a_{k1}/a_{11}$. We then add this equation and the kth equation. The resulting new equation will have had x_1 eliminated from it. The denominator (a_{11}) is called the *pivot*.

We will now have reduced the original system to n equations. All the equations (except the first which will have remained unchanged) will have had x_1 eliminated from them.

Linear and Higher-power Equations 111

In the second stage of the first part we replace all the equations (7.3'), (7.4'), etc. by new equations in which x_2 has been eliminated. We do this by using equation (7.2') in combination with equation (7.3'), then with equation (7.4'), and so on for all the remaining equations. In each case we multiply (7.2') by the negative of the quotient of the new coefficients of x_2 and add. In each case the resulting new equation will have had x_2 eliminated from it.

At the end of the second stage of the first part the first equation will still have remained unchanged. The second will have had x_1 eliminated from it. All the remaining equations will have had x_1 and x_2 eliminated from them.

The third and subsequent stages follow along exactly the same lines. We shall arrive finally at a set of equations in triangular form. In the second part we solve these *backwards* – that is, starting from the last equation, which will give x_n immediately, then substituting this value into the second from the last equation to give the value for x_{n-1}, and so on all the way back to the first equation which will give the value of x_1. Hence the method of back substitution will give all the values of $x_1, x_2 \ldots x_n$.

This Gaussian elimination technique proceeds by regular steps in which variables are successively reduced from the equations. Because of this it is frequently referred to by the phrase 'the successive elimination of variables'. This technique is obviously in a form which makes it suitable for computerisation. Most computer programs for solving linear equations do in fact use this approach. Such programs, though simple, are rather long and are not given here. Computer programs for the solution of non-linear equations are given later in this chapter.

7.4 Matrix inversion method

The general system of n linear equations in n unknowns, that is:

$$a_{11}x_1 + a_{12}x_2 + \ldots + a_{1n}x_n = b_1$$
$$a_{21}x_1 + a_{22}x_2 + \ldots + a_{2n}x_n = b_2$$
$$\ldots\ldots\ldots\ldots\ldots\ldots\ldots\ldots\ldots\ldots\ldots$$
$$a_{n1}x_1 + a_{n2}x_2 + \ldots + a_{nn}x_n = b_n$$

can be written as $AX = B$, where

$$A = \begin{pmatrix} a_{11} & a_{12} & \ldots & a_{1n} \\ a_{21} & a_{22} & \ldots & a_{2n} \\ \ldots & \ldots & \ldots & \ldots \\ a_{n1} & a_{n2} & \ldots & a_{nn} \end{pmatrix} \quad X = \begin{pmatrix} x_1 \\ x_2 \\ \ldots \\ x_n \end{pmatrix} \quad B = \begin{pmatrix} b_1 \\ b_2 \\ \ldots \\ b_n \end{pmatrix}$$

Provided the inverse of A (i.e., A^{-1}) is known, it is easy to solve these linear equations. For the equation can be written:

$$X = A^{-1}B.$$

7.5 Iterative method solution of equations

All that is required now is the matrix multiplication of A^{-1} by B and we can read off the values for $x_1 x_2 \ldots x_n$. Ways for finding the inverse of a matrix were discussed in Chapter 5.

7.5 Iterative method solution of equations

Equations of higher degree than the second are not easy to solve. Some business problems, however, require a solution. For such problems finding a very precise solution, fortunately, is not important. For example, provided we can find the roots (i.e., the values which satisfy the equations) of a cubic such as,

$$x^3 - 10.7x^2 + 32.4x - 25.2 = 0$$

fairly accurately, then we shall be satisfied.

Computers are particularly keen on this sort of work. If we can set up a program which will run automatically in a loop time and time again and always getting nearer to the solution, we shall have solved the problem in practical terms. Such a method is of course, an *iterative process*. We start off by applying one iterative approach to a quadratic, then to our cubic and finally to an equation with powers of four. There are other iterative methods to the solutions of equations than the one we shall be exploring, and some of these we mention briefly at the end.

Consider the equation:

$$x^2 - 2.8x + 0.75 = 0 \tag{7.4}$$

because this is a quadratic and of the form $ax^2 + 2bx + c = 0$ there is a formula for the solution. This is:

$$x = \frac{-b \pm \sqrt{b^2 - ac}}{a}$$

which gives the roots to be 2.5, 0.3.

We first demonstrate how an iterative method can solve this equation, and then show how this same process can be used for those equations in higher powers of x than two for which no simple formula for their solution exists.

Equation (7.4) can be rewritten in the form:

$$x = \frac{x^2}{2.8} + \frac{0.75}{2.8}$$

Let us suppose that we start off with (any) value for x and find the value of the right-hand side of this equation. This value is going to be our 'new' value for x. Symbolically:

$$x_{r+1} = \frac{x_r^2}{2.8} + \frac{0.75}{2.8} \quad \text{for} \quad r = 1, 2 \ldots, \text{etc.}$$

Clearly we want the value of x_{r+1} to equal x_r (or at least match it very closely). If we can find such a value then we have found one root of the equation.

Linear and Higher-power Equations

If – and only if – repeated use of this iterative process brings the difference between x_{r+1} and x_r ever closer then we shall have succeeded.

The program to do this (in BBC BASIC and using capital X and Y) is as in Figure 7.1. It will repeatedly find the new value of x_{r+1} ($=$ Y in the program). In line 80 we test whether the difference is greater than Q. In line 30 we have already defined Q to be a very small number.($1E - 6 = 0.000001$). When we are as close as this we say that for all practical purposes we have solved the equation.

```
10   REM   IT PROCESS
20   INPUT   'X1'; X
30   R = 1: Q = 1E – 6
40   R = R + 1
50   Y = X
60   X = X * X/2.8 + 0.75/2.8
70   PRINT R, X
80   IF ABS (Y–X) > Q THEN GOTO 40
90   PRINT:PRINT 'ROOT IS'; X
```

Fig. 7.1

If we run the program the computer will ask for the value of X1. If we start with 0 and then go through the digits the computer will print out a series of lines – such as for (say) X = 2 (see Figure 7.2) and then stop, printing out 'ROOT IS', and adding the result, that is 0.300000078. These results for different values of X are as in Figure 7.3.

1.	1.69642857
2.	1.29566782
3.	0.867412536
4.	0.536573039
5.	0.370682366
6.	0.316930506
7.	0.303730338
.	
.	
14.	0.300000078

Fig. 7.2

X1	ROOT IS:
0	0.299999867
1	0.300000128
2	0.300000078
3	TOO BIG AT LINE 60
4	TOO BIG AT LINE 60

Fig. 7.3

Whatever values of X we choose beyond this the computer will print out TOO BIG. From the first two lines we might reasonably assume that a root is very close indeed to 0.3. If we put this value in for X the computer confirms this by saying ROOT IS 0.3.

It seems therefore that we have found one root (0.3), but are unable to find the other by this process. It is, however, now easy to find this other root. If we call it α, the equation must be:

$$(x - 0.3)(x - \alpha) = 0$$

that is

$$x^2 - (\alpha + 0.3)x + 0.3\alpha = 0$$

This equation must be the same as:

$$x^2 - 2.8x + 0.75 = 0$$

Equating the coefficients of x (or the constant terms) gives immediately that $\alpha = 2.5$, so we have found the roots are indeed 0.3 and 2.5.

This method is effective for higher powers of x than 2. We have shown it for a quadratic (for whose solution, as we say, a simple formula does exist) not only to show the method, but also to provide a graphical explanation of why the root 0.3 can be arrived at, whereas the other root cannot be directly (though it is easy to determine once the first root is known by further work). Figure 7.4 shows the graphs of the straight line $y = x$ (*OA* in the graph) and of the quadratic curve

$$y = \frac{x^2}{2.8} + \frac{0.75}{2.8} \qquad (7.2)$$

For $x = 0.3$ and for $x = 2.5$ we shall get the same values for y (0.3 and 2.5) as expected, since these are the roots of the equation $x = \dfrac{x^2}{2.8} + \dfrac{0.75}{2.8}$.

It is easy to construct roughly the quadratic curve 7.2. We already have two points on it, and a few more will give the shape of the curve fairly adequately. If, for instance, we put $x = 2$, we get

$$y = \frac{4}{2.8} + \frac{0.75}{2.8} = 1.4286 + 0.2679$$

$$= \text{(approx.) } 1.696$$

The point P_1 (2, 1.696) therefore lies on the curve.

In order to understand the way the iterative process searches for the root of the equation let us suppose we start with the value $x_r = 2$ (i.e., OQ_1 in Figure 7.4). The computer will calculate the value of y that is, 1.696 or $P_1 Q_1$ in Figure 7.4) and then put this figure in as the new x. This new x will be OQ_2 (because $OQ_2 = Q_2 R_2 = P_1 Q_1$). In the second iteration the computer will find

Linear and Higher-power Equations

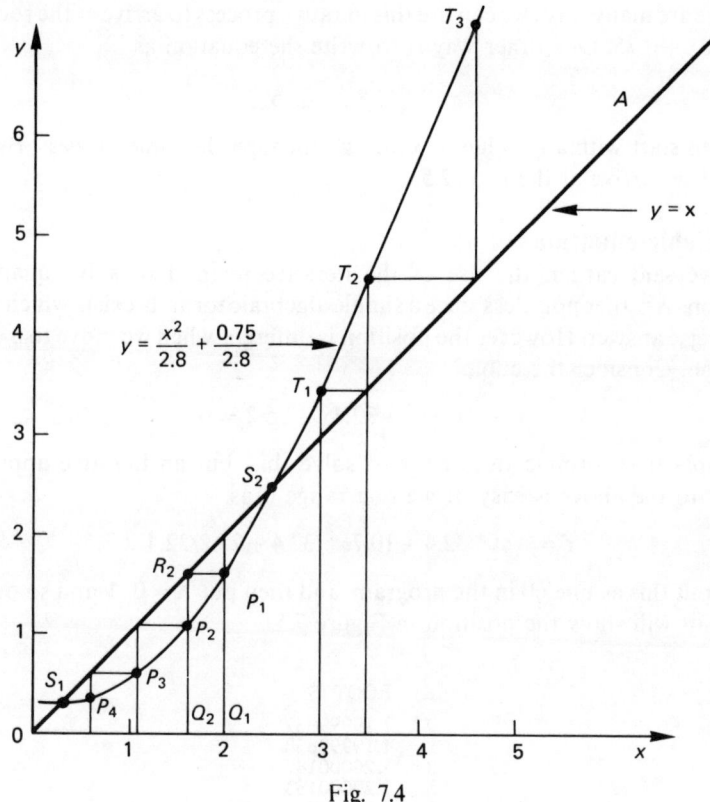

Fig. 7.4

the next new value of x (that is, the new value for y). This is P_2Q_2. After the next iteration the point P_3 will be arrived at.

Although up to now this process of getting nearer to (that is 'converging' to) the root at S_1 is slow, a few more iterations and it is easy to see that we shall soon be very close to S_1. This is the diagramatic illustration of the line-by-line results printed out by the computer showing how we arrive at the root S_1. If we had not put in the caveat, contained in lines 30 and 80 of the program (which stops the computer when it gets near enough to the answer), it would have worked away to the limit of the power of the machine, getting ever closer to the precise answer.

If we start with $x = 3$ and use the same process, the results will get further and further away. Diagramatically this is shown by the points T_1, T_2, T_3, etc. This is a 'divergent' process – we do not get nearer but further away from the root. Very soon the computer prints something like TOO BIG as it reaches the limit of its capacity.

There are many ways we can use this iterative process to arrive at the roots of $x^2 - 2.8x + 0.75$. One other way is to write the equation as

$$x = \sqrt{2.8x - 0.75}$$

and then start with any value of x and go through the same process. By this method we arrive at the root 2.5.

7.6 Cubic equations

As we said earlier, the use of the iterative method to solve quadratic equations is rather pointless since a simple algebraic formula exists which gives the precise answer. However the position is different when we move to a cubic equation. Consider the cubic

$$x^3 - 10.7x^2 + 32.4x - 25.2 = 0$$

A complicated formula does exist to solve this, but an iterative approach similar to the above is easy. If we re-arrange it as:

$$x = -x^3/32.4 + 10.7x^2/32.4 + 25.2/32.4$$

and input this as line 60 in the program, and then put X = 0, 1 and so on, the print out will show the position in Figure 7.5.

X1	ROOT IS:
0	1.19999837
1	1.19999858
2	1.20000147
3	1.20000153
4	5.99999838
5	5.99999872
6	6
7	6.00000158

Fig. 7.5

We can test whether the roots are indeed 1.2 and 6 by putting these in as values for x. In each case the print out confirms this. We therefore know that two out of the three roots are 1.2 and 6. If we suppose that the third root is α the equation must be:

$$(x - 1.2)(x - 6)(x - \alpha) = 0$$

that is, multiplying out

$$x^3 - (7.2 + \alpha)x^2 + (7.2\alpha + 7.2)x - 7.2\alpha = 0$$

This equation is the same as:

$$x^3 - 10.7x^2 + 32.4x - 25.2 = 0$$

Equating the coefficients of x^2 (or of x, or the constant term) gives immediately that $\alpha = 3.5$.

The roots are therefore 1.2, 3.5 and 6.0.

Linear and Higher-power Equations   117

7.7 Power of four

Consider the equation

$$x^4 - 19x^3 + 113x^2 - 245x + 150 = 0$$

No simple formula exists to find the roots of this equation. However if we write:

$$x = x^4/245 - 19x^3/245 + 113x^2/245 + 150/245$$

and input this as a new line 60 in the program, and then try, more or less haphazardly, the values of 0, 1, 2, 4, 5, 6, 9, 12 for x, the print out will show the position in Figure 7.6.

X	ROOT IS
0	0.99999817
1	1.
2	1.00000183
4	4.99999514
5	5.
6	5.00000467
9	5.00000515
12 ETC	TOO BIG

Fig. 7.6

For inputs of $x = 1$ and $x = 5$, the print out confirms that the roots are 1 and 5 respectively. We have therefore found two of the roots, but are unable – at any rate easily – to find the other two by this process. In fact, since the other roots happen to be simple whole numbers (3 and 10) we shall hit on the exact right answers by trying these figures, and the computer program will confirm it. (A better constructed example would not do this!) If we call these roots α and β, the equation must be

$(x - 1)(x - 5)(x - \alpha)(x - \beta) = 0$, that is

$$x^4 - (6 + \alpha + \beta)x^3 + (5 + 6\alpha + 6\beta + \alpha\beta)x^2 - (5\alpha + 5\beta + 6\alpha\beta)x + 5\alpha\beta = 0$$

This equation is the same as

$$x^4 - 19x^3 + 113x^2 - 245x + 150 = 0$$

Equating the coefficients of x^3, x^2, x and the constant term gives:

$$6 + \alpha + \beta = 19$$
$$5 + 6\alpha + 6\beta + \alpha\beta = 113$$
$$5\alpha + 5\beta + 6\alpha\beta = 245$$
$$5\alpha\beta = 150$$

The last of these equations gives $\beta = 30/\alpha$. Substituting this value in any of the other equations gives a quadratic in α. If we choose the first, the equation is $\alpha^2 - 13\alpha + 30 = 0$. The formula for the solution of a quadratic gives the values of 3 and 10.

The roots of the original equation are therefore 1, 5, 3 and 10.

7.8 Other iterative methods

One of the disadvantages of the preceding method is that it frequently takes some time to converge to the answer required – that is, to the root. A method which is generally more rapid – usually attributed to Newton – relies on the fact that the tangent to a curve at a point, and the curve itself, normally meet the x axis fairly near to each other. The slope of the tangent is the *derivate*. Provided we can differentiate the function – and this is no problem with a cubic, power of four, or indeed any polynomial – we can move fairly quickly to the root by making the new value of x where the tangent cuts the horizontal axis. Figure 7.7 shows the position.

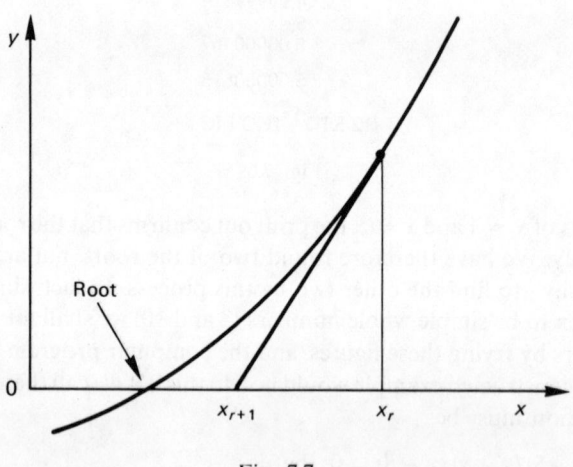

Fig. 7.7

Another approach relies on the fact that if a function has a root lying between two values of x, substituting in these values of x will give opposite signs for the value of the function. If we can find two such values by trial and error, then bisecting these values must give a value nearer to the root. Repeating the process will reduce the interval from $\frac{1}{2}$ to $\frac{1}{4}, \frac{1}{8}, \frac{1}{16}$, etc. of its original difference. This approach is called the *interval bisection method*. In effect, it uses the techniques of the binary search, which is eminently suitable for computer use. A more general example of binary search arises when, for example, looking for a word in a dictionary or file. Searching sequentially (word after word alphabetically) is not the most efficient way. A much quicker way is to start in the middle, decide which half the word is in, take this half, then halve it again, and so on.

Linear and Higher-power Equations 119

7.9 Exercises
Worked answers are provided in Appendix C to question(s) marked with an asterisk.

1. Solve the equations:
$$3x - 2y = 7$$
$$x + 2y = 1$$
   by (a) a traditional commonsense approach;
   (b) the matrix inversion method;
   (c) the use of the Gaussian elimination technique;
   (d) Cramer's method.

2. Solve the equations:
$$2x + 5y = 8$$
$$6x - 4y = -14$$
   by (a) a traditional commonsense approach;
   (b) the matrix inversion method;
   (c) the use of the Gaussian elimination method;
   (d) Cramer's method.

3. Solve the equations:
$$2x - 4y = 10$$
$$-3x + 6y = -10$$
   by any method you like, or can.

4. Solve for x in each of these equations:
   (a) $6x - 8 + x + 5 = 2x + 12 - 5x$
   (b) $2x - \dfrac{2}{3} = \dfrac{1}{2}x + 4$
   (c) $3x - 4 - x = 5x + 8 - 3x - 13$

5. Find the values for x which satisfy the equation $2x - 3y = 15$, if
   (a) $y = 1$ (b) $y = 2$ (c) $y = 3$

   Plot the graph of the equation.

6. Solve the equations:
$$x + y + z = 6$$
$$2x - y + 2z = 6$$
$$x - z = -2$$
   by (a) a traditional commonsense approach;
   (b) the matrix inversion method;
   (c) the Gaussian elimination method.

7. Solve the equations:

$$2x + 2y + z = 3$$
$$2x - 2y + z = -1$$
$$x + y - 2z = 4$$

by (a) a traditional commonsense approach;
   (b) the matrix inversion method;
   (c) the Gaussian elimination method.

8. Solve the equations:

$$2x + 2y - 4z = 6$$
$$2x - 3y + z = 1$$
$$9x - 9y - 9z = 36$$

by (a) a traditional commonsense approach;
   (b) the matrix inversion method;
   (c) the Gaussian elimination method.

9. Produce computer programs in BASIC (or any other high-level language you like) for:

(a) *Finding the root of a function by the method of bisections* – The program should be based on the fact, noted in the text, that if there are two points with x coordinates of (say) x_n and x_{n+1}, for which the values of the function $f(x_n)$ and $f(x_{n+1})$ have opposite signs, then there must be (at least one) root between x_n and x_{n+1}. Use IF, THEN, ELSE approach. If the values $f(x_n), f(x_{n+1})$ have opposite signs then try $f\left(\dfrac{x_n + x_{n+1}}{2}\right)$: if this has an opposite sign to (say) $f(x_n)$, then bisect these two functions and so on.

(b) *Solving the quadratic $ax^2 + 2bx + c = 0$* – This needs only a very simple program which merely incorporates the formula:

$$x = \frac{-b \pm \sqrt{b^2 - ac}}{a}$$

10.* (a) St Swithin's hospital has three departments which service the main operating departments of the hospital. The three are maintenance, laundry and catering. In each case, part of the services provided is within the department itself and/or to the other service departments. Past records show the proportions of such services to be as follows:

    Maintenance: 5% to itself, 25% and 10% to laundry and catering respectively.

Linear and Higher-power Equations

Laundry: 5% to maintenance and 10% to catering.
Catering: 5% to itself, 10% and 15% to maintenance and laundry respectively.

In a given period, direct costs allocated to the operating of each of these departments are shown as:

£
Maintenance: 56,000
Laundry: 250,000
Catering: 35,000

You are required to:
(i) draw up the cost equations to derive the gross operating cost of each of the service departments;
(ii) demonstrate, both algebraically and numerically, the matrix algebra steps which would lead to obtaining the inverse matrix to solve the unknown values in the equations of (i) above. (*Do not attempt to invert the matrix.*)

(b) You are given the following as the inverse matrix for the problem of (a) above:

1.083 0.067 0.125   (Figures rounded to
0.293 1.034 0.194   three decimal places)
0.145 0.116 1.086

You are required, using the inverse matrix, to obtain the gross cost of operation of each service department. Workings must be shown.
(*Institute of Cost and Management Accountants*, November 1983)

11. Invert the matrix obtained in Question 10(a)(i) above yourself by one of the methods discussed in Chapter 5.

Chapter 8
Graphs

8.1 Introduction

In Chapter 4 we said that if R represents the set of *all* real numbers, then R^2 will be the set of all ordered pairs of all real numbers. The visual representation of this was all the points in a two-dimensional Cartesian plane – i.e., the plane itself (this page is part of one such plane).

It may be that we wish to concern ourselves not with the whole plane, but with certain points on the plane, that is with certain ordered pairs only. Frequently the way that these particular ordered pairs are defined is by some *equation*. The equation we used was $x^2 + y^2 = 25$. Only certain points on the plane *satisfy this equation*. These points all lie on a circle. All other points not satisfying this equation (such as, for example, the point (6, 7), since $6^2 + 7^2 \neq 25$) will not be on this circle.

Lines formed by all points on a plane which satisfy an equation are called *graphs*. In this chapter we discuss mainly *two*-dimensional plane geometry, though its extension into three- or four-dimensional geometry (though not its visual representation!) is fairly straightforward.

In this last paragraph, we have taken it for granted that the reader remembers the basic simple facts about graphs – for example what the origin, co-ordinates, etc. mean. It may be that this is not the case especially if modern rather than traditional maths formed the main part of the curriculum. A restatement of the basic points may therefore be helpful.

8.2 Axes and co-ordinates

In order to be able to refer to points on a plane the first step is to fix (fairly arbitrarily) two *axes* at right angles to each other. These are usually the 'x' or horizontal axis, and the 'y' or vertical axis, as in Figure 8.1.

The position of any point P (say) on this plane is determined by its *co-ordinates*.

These are:

1. the x co-ordinate, which is its distance along the x axis (the distance OR or SP in Figure 8.1) measured from left to right and

Graphs

2. the y co-ordinate, which is its distance along the y axis (the distance OS or RP in Figure 8.1, measured upwards.

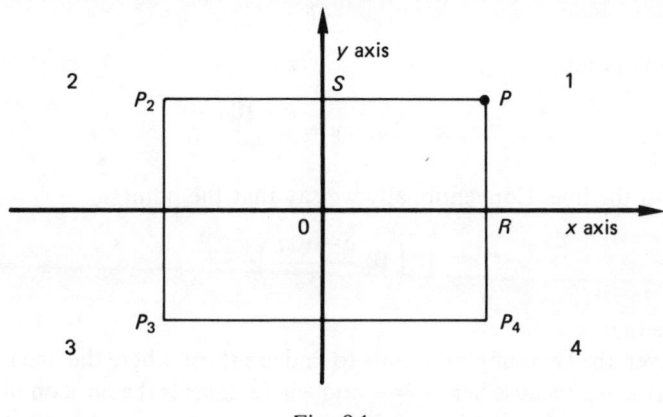

Fig. 8.1

Thus if $OR = 6$ and $RP = 4$, P is referred to as the point $(6, 4)$.

If we suppose that the point P moves along PS towards S, the distance SP will get less and less. In Figure 8.1, it will decrease till at the point S it will be $\dot{0}$. The co-ordinates of S are, therefore, $(0, 4)$. If P still continues in the same direction, the distance SP will be negative and will steadily increase in size. At P_2, the 'reflection' of P in OS (SP_2) will be -6 and the co-ordinates of P_2 will be $(-6, 4)$. Similarly the co-ordinates of P_3 will be $(-6, -4)$, and of P_4 $(6, -4)$. To sum up, points to the *right* of the vertical axis (OS) and above the horizontal axis (OR) – i.e., the quadrant marked 1 in Figure 8.1 – have both co-ordinates positive. Points in quadrant 2 have a negative x co-ordinate and a positive y. In quadrant 3 both co-ordinates are negative: in 4 the x co-ordinate is positive, the y negative.

8.3 Graph of a linear equation

The general form of the linear equation in x and y is usually written as:

$$ax + by = c \qquad (8.1)$$

though there are other forms, as we shall see. The letters a, b, c are *constants*. By this we mean that in any actual problem or application the figures they stand for would be known and would remain the same (or *constant*) throughout.

The graph of this equation is a straight line. To construct or draw this line, we need only to find *two* points on it. When we join them, this will give us the *straight line*. If any value for x is chosen and put into this equation, we can find from the equation the corresponding value for y (i.e., the y co-ordinate of the point). Thus, if we put $x = 10$ in equation (8.1), then:

$$a \times 10 + by = c$$

giving

$$y = \frac{c - 10a}{b}$$

so that the point

$$x = 10, \quad y = \frac{c - 10a}{b}$$

will lie on the line. Conventionally we say that the point

$$\left(10, \frac{c - 10a}{b}\right)$$

is on the line.

However, the two simplest points to find are those where the line meets the vertical (y axis), because here $x = 0$, and where it meets the horizontal (x axis), because here $y = 0$ (see Figure 8.2).

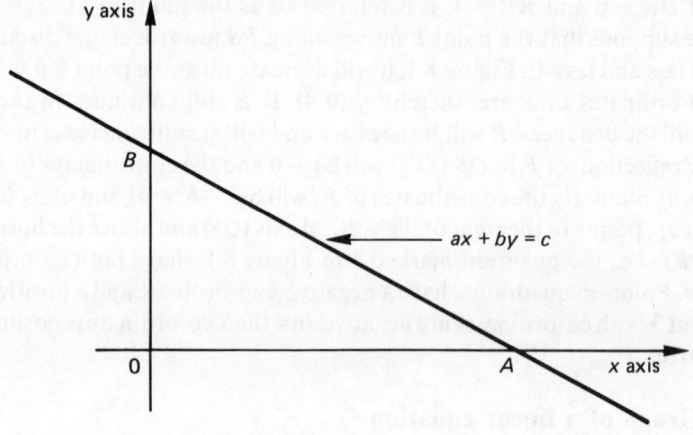

Fig. 8.2

For the first point (B in Figure 8.2) we put $x = 0$ in the equation $ax + by = c$, giving:

$$by = c$$

or

$$y = c/b$$

(provided $b \neq 0$: here and onwards we make the assumption that such a denominator is not zero). So B is the point $(0, c/b)$, and the distance OB is c/b.

Graphs

For the second point (*A* in Figure 8.2) we put $y = 0$ in the equation $ax + by = c$, giving:

$$ax = c, \quad \text{i.e.}$$

$$x = c/a$$

so that *A* is the point $(c/a, 0)$ and *OA* is c/a.

The gradient of the line, usually called the *slope*, is measured by OB/AO.

It is important to take account of the sign. *AO* is drawn in the negative direction – i.e. in the opposite direction to the positive direction (from left to right) along the *x* axis. The slope is therefore:

$$\frac{OB}{-OA} = \frac{c/b}{-c/a} = -\frac{a}{b}$$

8.4 Points lying on either side of a line

We noted earlier in this chapter that there are general forms for the linear equation in *x* and *y* other than that one which we have used so far, which is:

$$ax + by = c \tag{8.1}$$

One of these is

$$y = mx + c \tag{8.2}$$

It is easy to change (8.1) into the form used in (8.2). If, in (8.1), we move ax across to the other side of the equation and then divide both sides by b, we have:

$$y = -\frac{a}{b}x + \frac{c}{b}$$

which is in the form of (8.2) where $m = -\frac{a}{b}$ and a new '*c*' equal to $\frac{c}{b}$.

This form is particularly useful since $-\frac{a}{b}$ is the slope and $\frac{c}{b}$ is the intercept on the *y* axis. Hence we can use the form $y = mx + c$ directly if we know the slope (*m*) and the intercept on the vertical or *y* axis (*c*).

Costing provides one good example for the use of this form. The cost of producing a number of units of a product has two parts. First, those that do not vary with throughput (referred to as 'fixed costs'), and secondly those that do vary with throughput ('variable' costs). The assumption is often made that these variable costs are *directly* variable with throughput. In other words if we double the throughput we double the variable costs. If we call fixed costs '*a*' and variable costs per unit '*b*', then total costs will be as we noted in Chapter 7, $y = a + bx$, where the *x* axis indicates units of throughput and the *y* axis units of cost in money terms. The total cost function is therefore a *straight line* cutting the *y* axis at the point (o, a), and with a slope *b*.

If variable costs are not directly proportional to throughput, then one or more additional terms will be required. The expression for total costs most used in such circumstances is the second order equation:

$$y = a + bx + cx^2$$

If *diseconomies of scale* exist, then c will be positive. Average cost (AC) will be found by dividing through by x. Hence $AC = y/x =$ (in this case) $\dfrac{a}{x} + b + cx$.
Another form of the linear equation much used is:

$$ax + by + c = 0 \tag{8.3}$$

This is obtained even more easily from (8.1) by simply moving c across to the left-hand side and changing its sign. The 'new' c in this version of the general form is equal to the $-c$ of (8.1). It is important to understand that all these forms are merely variations on the same thing.

This last form (8.3) is particularly helpful when we are considering points *not on the line*. For all those points lying on the line, by definition, the value of the expression $ax + by + c$ must be equal to 0. But what will be the values for this expression for points either side of the line? They cannot be zero. What will their value be?

The best way to approach this problem is by an example. Consider the line $3x - 5y + 15 = 0$. Putting $x = 0$ we get $y = 3$, and similarly for $y = 0$, $x = -5$, so that the line passes through the points $(0, 3)$ and $(-5, 0)$, and can be easily constructed by joining these two points. Its graph is shown in Figure 8.3.

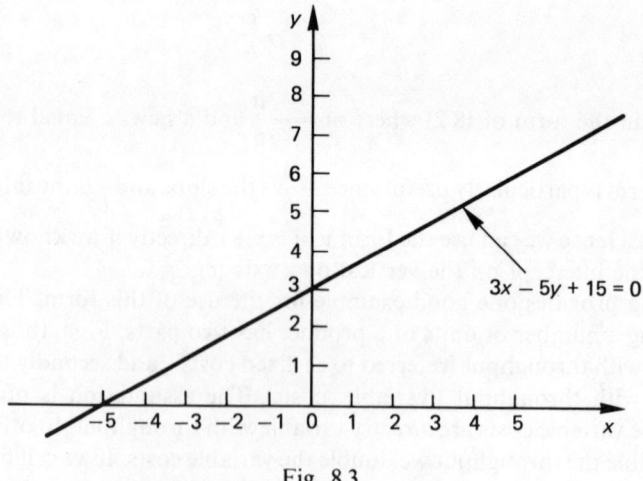

Fig. 8.3

We now calculate the values of the expression $3x - 5y + 15$ for points not on the line. The values for some of these points are shown below in Table 8.1.

Graphs

TABLE 8.1

Point	Value of: $3x - 5y + 15$	Point	Value of $3x - 5y + 15$
(4, 2)	17	(5, 2)	20
(4, 3)	12	(3, 2)	14
(4, 4)	7	(2, 2)	11
(4, 5)	2	(1, 2)	8
(4, 6)	−3	(0, 2)	5
(4, 7)	−8	(−1, 2)	2
(4, 8)	−13	(−2, 2)	−1
(4, 9)	−18	(−3, 2)	−4

Imagining that these points are put in on the graph in Figure 8.3, we note that the sign of $ax + by + c$ is the same for all points on one side of the line, and changes when we cross the line. Although, as usual, we do not prove it, it is always true that for all points on one side of any line $ax + by + c = 0$, the value of $ax + by + c$ is *positive* and for all points the other side of the line it is *negative*.

In our example, for points to the right of the line (that is, nearer to the origin), the value of the expression is *positive*. Put in the form of an algebraic statement:

$$3x - 5y + 15 > 0$$

For all points on the other side of the line (that is, further away than the origin), the expression is *negative*. For these points:

$$3x - 5y + 15 < 0$$

It is also intuitively obvious – though the reader may well wish to show it by taking several sample points – that the value of the expression increases in *absolute value* the further away the point is from the line. For points far down in the bottom right-hand corner the value for the expression will be large and positive; for points far up in the top left-hand corner the values will be large and negative.

All these results will be of importance to us later in Chapter 12.

8.5 Non-linear graphs

The graphs of non-linear equations (i.e., equations containing higher powers of x and y than the first) are *curves*. Figure 8.4 shows the graph of part of a typical curve.

The straight line through any point on a curve which has the same gradient as the curve at that point is called the *tangent at that point*. In Figure 8.4 *SP* is the tangent at *P*. The slope (or gradient) of this tangent is clearly *TP/ST*, so that this is the measure of the slope of the curve at *P*.

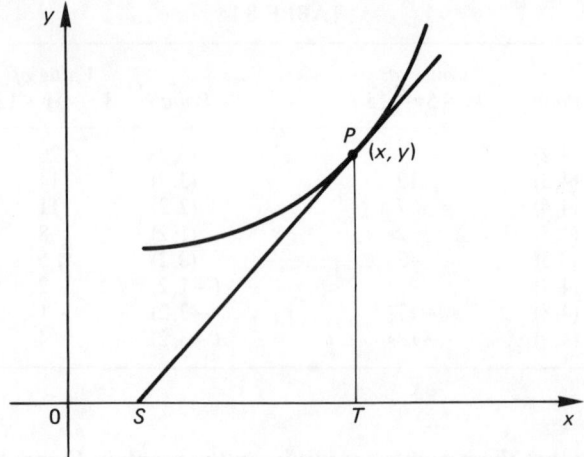

Fig. 8.4

One way of looking at the slope of a curve is to regard it as the rate at which the y co-ordinate is increasing relative to a slight increase in the x co-ordinate. This can best be seen by drawing in two points on the curve close to each other and comparing the increase in the co-ordinates. This approach will be followed up in the next chapter on the calculus.

In school and college textbooks on co-ordinate geometry the properties of a number of non-linear graphs are considered. Second order equations (i.e., equations with x^2, y^2, xy, etc. but no higher indices) receive a great deal of attention. The general equation of the second order in x and y is: $ax^2 + by^2 + cxy + dx + ey + f = 0$.

The graph of this is called a *conic*. (The term 'conic' is derived from conic section: all conics can be obtained by slicing through a suitable cone.) Names are given to certain special types of conics. For example ellipses (the path of the earth round the sun is an ellipse), parabolas (the path of comets), hyperbolas and circles are special cases of a conic.

Ellipses and parabolas are of interest to physicists, mathematicians and engineers. For management purposes hyperbolas and circles are the only conics of much importance. We conclude this chapter by looking first at circles and hyperbolas and then at the exponential function (and its application in the *learning curve*). All these have management applications.

8.6 The circle

Figure 8.5 shows a circle with its centre at O, the origin. By the definition of a circle, any point on it will be the same distance from the centre. (A circle is drawn with a compass by using this fact.) Let us say that this distance is r.

If we can find the equation for x and y which holds true for the co-ordinates of any point on the circle, then we shall have found the equation to the circle.

Graphs

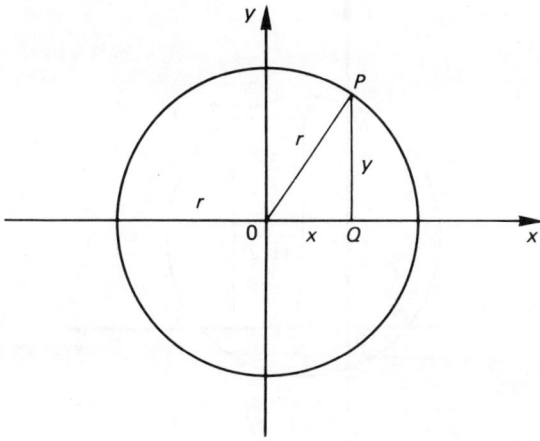

Fig. 8.5

In Figure 8.5, if P is any such point with co-ordinates (x, y), then OQ is x and QP is y. The triangle OPQ has a right angle at Q. Therefore, by Pythagoras's theorem (for the square on the hypotenuse):

$$OQ^2 + QP^2 = OP^2$$

But OP is the radius (r), which is constant, and $OQ = x$, $QP = y$. So that:

$$x^2 + y^2 = r^2$$

which is the equation to the circle, centre O with radius r.

For a circle whose centre is not at the origin we still use the same approach. In the example shown in Figure 8.6 we show a circle whose centre is at the point C $(1, 2)$ and whose radius (r) is equal to 3.

The distance of a point P on the circle, whose co-ordinates are (x, y), from C will be 3, i.e. $CP = 3$.

The distance $MP = NP - NM = x - 1$.

Similarly $CM = y - 2$.

By Pythagoras's theorem $MP^2 + CM^2 = CP^2$, i.e. $(x-1)^2 + (y-2)^2 = 9$ or $x^2 - 2x + 1 + y^2 - 4y + 4 = 9$, giving $x^2 + y^2 - 2x - 4y - 4 = 0$, which is the equation to the circle centre $(1, 2)$ radius 3.

The points where this circle intersects the x axis (RS in Figure 8.6) are found by putting $y = 0$ in the above equation. This gives:

$$x^2 - 2x - 4 = 0$$

Solving this equation gives $x = -1.24$, or 3.24, so that the co-ordinates of the points of intersection are $(-1.24, 0)$ and $(3.24, 0)$. Using the same approach, it is evident that the equation to a circle with centre at (a, b) and of radius r is $(x - a)^2 + (y - b)^2 = r^2$.

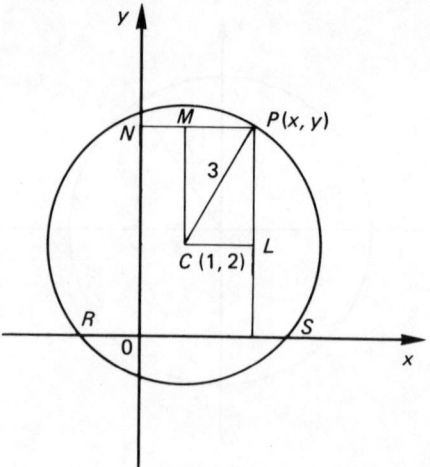

Fig. 8.6

8.7 The hyperbola

Another special case of the general conic is the hyperbola. A standard form for the hyperbola is:

$$\frac{x^2}{a^2} - \frac{y^2}{b^2} = 1$$

The particular type of hyperbola which is most used in management studies is the *rectangular hyperbola*. Here $a = b$ so that a standard form for the rectangular hyperbola is:

$$x^2 - y^2 = a^2, \quad \text{that is}$$
$$(x - y)(x + y) = a^2$$

It is possible to transform this equation. The lines $x + y = 0$ and $x - y = 0$ are the lines which bisect the normal axes $x = 0$, and $y = 0$. They, too, are at right angles to each other. What will be the new equation for our rectangular hyperbola if these two lines are regarded as our new axes?

The co-ordinates of $P(x, y)$, relative to the new axes will be (u, v), say, where $u = OM$ and $v = MP$. To find the values of these we need to note that all the triangles on this diagram have two angles at 45° and one at 90°. By Pythagoras's theorem, the ratio of the sides is therefore as shown below:

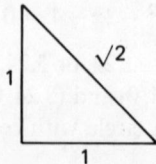

Graphs

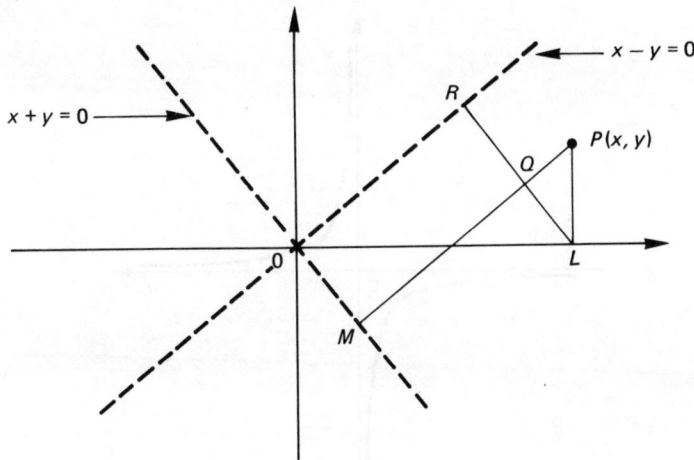

Fig. 8.7

Now
$$MP = MQ + QP = OR + QP = \frac{OL}{\sqrt{2}} + \frac{LP}{\sqrt{2}}$$
$$= \frac{x+y}{\sqrt{2}} = v, \quad \text{by definition}$$

Similarly
$$OM = \frac{x-y}{\sqrt{2}} = u, \quad \text{by definition}$$

But the relationship that we have established for P is
$$(x-y)(x+y) = a^2$$
$$u\sqrt{2} \cdot v\sqrt{2} = a^2$$
$$uv = \frac{a^2}{2}$$

(Or, more simply)
$$uv = k \quad \text{(where } k \text{ is a constant)}$$

This is the equation to a rectangular hyperbola relative to new axes. A typical rectangular hyperbola (the exact shape will depend on the constant) is shown in Figure 8.8.

One good example of the use of a rectangular hyperbola arises when a business sells its product in a market when the total sales revenue remains constant. If unit price is p, and the quantity sold is q, then total revenue is given by
$$pq = \text{constant}$$

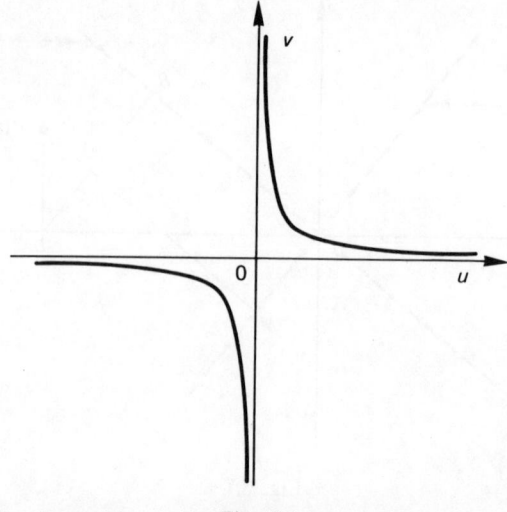

Fig. 8.8

8.8 The exponential function

Many problems in business studies can be expressed in functions of the first degree in x and y (linear equations) or the second degree (conics). One further function which is important to us has x as the exponent. The equation to this function – understandably called the 'exponential' – is:

$$y = a^x$$

This equation was discussed in Chapter 3. It will be recalled that we can write $x = \log_a y$, that a is referred to as the base, and that two important values for the base are 10 and the irrational number e.

The graph of a typical exponential function is shown in Figure 8.9.

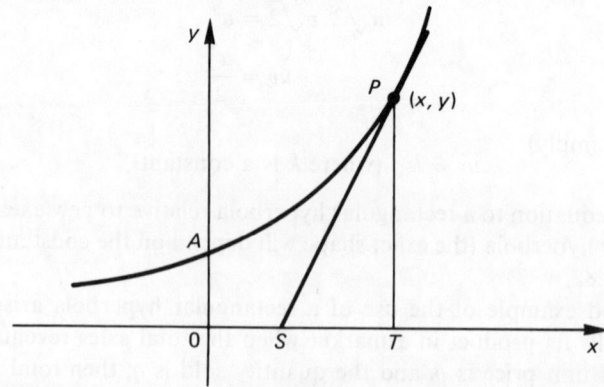

Fig. 8.9

Graphs

When the value of the base a is 10 the principal use of this function is by way of common logarithms and we noted that logarithm tables have been widely used for arithmetical calculations.

When the value of the base is e, the exponential has the unique property that the *slope at any point on its curve is equal to the value of the y co-ordinate at that point*.

It will be remembered that the slope of a curve at any point P is measured by the ratio TP/ST (where SP is the tangent at P). Since TP is y, this means that TP/ST is equal to TP. It follows that ST is of unit length. If, for example, we double the value of y (i.e., move to a new point where the new 'TP' is twice the old value), then the gradient at this new point will be *double* the old gradient.

We have already seen that the slope of a curve at any point is a measure of the *rate of growth of the curve*. The significant thing about this curve is that its rate of growth at any point is equal to the value of the y co-ordinate at that point on the curve.

If we make x the time base t, so that steady movement along the horizontal axis represents the passing of time, the exponential curve represents normal uninhibited growth. A good economic example of this is the overall rate of growth in a developing country. At the start, the few factories in existence will mean a slow rate. When more factories are available, more equipment for new factories can be produced from these. The rate of growth should, therefore, rise directly in proportion to the number of factories already in existence.

The population of a country (or the 'germ' population in a culture) tends to grow according to an exponential curve. The more persons there are, the greater will be the number of births. Where there are constraints – e.g., the inability of the country to support the people, inhibiting government control or shortage of food in the culture – this curve will not apply.

8.9 Graphs and business problems

The aim of this book is to discuss those mathematical techniques which will be useful in solving problems in businesses. Graphs are an example of such a technique. If we can formulate a model which represents graphically the behaviour of a business (or at least some of the *key variables* forming part of it) then we are someway towards understanding how the business works.

We have already seen some examples of this. Thus either the line $y = a + bx$ or the curve $y = a + bx + cx^2$ are often used for *total costs*. The rectangular hyperbola $pq = $ constant shows price against quantity, if total market revenue is constant.

There are other common graphical examples, some of which we have not discussed. One of these is a *periodic* function. The special characteristic of the graph of this function is that its shape is a series of repeated curves. Figure 8.10 shows a typical periodic function.

Many business-related phenomena exhibit this pattern of behaviour. Seasonal figures of a business (such as those for sales or production) may well

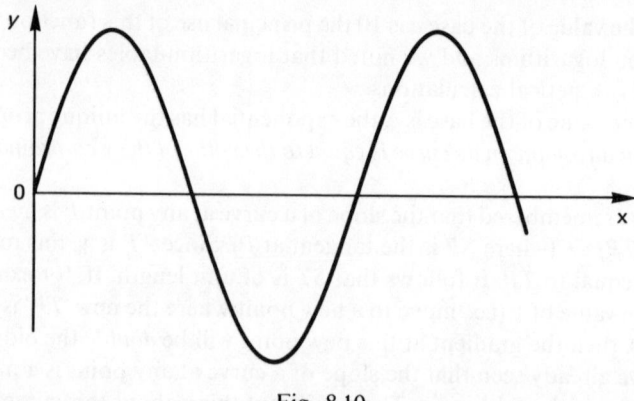

Fig. 8.10

be one example. Agricultural figures, commodity prices, even absenteeism and accident rates will all probably conform to this pattern. An example of a periodic function is given in Question 13 at the end of this chapter.

In Figure 8.11 we show another typical business behaviour pattern. This curve rises steeply at the start, then flattens out. It approaches a horizontal line, but never quite reaches it. Because of this we say that the curve is 'asymptotic' to this line. A useful application of this curve is when the vertical axis represents the work or behaviour rate at any particular time, and time is shown along the horizontal axis.

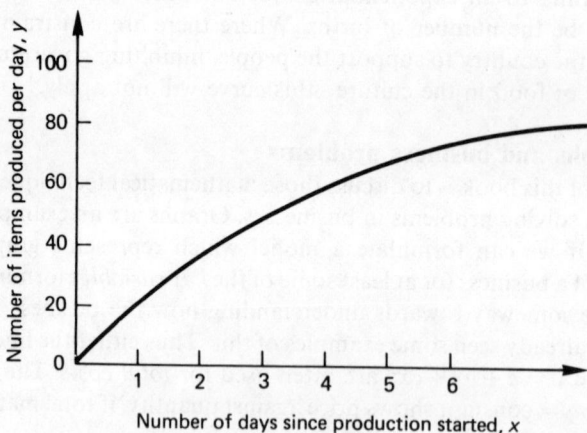

Fig. 8.11

Suppose we start production of a new microchip assembly. Production will start slowly but will then rise as operators become accustomed to their work, adapt their work patterns to cope more easily with the problems, and iron out

Graphs

any minor 'bugs' in the system. The curve most suitable for representing this *learning process* is an exponential function:

$$y = a(1 - e^{-bx})$$

where a and b are constants.

One personal example of the learning curve followed by an application to industrial pricing now follows. In this first example we are concerned with costs. Since costs will decrease with greater production efficiency the curve starts with high (cost) values which then steadily decrease. It can be seen as a 'mirror image' of the learning curve in Figure 8.11.

DOUBLE GLAZING

Having read (in 'Which', August 1982) that it is cheaper to install double glazing oneself I decide to 'have a go' and do all the windows in my house, which are all of standard size. I realise that the first will take some time and then each successive one will be simpler and quicker. Figure 8.12 shows how it actually worked out.

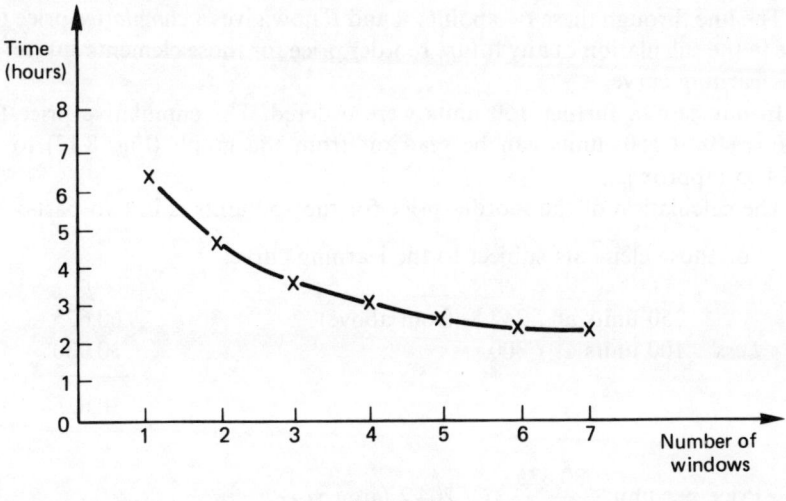

Fig. 8.12

Plotted on normal graph paper the learning curve is a complex curve, and extrapolation is not easy. If, however, we use log log paper, where the measurements along the axes are in a logarithmic sequence already, the curve appears as a *straight line*. This means that extrapolation can be easily and accurately done.

Figure 8.13 uses log log graph paper. The special property that this paper possesses can be seen by looking at the measurements along the vertical axis. If we double 100, we have 200, double 200 we have 400, and so on. The distances along the axis on this paper from 100 to 200, from 200 to 400, etc. are all the same. Proceeding in the opposite direction – i.e., downwards – we would find (if space had been available to show it) that 50 to 100 is also the same length. It is apparent that the '0' line is at 'infinity'. A straight line, such as *A–B* never, therefore, cuts the 'axis' at a finite point.

BUYING OUT

Suppose that a factory subcontracts out an order for 100 parts, machined to specification. The agreed cost price is £720 per unit. Of this total price £420 is made up of raw materials and other items not subject to the learning curve. The remaining £300 is for direct labour. Suppose, further, that in this particular industry an 85% learning curve has been found, from past experience, to apply. What price should be paid for a second order of (say) 150 units?

Two points need to be plotted on the log log graph paper to determine the straight line that is the learning 'curve'. First, the point *A* corresponding to the original 100 units at £300 for those elements subject to the learning curve. Second, the point *B* corresponding to double the original number of units (i.e., 200 units) at 85% of that price (i.e., 85% of £300 = £255).

The line through these two points *A* and *B* now gives a *cumulative* price for use in the calculation of any future reorder price for those elements subject to the learning curve.

In our case, a further 150 units were ordered. The cumulative price for 250 = (100 + 150) units can be read off from the graph (Fig. 8.13) to be £242.5 (approx.).

The calculation of the reorder price for the 150 units is in two parts:

1. For those elements subject to the learning curve:

		£
250 units at £242.5 (from above)		60,625
Less 100 units at £300		30,000
		30,625

Price per unit $\dfrac{£30,625}{150}$ = £204.2 (approx.)

2. Price per unit of those items subject to
learning curve (from 1 above)   204.2
Add   Items not subject to learning curve   420.0

New price   624.2

There are two things to bear in mind when using the learning curve. First, it applies only to certain costs, principally direct labour and labour-related costs. Raw materials are not usually subject to the learning curve, though their cost may reduce a little as waste is eliminated, or better purchasing is made. Selling

Graphs

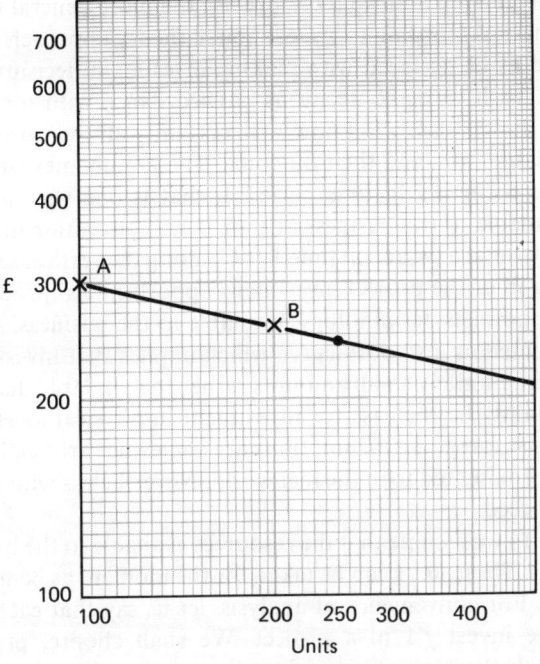

Fig. 8.13

and administrative expenses are often *allocated* on a direct labour basis, but they are not likely actually to be subject to the learning-curve reduction.

Second, in the calculation of those costs subject to the learning curve, it is not the cumulative figure that is required but the increase in the new cumulative figure over the previous one.

The learning curve can be used repeatedly in this fashion to give the price or cost of new orders. If substantial times elapse between orders, so that some of the 'learnt' skill is forgotten each time, then the curve will consist of a series of *stepped parallel lines*, each new line starting some way above the end of the previous one.

In America particularly the learning curve has been used fairly extensively in subcontracting work in a number of major industries. The rate varies with the industry. In the aircraft industry, for instance, an 80% rate has been found to be appropriate.

8.10 Economic approach to investment, financing and dividend decision making

PROJECT INVESTMENT OPPORTUNITIES

The term used for a business by economists in this sort of analysis is a 'firm'. For accountants, however, a firm has the special connotation of a business which is a partnership and we therefore use the more general term 'business' throughout here. According to the economic theory approach a business will be considered as being faced with making investment decisions as at time 0 (now). It will wish to make decisions in accordance with the wishes of the owners (shareholders). It will know with certainty all the various investment projects available to it, and their outcome. These outcomes will be restricted here to those occurring in time period 1, that is one year from now. The resources available to the business at time 0 are conventionally shown as C_0 along the horizontal axis, and at time 1 as C_1 along the vertical axis. Figure 8.14 shows the graph of the project investment opportunities (production opportunities in economists' terminology) available to the business. As can be seen this is represented as a smooth curve which implies that investment projects are available in infinitely small amounts rather than in large discrete amounts as would certainly be the case for investment in physical assets.

Indeed all the foregoing assumptions are somewhat artificial and are likely only to be approximated to in the real world. Nevertheless with this important caveat we proceed.

In Figure 8.14 we assume that the resources available to the business in time 0 are 100 units. These units can be taken for the moment as being expressed in money terms. For convenience of analysis, let us say that each unit is £1.

Suppose we invest £1 in a project. We shall choose, presumably, the investment with the highest rate of transformation. In Figure 8.14 this £1

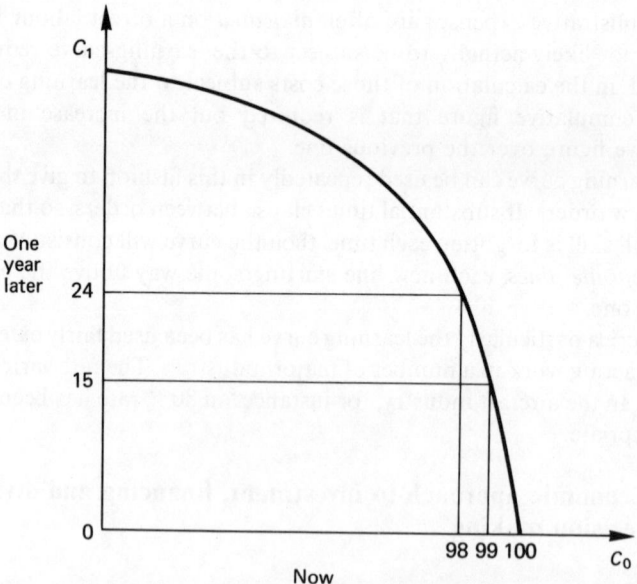

Fig. 8.14

investment in Year 0 will be worth £15 in Year 1. The next £1 invested will be worth an additional £9 in Year 1. As we go further along the curve the marginal return on each investment drops. The tangent to the curve at any point represents this *marginal rate*. At what point should the business decide to forego all further projects because of this low return?

In order to determine this we need to recall that one of the basic assumptions made in this analysis is that decisions are made by management in accordance with the wishes of shareholders. All company law in the UK is based on this principle of shareholder control. We may, however, particularly question the correctness of this assumption. In any large corporation the dichotomy between management and shareholders is great. Shareholders – other than large institutional investors, who may demand a seat on the Board – have no day-to-day control over management decisions. According to agency theory, management should act as agents for the owners: all available evidence indicates that in fact they are often more motivated by personal, selfish considerations. Nevertheless we again proceed with the analysis and consider the position from the shareholders' point of view.

A shareholder's preference for resources as between time 0 and time 1 can be represented by an indifference curve. Figure 8.15 shows two such indifference curves for a particular shareholder.

If two points, say, X_a and X_b are on a particular indifference curve this means that the shareholder is equally satisfied (i.e., indifferent as between) the resources in time 0 and time 1 as represented by these two points. Of course he

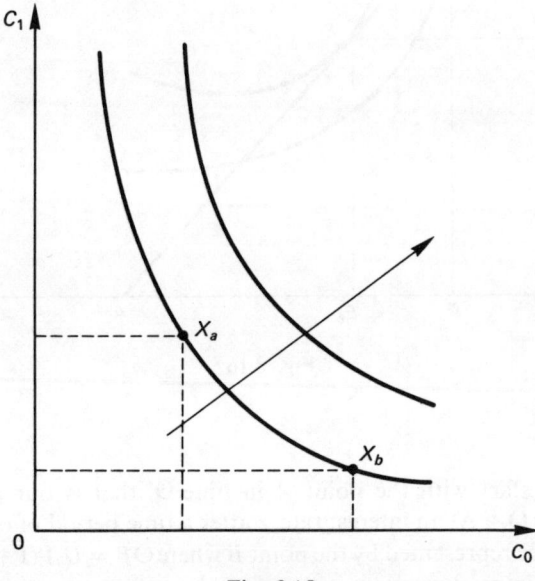

Fig. 8.15

would prefer more satisfaction in total if that were possible. In that case he would move to a higher indifference curve (i.e., one giving more utility or satisfaction) in the general direction given by the arrow. One such higher indifference curve is shown in Figure 8.15 to the right and above.

THE MARKET INTEREST RATE

So far we have considered two types of curves representing *investment opportunities* from the point of view of the business, and *indifference curves* which relate to the satisfaction preferences of a shareholder. Both the business and any shareholder are operating, however, against the background of the general capital market situation. In a capital market which is perfect – another assumption we need to make – there is an interest rate r (say) which transforms resources in time 0 into time 1 or vice-versa. Figure 8.16 shows the capital market interest rate line. The slope of this line is easy to determine.

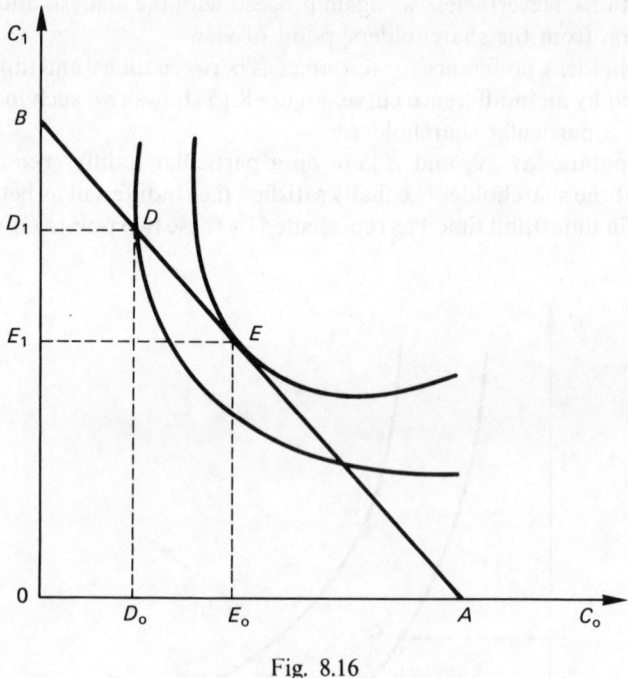

Fig. 8.16

Assume we start with the point A in time O, that is our resources are represented by OA. At an interest rate, r after a time period of one year these resources will be represented by the point B where $OB = OA(1+r)$. The slope of the market interest rate line is therefore:

Graphs

$$-\frac{OB}{OA} \quad \text{(the negative sign because the line slopes backwards)}$$

$$= -\frac{OA(1+r)}{OA}$$

$$= -(1+r)$$

Figure 8.16 also shows two of an individual's indifference curves. Consider the position of the individual who is at point D. This individual can increase his utility (i.e., move to a higher indifference curve) by a capital market transaction. He borrows (and hence gives up in time 1) $D_1 E_1$ units, and gets (in time 0) $D_0 E_0$. His position has now moved from D to E. The point of tangency with one of his indifference curves is clearly his point of maximum utility.

FISHER'S SEPARATION THEOREM

Consider three possible investment/dividend policies. By implication we are saying that if a business does not invest its available cash resources in some form (either as projects or using its spare money on the capital market), then it will be paying it out as dividends.

Suppose the business is considering the three possible investment/dividend policies shown in Figure 8.17. The corresponding market interest lines (all these lines will have the same constant slope) are drawn through these three points. If P is adopted then the wealth in time 0 will increase from F to P_0; if Q

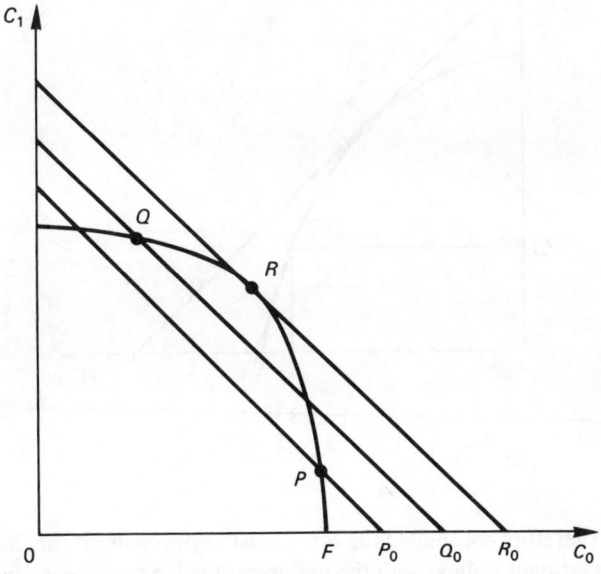

Fig. 8.17

from F to Q_0, and if R from F to R_0. Maximum wealth will be achieved by adopting investment policy R.

OPTIMAL INVESTMENT POLICY

It is also fairly obvious that an investment policy represented by the point R is the best investment policy for the business to pursue if it requires a rate of return which is greater than the market interest rate. We can show this, more formally, by considering a series of small investment projects each requiring a small outlay of resources in time 0.

In Figure 8.18 a business, starting with E units of resources in time 0, and considering only the first project with the highest rate of return, will give up EE_1 resources in time 0 to receive E_1R_1 resources in time 1. The slope of ER_1 will be $-(1+r_1)$ where r_1 is the return on the project investment. Continuing along the 'curve' of investment project possibilities we shall reach the point R where the marginal investment return equals the market interest rate. Looked at the other way round, we will continue investment until the discounted value of the investment in Year 1 exceeds the investment in Year 0. This is the *net present value*. Hence the policy of continuing investment so long as the net present value is positive is the policy which will arrive at the point R.

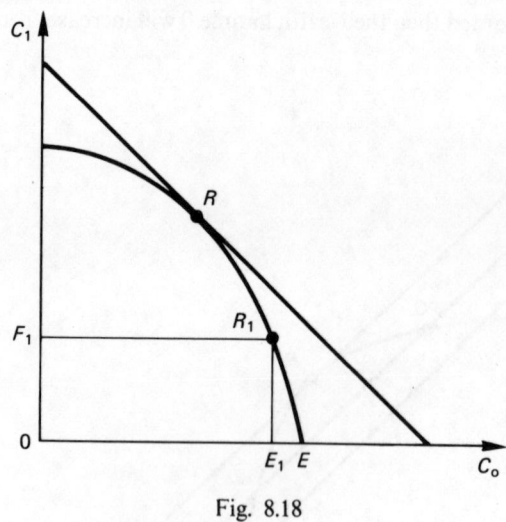

Fig. 8.18

We have therefore reconciled the economist's approach (to the theory of the firm and investment policy) with the net present value approach, (favoured by financial investment analysts). Question 21 of Chapter 9 uses this approach.

Graphs

8.11 Exercises

Worked answers are provided in Appendix C to question(s) marked with an asterisk.

1. Draw the curve:
$$\frac{x^2}{36} + \frac{y^2}{32} = 1$$

   This is an *ellipse*. The points $(2,0)$, $(-2,0)$, A and B (say) are its foci. If P is any point on the curve then the sum of the distances $AP + PB$ is always constant. This is a characteristic of any ellipse.

2. Draw the curve:
$$y^2 = 8x$$

   This is a *parabola*. Its focus is the point $(2,0)$ and its axis is the x axis. If we imagine a reflecting surface made by rotating a parabola about its axis, then rays of light or radio waves parallel to its axis would all be reflected and converge straight to the focus. The mirrors in reflecting telescopes and the Jodrell Bank radio telescope are parabolic in shape.

3. The volume of production, y, of a new item x months after the beginning of a production run is given by the equation
$$y = 100(1 - e^{-0.2x})$$

   Draw this curve (which is an example of the *learning curve*). Note that the maximum value cannot exceed 100 units per month. From the graph, what will be the production for month 5?

4. Maximize the expression $10x + 4y$ given the following constraints:
$$x \geqslant 0 \quad y \geqslant 0$$
$$x \leqslant 10 \quad y \leqslant 11$$
$$3x + 2y \leqslant 40$$

5. Draw the graph of the function $y = a^x$, when:

   (i) $a = 2$, and
   (ii) $a = 3$

6. A company's sales S are given by the formula
$$S = 5000 + 10,000(1 - e^{-0.02x})$$

   where $x =$ the advertising effort, in £000. Draw the graph of this function.

7. If $y = 10x^2$ plot $\log_e y$ against x.

8. Draw the hyperbola:
$$\frac{x^2}{16} - \frac{y^2}{4} = 1$$

As in the case of an ellipse the curve is *symmetric*, with two foci. However the hyperbola has two non-intersecting branches.

9. The costs (y) of manufacture of a product at three levels of output (x) have been found to be as follows:

Output (units)	20	25	30
Total cost (£)	1,800	2,225	2,700

You are told that the relationship between total costs and output can usually be expressed fairly accurately by a simple quadratic equation. You are asked to

(a) find this total cost function; and
(b) sketch the graph of this total cost function over the range 20–30 units.

10. For the last question you are advised that revenue is £89.5 per unit. By drawing the line $y = 89.5x$ and finding the points where it cuts the curve show that there are two break even points at which revenue equals costs.

Confirm your efforts by substituting the value of 89.5x for y into the equation to the cost function and solving for x.

11. The demand for the number of units (y) of a product at a price (x) is given by the equation (a rectangular hyperbola):

$$xy = 6$$

The supply for these units is given by the equation:

$$y = 3(x-1)$$

Show that supply and demand will be equal when the price is 2, and the number of units is 3.

12*. You have been asked about the application of the learning curve as a management accounting technique.

You are required to:

(a) define the learning curve;
(b) explain the theory of learning curves;
(c) indicate the areas where learning curves may assist in management accounting;
(d) illustrate the use of learning curves for calculating the expected average unit cost of making:
 (i) 4 machines; and
 (ii) 8 machines,
 using the data given below.

Data:
Direct labour needed to make the first machine—1,000 hours
Learning curve—80%
Direct labour cost—£3 per hour
Direct materials cost—£1,800 per machine
Fixed cost for either size order—£8,000
(*Institute of Cost and Management Accountants,* May 1984)

13. V. Redbrick plc is in the business of supplying distance learning packages. Sales in this type of business follow a cyclical pattern. Last year's monthly sales followed this pattern:

Month	Jan.	Feb.	Mar.	Apr.	May	Jun.
Sales	1,500	1,600	1,673	1,700	1,673	1,600
Month	Jul.	Aug.	Sept.	Oct.	Nov.	Dec.
Sales	1,500	1,400	1,327	1,300	1,327	1,400

You are required to show that monthly sales (y) are given by:

$$y = 1,500 + 200 \sin \theta$$

where $\theta = 0$, $\pi/6$, $\pi/3$ etc., corresponding to the months January, February, March, etc.

The study of periodic function often involves trigonometry. Indeed it can be shown that *any* periodic function can be expressed as the sum of sine and cosine functions.

Elementary books on trigonometry give the following definitions:

$$\sin \theta = \frac{\text{opposite side}}{\text{hypotenuse}} = \frac{AB}{OB}$$

$$\cos \theta = \frac{\text{adjacent side}}{\text{hypotenuse}} = \frac{OA}{OB}$$

$$\tan \theta = \frac{\text{opposite side}}{\text{adjacent side}} = \frac{AB}{OA} = \frac{\sin \theta}{\cos \theta}$$

(where θ is the angle AOB in a right angled triangle, with $OAB = 90°$)

A short extract from a set of trigonometric tables is as follows:

θ	0	$\pi/6$	$\pi/4$	$\pi/3$	$\pi/2$	π	$3\pi/2$	2π
$\sin \theta$	0	0.50	0.71	0.87	1	0	−1	0
$\cos \theta$	1	0.87	0.71	0.50	0	−1	0	1

In the above table the value of θ is expressed in *radians*. 2π radians equal 360°.

Chapter 9
Calculus

9.1 Differentiation

For explaining how total cost (y) varies with throughput (x) we have generally used two basic functions. They are the straight line:

$$y = a + bx \tag{9.1}$$

and an equation of the second degree:

$$y = a + bx + c^2 \tag{9.2}$$

If all variable costs are simply and directly variable with output we will use equation (9.1). If not equation (9.2) may give a better and more accurate representation. It all depends on the particular business situation.

In this section we are concerned with *growth* – that is, with the rate at which the y co-ordinate changes relative to the x co-ordinate. If our cost function is a straight line as in equation (9.1) then the growth rate will be constant for all values. This rate will be the slope of the line. Figure 9.1 shows the graph of this line $y = a + bx$.

If we choose two points P, Q on the line whose co-ordinates are $(x_1 y_1)$ and $(x_2 y_2)$ respectively, the slope, or gradient from Figure 9.1 is:

$$\frac{MQ}{PM} = \frac{y_2 - y_1}{x_2 - x_1}$$

and obviously this is the same whatever two points on the line are chosen. In Chapter 8, we showed that this slope will be equal to b.

If cost behaviour is better represented by some function of the second order, as for example that in equation (9.2), the graph will be some curve as shown in Figure 9.2; the exact shape of this curve will depend on the values of the constants a, b, c.

The slope of this curve varies. If we wish to determine its value at a point $P(x, y)$, we start by choosing a neighbouring point Q close to P. To emphasise that only a small change in the values for x and y is being considered we call the co-ordinates of $Q(x + \delta x, y + \delta y)$. The symbols $\delta x, \delta y$ do not mean $\delta \times x$ or $\delta \times y$. They merely represent the slight increases in value of x and y moving from P to Q. (Read these symbols as 'delta x' and 'delta y'.)

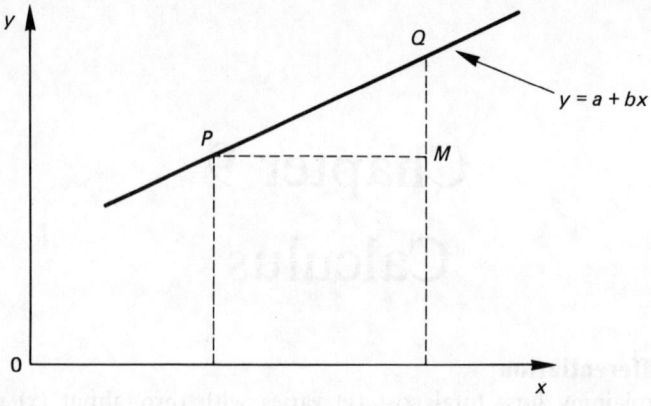

Fig. 9.1

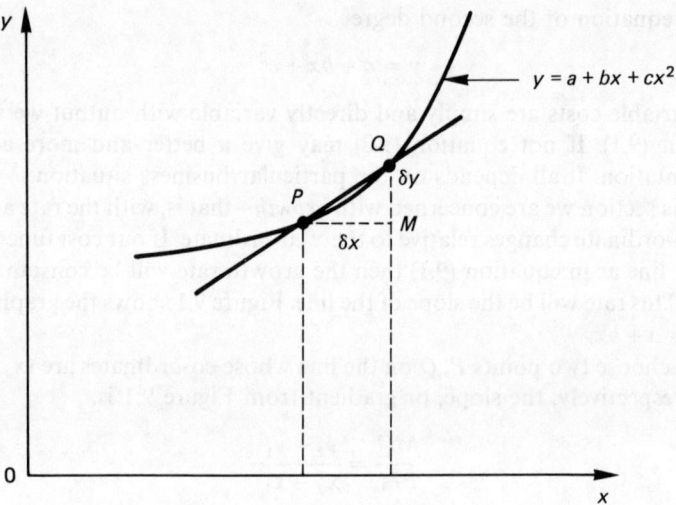

Fig. 9.2

Since P and Q are both on the curve $y = a + bx + cx^2$, then $P(x, y)$ and $Q(x + \delta x, y + \delta y)$ must both satisfy this equation. This may seem confusing since apparently we are here (as elsewhere) using x and y both in the general sense of all values which satisfy the equation and in the specific sense of one set – i.e., the co-ordinates of P. The reader may wish to make (x_1, y_1) the value of P's co-ordinates from now on. The final result will be the same. We follow the normal (if rather confusing) convention and use x, y in the double sense.

Calculus

The slope or gradient of *PQ* in Figure 9.2 will be:

$$\frac{MQ}{PM} = \frac{\delta y}{\delta x}$$

If we imagine that we choose a succession of points *Q* all of which are on the curve, but which move closer and closer to *P*, we see that as $\delta x \to 0$ (read 'delta x tends to zero'), the slope of *PQ* becomes closer and closer to the slope of the tangent at *P*.

We define the limit of the ratio of $\delta y/\delta x$ as $\delta x \to 0$ as the *derivative* of *y* with respect to *x* (or alternatively, the differential coefficient of *y* with respect to *x*) and we call it dy/dx (read '*dy* by *dx*').

Hence

$$\frac{dy}{dx} = \lim_{\delta x \to 0} \frac{\delta y}{\delta x}$$

Note that dy/dx is not dy divided by dx, but just a symbol for the limiting value of the ratio $\delta y/\delta x$.

The process of finding the value for dy/dx for any curve is called *differentiation*.

In the case of the straight line $y = a + bx$ we know that the value of the slope is constant for all values of *x* or *y*, and is equal to *b*.

In the case of the curve $y = a + bx + cx^2$ to find the value of dy/dx we need to proceed as follows:

Since $P(x, y)$ and $Q(x + \delta x, y + \delta y)$ are both on this curve we have,

For *P* $\qquad y = a + bx + cx^2$

For *Q* $\qquad (y + \delta y) = a + b(x + \delta x) + c(x + \delta x)^2$

i.e., $\qquad y + \delta y = a + bx + b\,\delta x + cx^2 + 2cx\,\delta x + c\,\delta x^2$

Subtracting the first equation in this group from the last gives:

$$\delta y = b\,\delta x + 2cx\,\delta x + c\,\delta x^2$$

Dividing through by δx we have:

$$\frac{\delta y}{\delta x} = b + 2cx + c\,\delta x$$

In the limit as $\delta x \to 0$, $c\,\delta x$ will tend to zero, and so we have,

$$\frac{dy}{dx} = \lim_{\delta x \to 0} \left(\frac{\delta y}{\delta x}\right) = b + 2cx$$

Since dy/dx is the slope of the curve, it follows that the slope of this curve varies with (or can be said to be a function of) the *position of the point itself*.

Thus:

if $x = 0$, the slope is b

if $x = 1$, the slope is $b + 2c$, and so on.

To sum up, for the curve $y = a + bx + cx^2$

$$\frac{dy}{dx} = b + 2cx$$

We line up the derivative in this way to show that it can also be arrived at by taking the sum of the derivatives of the three terms one after another. That is, for:

$$y = a + bx + cx^2$$

$$\frac{dy}{dx} = \frac{d(a)}{dx} + \frac{d(bx)}{dx} + \frac{d(cx^2)}{dx}$$

This derivative of a, that is $d(a)/dx$, is obviously zero since the rate of change of something which is constant must be nil. Similarly the rate of change of bx (i.e., $d(bx)/dx$) is b since we know that the graph of $y = bx$ is a straight line of constant gradient b. The derivative of cx^2 (i.e., $d(cx^2)/dx$) is not so obvious. It can best be seen as an example of the more general rule that if:

$$y = x^n, \text{ then } \frac{dy}{dx} = nx^{n-1}$$

In Question 19 at the end of this chapter we indicate how to prove this rule for values of n which are integers. In fact this rule applies for all values of n. For example if n equals $\frac{1}{2}$ so that $y = x^{\frac{1}{2}} = \sqrt{x}$, then

$$\frac{dy}{dx} = \tfrac{1}{2} x^{\frac{1}{2}-1} = \tfrac{1}{2} x^{-\frac{1}{2}} = \frac{1}{2\sqrt{x}}$$

Similarly if:

$$y = \frac{1}{x} = x^{-1}, \text{ then } \frac{dy}{dx} = -1x^{-2} = -\frac{1}{x^2}$$

However at this point we merely state this general rule with the comment that this is probably the most used (and useful) rule in calculus.

9.2 The product rule

In the case of simple polynomials – i.e., functions of the form:

$$y = a + bx + cx^2 + dx^3 + \ldots +$$

it is easy to find the derivative. This is done by differentiating each term one after another. In more complicated functions of x it is not always so simple. There are however certain approaches which can be a great help. The first of

Calculus

these can be used when the complicated function is itself the product of two functions of x, for example if $y = (x+4)(x^3+3)$.

Before differentiating this, we consider first the more general situation, that is how to differentiate $y = uv$ when u and v are both functions of x.

To find this derivative we proceed along the same lines as before, and consider a small change in y to $y + \delta y$, and the corresponding small changes of u and v into (say) $u + \delta u$ and $v + \delta v$. We therefore have:

$$y + \delta y = (u + \delta u)(v + \delta v)$$
$$y + \delta y = uv + v\,\delta u + u\,\delta v + \delta u\,\delta v$$

Subtracting $y = uv$ from this gives:

$$\delta y = v\,\delta u + u\,\delta v + \delta u\,\delta v$$

Dividing through by δx:

$$\frac{\delta y}{\delta x} = v\frac{\delta u}{\delta x} + u\frac{\delta v}{\delta x} + \frac{\delta u\,\delta v}{\delta x}$$

As

$$\delta x \to 0, \quad \frac{dy}{dx} = \lim_{\delta x \to 0}\frac{\delta y}{\delta x} = v\frac{du}{dx} + u\frac{dv}{dx}$$

since

$$\frac{\delta u\,\delta v}{\delta x} = \delta u \cdot \frac{\delta v}{\delta x}$$

which $\to 0$ as δu, too, tends to zero.

This gives the *product rule*:

$$\frac{dy}{dx} = v\frac{du}{dx} + u\frac{dv}{dx}$$

To use this rule in differentiating, for example, the function $y = (x+4)(x^3+3)$, we write:

$$u = x+4, \quad \text{giving } \frac{du}{dx} = 1$$

$$v = x^3+3, \quad \text{giving } \frac{dv}{dx} = 3x^2$$

Hence

$$\frac{dy}{dx} = v\frac{du}{dx} + u\frac{dv}{dx}$$

$$= (x^3+3) + (x+4)3x^2$$

$$= 4x^3 + 12x^2 + 3$$

In this particular case we could have found the derivative by first multiplying out the brackets, but this method is very useful for more difficult functions.

9.3 The quotient rule

The corresponding rule when, instead of having two functions multiplied together, we have one divided by the other, that is, when

$$y = \frac{u}{v}$$

$$\frac{dy}{dx} = \frac{1}{v^2}\left(v\frac{du}{dx} - u\frac{dv}{dx}\right)$$

and this 'quotient' rule is proved by the same method as for the product rule above.

Thus if

$$y = \frac{x+3}{2x}$$

we let

$$u = x+3 \qquad \frac{du}{dx} = 1$$

$$v = 2x \qquad \frac{dv}{dx} = 2$$

$$\frac{dy}{dx} = \frac{1}{(2x)^2}[2x - (x+3)2] = \frac{1}{4x^2}(-6)$$

$$\frac{dy}{dx} = \frac{-3}{2x^2}$$

9.4 The function of a function (or chain) rule

The third helpful rule for differentiating comparatively complex functions is used when y is a function of u, and u is a function of x. The rule is:

$$\frac{dy}{dx} = \frac{dy}{du} \times \frac{du}{dx}$$

The proof both of this rule, and that of the following rule, also follow along similar lines. This rule can be a powerful aid. If we wish to find the value of dy/dx for $y = (2+3x)^5$, we put $u = (2+3x)$ so that

$$y = u^5 \qquad \frac{dy}{du} = 5u^4$$

but $\frac{du}{dx} = 3$, so the rule

$$\frac{dy}{dx} = \frac{dy}{du} \cdot \frac{du}{dx} \quad \text{gives} \quad \frac{dy}{dx} = 5u^4 \cdot 3$$

Calculus

So that $\dfrac{dy}{dx} = 15(2+3x)^4$.

Our final rule is that:
$$\frac{dy}{dx} = \frac{1}{\dfrac{dx}{dy}}$$

It may seem that this is not a calculus rule but a simple and well-known algebraic statement. However it must be remembered that dy/dx and dx/dy do not mean dy or dx divided by dx, dy respectively (though often they can be *manipulated* as if this is the case). They are the *limiting values* of the ratios $\delta y/\delta x$ and $\delta x/\delta y$ as $\delta x \to 0$ and $\delta y \to 0$.

9.5 The exponential and logarithmic functions

In Chapter 8 we discussed the general form of the group of exponential curves $y = a^x$ and in particular that exponential which has as its base the irrational number e (i.e., $y = e^x$).

We also noted that the unique property of this latter curve was that the slope at any point on it was equal to the value of the y co-ordinate at that point. Now dy/dx is the slope of a curve, so that for the function $y = e^x$, $dy/dx = y = e^x$.

It will also be recalled from Chapter 3 that if $y = e^x$, then $x = \log_e y$. To find the derivative of a logarithm we use this fact. Thus if $x = \log y$ (dropping the base e as what is intended is clear from the text):

$$\frac{dx}{dy} = \frac{1}{\dfrac{dy}{dx}} \quad \text{but} \quad \frac{dy}{dx} = y$$

so

$$\frac{dx}{dy} = \frac{1}{y} \quad \text{for} \quad x = \log y$$

Since it is *conventional* to write functions in the form $y = f(x)$ rather than $x = f(y)$, we put y in the place of x, and x in the place of y, when we have for the function

$$y = \log x, \quad \frac{dy}{dx} = \frac{1}{x}$$

Compare this with the derivative of $y = e^x$ above.

9.6 Maxima and minima

The most important characteristic of the derivative dy/dx of a function of x is that its value gives the slope of the graph of the function at any point on it. A number of applications are given later in this chapter.

However one particular value of the slope requires special consideration. When its value is zero, the tangent at that point will be horizontal, that is parallel to the x axis. Hopefully it is again intuitively obvious that when this occurs the value of the function is stationary, and this will mean that it is at a maximum or minimum. An example will help explain.

We have used the second order equation $y = a + bx + cx^2$ as a useful model in giving total costs (y) for a throughput (x) in certain practical circumstances. A positive value for c will imply that *diseconomies of scale*, even if small, do exist. Let us be even more precise and give values to each of a, b and c. Using *TC* (Total Costs) as a replacement for y for ease of identification, consider the equation:

$$TC = 50 + 2x + \tfrac{1}{2}x^2 \tag{9.3}$$

The graph of this function over the range of throughput units from 0 to 25 is shown in Figure 9.3. To find the average cost per unit of output x we divide

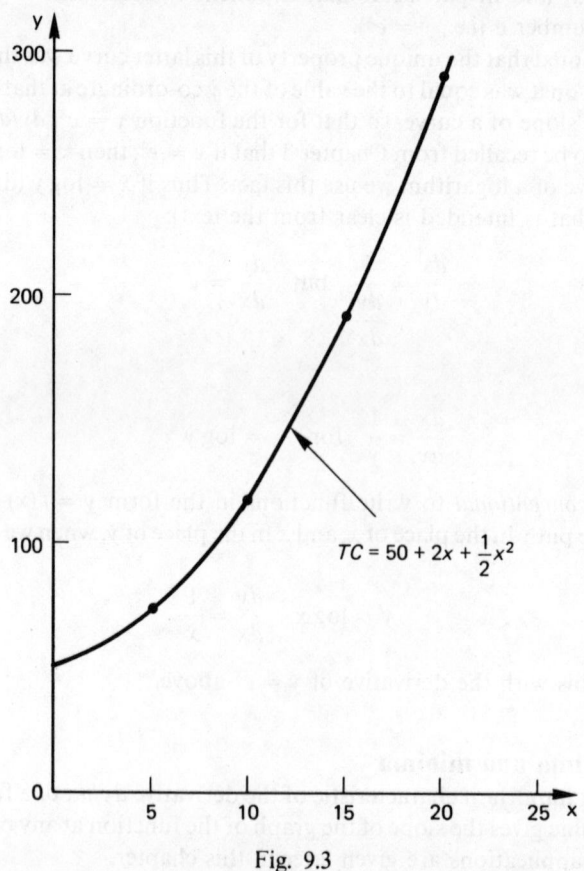

Fig. 9.3

Calculus

equation (9.3) through by x. If we denote average cost per unit by AC (so that $TC/x = AC$) we have

$$AC = \frac{50}{x} + 2 + \tfrac{1}{2}x \tag{9.4}$$

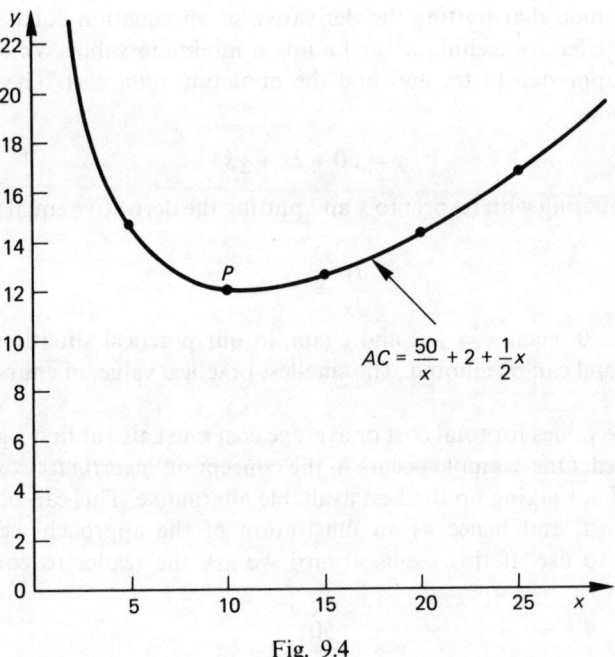

Fig. 9.4

The graph of this function over the same range for x is shown in Figure 9.4. Average cost (AC) starts very high for small values of x (this is because the fixed costs of 50 must be carried by only a few units), then decreases till it reaches some point P, then increases again as throughput rises. To find the exact position of P (this point of minimum average cost) we use the fact that at P the slope will be *zero*. The slope, however, can easily be found by differentiating (9.4) with respect to x. That is:

$$\frac{d(AC)}{dx} = -\frac{50}{x^2} + 0 + \tfrac{1}{2} \tag{9.5}$$

using the general rule for differentiating $y = x^n$, and applying it to each term in this equation. At P we will have:

$$0 = -\frac{50}{x^2} + \tfrac{1}{2}$$

giving

$$x = \sqrt{100} = \pm 10$$

We ignore the solution $x = -10$ (which, although it satisfies the equation, is a nonsense in this particular practical problem, since we cannot have a 'negative' output). It follows that for $x = 10$ *average cost is a minimum*. Putting this value for x into equation (9.4) we have as the minimum value for average cost $\frac{50}{10} + 2 + \frac{1}{2} \times 10$, i.e. 12.

We conclude that putting the derivative of an equation equal to zero is usually an effective technique for finding a minimum value. We could have used this approach to try and find the minimum *total* cost. The total cost equation was:

$$y = 50 + 2x + \tfrac{1}{2}x^2$$

Differentiating with respect to x and putting the derivative equal to zero we have

$$\frac{dy}{dx} = 2 + x \tag{9.6}$$

This equals 0 when $x = -2$ and again, in our practical situation, this is a nonsense, and can be ignored. The smallest practical value, of course, is when $x = 0$.

Negative values for total cost or average cost must also at first sight appear to be absurd. One example occurs in the concept of *opportunity cost* – that is, the cost of not taking up the best available alternative. This can be seen as a negative cost, and hence as an illustration of the approach we are now proposing to use. If this seems absurd we ask the reader to consider the equation:

$$y = -\frac{50}{x} - 2 - \tfrac{1}{2}x \tag{9.7}$$

and to look at it just as a function whose graph we wish to explore (and not for any possible practical application). Its graph is shown in Figure 9.5.

If the equation for the average cost curve (9.4) is shortened to $y = f(x)$ (read, y is a function of x), then this equation is $y = -f(x)$. Its graph is the mirror image of that in Figure 9.4 with the x axis as the mirror. The point P' in Figure 9.5 is now the maximum (as opposed to the minimum) value of the function. It is evident that at P', too, the derivative is equal to zero. Both at maximum and minimum points dy/dx is zero. But if we do not have the graphs of a particular function drawn for us, how can we determine whether the point where the derivative is equal to zero is a maximum or minimum? We now show how this is possible merely by analysing the equation.

We saw that in Figure 9.4 the value of the derivative (graphically the slope of the tangent) started negative, became zero at P, and then became positive as x increased. If we imagine that we draw a graph of the *derivative* (as y) against x, the value of y will start negative, and become zero and then positive. The slope of the derivative curve will therefore be upwards (that is positive) around the point P.

Calculus

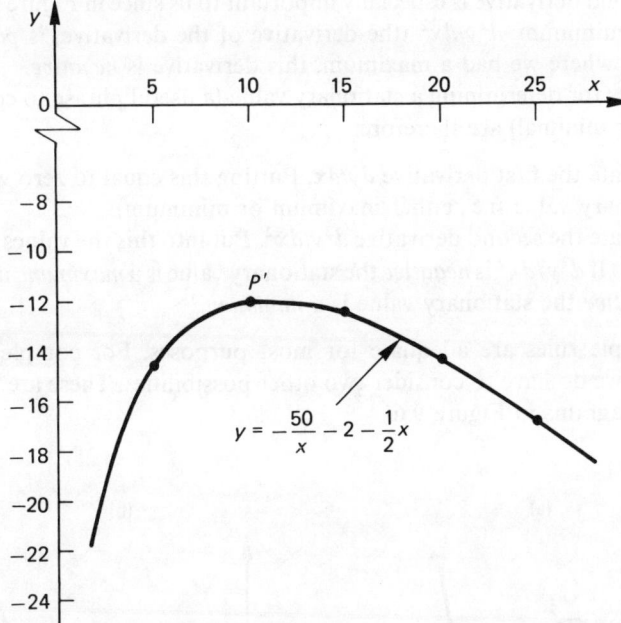

Fig. 9.5

In Figure 9.5 the derivative will start positive, become zero and then be negative. The slope of this *derivative* curve will therefore be downwards (that is, it has a negative value) around the point *P*. But the slope of any curve investigated is measured by its derivative. In Figures 9.4 and 9.5 the curves investigated are themselves derivatives. *Their* slope is therefore the derivative of a derivative. We call this the second derivative. In general if the original function is $y = f(x)$ the first derivative is:

$$\frac{dy}{dx} = \frac{df(x)}{dx}$$

The second derivative, that is:

$$\frac{d\left(\dfrac{dy}{dx}\right)}{dx}$$

is conventionally written as $\dfrac{d^2y}{dx^2}$.

Higher derivatives are obtained in the same way, that is by differentiating each lower derivative; the *n*th derivative is conventionally written $d^n y/dx^n$.

The second derivative is especially important to us since in Figure 9.4 where we had a minimum, d^2y/dx^2 (the derivative of the derivative) is *positive*. In Figure 9.5 where we had a maximum, this derivative is *negative*.

The rules for determining a stationary value (a useful phrase to cover both maxima or minima!) are therefore:

1. Calculate the first derivative dy/dx. Putting this equal to zero will give a stationary value (i.e., either maximum or minimum).
2. Calculate the second derivative d^2y/dx^2. Put into this the values obtained from 1. If d^2y/dx^2 is *negative* the stationary value is a *maximum*; if d^2y/dx^2 is *positive* the stationary value is a *minimum*.

These simple rules are adequate for most purposes. For completeness of treatment we do have to consider two other possibilities. These are shown in the two diagrams in Figure 9.6.

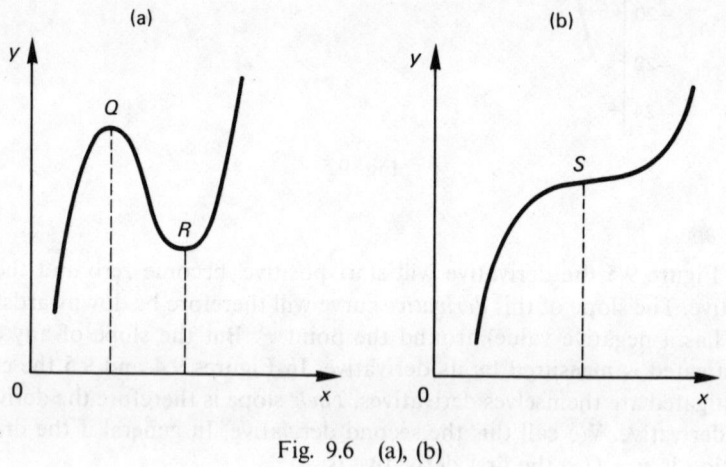

Fig. 9.6 (a), (b)

The function in Figure 9.6(a) has no absolute maximum or minimum since y increases indefinitely as x increases, and decreases indefinitely as x decreases. However dy/dx will be zero at the points Q and R. These are referred to as the *local* maximum and minimum.

The graph of the function in Figure 9.6(b) has no local maximum or minimum value but has a point where dy/dx is zero. This point S is called a *point of inflexion* and d^2y/dx^2 will be zero there, too. To determine the exact shape of a curve when both dy/dx and d^2y/dx^2 are zero it is best to sketch the curve around that point. This means using neighbouring values for x, finding the corresponding values for y from the equation, and then putting in these points on the graph.

9.7 Differentials

Earlier we defined the derivative dy/dx as the limiting value for the ratio of the rate of change of y divided by the rate of change of x, that is:

$$\frac{dy}{dx} = \lim_{\delta x \to 0} \frac{\delta y}{\delta x}$$

or

$$\delta y = \frac{dy}{dx} \delta x \quad \text{(as } \delta x \to 0\text{)}$$

Conventionally we write,

$$dy = \frac{dy}{dx} dx$$

where dy is called the differential of y and dx the differential of x. These are both *limiting values*. It would be easier if we used δy and δx, because dy and dx look as if they are the top and bottom of dy/dx. This is not so; dy/dx is the limiting value of the ratio $\delta y/\delta x$.

The benefit arising from this approach is that we can now calculate the ratio of *relative rates of change* at a point. Thus for our total cost curve:

$$y = 50 + 2x + \tfrac{1}{2}x^2$$

$$\frac{dy}{dx} = 2 + x$$

At $x = 1$, $dy = (dy/dx) dx = (2 + x) dx = 3 dx$ implying that for this value of x the relative increase in y is three times that for x. (In Figure 9.3, remember that units are shown as different lengths along the two axes.)

If, on the other hand, $x = 100$, $dy = (2 + x) dx = 102 dx$, showing that the relative rate of change of y at this point is now 102 times that of x. Obviously for this function the change in y resulting from a small change in x becomes larger as x increases in value.

9.8 The integral calculus

In a previous section we discussed the total cost curve. (In that section we divided the equation through by x to find average cost, and then we considered the maxima and minima of this average cost curve.) Here we are not concerned with the average cost curve but with the *total cost curve itself*. Its equation is:

$$y = 50 + 2x + \tfrac{1}{2}x^2 \tag{9.3}$$

Its derivative is:

$$\frac{dy}{dx} = 2 + x \tag{9.6}$$

and this derivative was easily obtained from equation (9.3) by using those rules for differentiation that we have established already.

The problem we are concerned with here is a different one, too. It is this. How would we find the original equation if we knew only its derivative to be that shown in (9.6)?

In this example, given our knowledge of differential calculus, we could probably work backwards, and might arrive at

$$y = 2x + \tfrac{1}{2}x^2$$

for the original equation.

Hopefully we would also realise that this equation might well include a constant term which would disappear on differentiating. It would not be possible to determine the value of this constant just from knowing the equation to its derivative. For example $y = 2x + \tfrac{1}{2}x^2$, $y = 10 + 2x + \tfrac{1}{2}x^2$ and $y = 4.29 + 2x + \tfrac{1}{2}x^2$, all have the same derivative.

To the question as to what this original equation was, given only the equation to its derivative, our answer must be:

$$y = C + 2x + \tfrac{1}{2}x^2$$

where C is an arbitrary constant (called the *constant of integration*), whose value can be determined only when additional information is made available. Thus if we are also told that $y = 14$, when $x = 2$, then C must be 8.

This process of going back to the original function is called *integration*. It will be obvious that it is, in effect, the reverse of differentiation.

In our case, the derivative was $2 + x$. In the integral calculus the higher powers of x are often put first. We therefore write this as $x + 2$. We then say that

$$\int (x+2)dx = \tfrac{1}{2}x^2 + 2x + C$$

where $\int$ is the integral sign. The result is called the *indefinite integral* of $(x + 2)$. The adjective indefinite is used to draw attention to the fact that the value of the constant has not been determined. The term dx is used to draw attention to the fact that x is the variable being considered.

Because integration is the reverse of differentiation it is often possible, as we have seen, to find directly the integral of a function. Generally, however, integration is more difficult than differentiation, and in some cases may require the exercise of considerable mathematical ingenuity.

9.9 Simple integration techniques

Earlier in this chapter we saw that the most useful rule in differentiating was that for the function

$$y = x^n$$

whose derivative was

$$\frac{dy}{dx} = nx^{n-1}$$

Calculus

Putting a 'new' $n = n+1$ or, more simply, by commonsense we can see that:

$$\int x^n dx = \frac{x^{n+1}}{n+1} + C \qquad (9.8)$$

and this is the most useful rule in the integral calculus. It is true for all values of n, except when $n = -1$. To find the value of the integral of x^{-1} we recall that:

$$\frac{dy}{dx} = \frac{1}{x} \quad \text{if } y = \log_e x$$

Hence:

$$\int \frac{1}{x} dx = \log_e x + C \qquad (9.9)$$

The exponential function e^x is easily integrated, For if:

$$y = e^x$$

$$\frac{dy}{dx} = e^x$$

Hence:

$$\int e^x dx = e^x + C \qquad (9.10)$$

Functions which merely consist of polynomials can easily be integrated by using the results of (9.8) and (9.9). For example the integral of

$$y = x^3 + 7x^2 + 2x + 9 + \frac{4}{x}$$

is

$$\int y \, dx = \frac{x^4}{4} + \frac{7}{3}x^3 + x^2 + 9x + 4\log_e x + C$$

For more complex functions there are two approaches which are often helpful.

The first can be derived from the product rule for differentiating – that is, if u and v are both functions of x:

$$\frac{d(uv)}{dx} = u\frac{dv}{dx} + v\frac{du}{dx}$$

Integrating both sides of this, we have:

$$\int \frac{d(uv)}{dx} dx = \int u \frac{dv}{dx} dx + \int v \frac{du}{dx} dx$$

$$uv = \int u \, dv + \int v \, du$$

$$\int u \, dv = uv - \int v \, du \qquad (9.11)$$

This rule is called *integration by parts*. It can be a very powerful tool for integrating some functions. Thus to find $\int xe^x \, dx$, we put:

$$x = u, \quad \text{so that } du = dx, \text{ and}$$

$$e^x dx = dv, \quad \text{so that } \frac{dv}{dx} = e^x \text{ and hence } v = e^x$$

Applying the rule in (9.11) above we have:

$$\int xe^x \, dx = xe^x - \int e^x \, dx = xe^x - e^x + C$$

The other useful technique for integration is called *integration by substitution*. An example will best explain this approach.

Thus, to find

$$\int (4x + 5)^8 \, dx$$

we write

$$u = 4x + 5, \text{ so that } du = 4dx$$

Then

$$\int (4x + 5)^8 \, dx = \int \frac{u^8}{4} \, du = \frac{u^9}{36} + C$$

$$= \frac{(4x + 5)^9}{36} + C$$

9.10 Calculation of the area under a curve

We know that the differential of any function is graphically represented by the slope of its curve at that point. The process of differentiating therefore means moving from the value of a function at any point to the value of its slope at that point. Integration is the reverse of differentiating; it therefore implies moving from the slope to the value of the function itself.

But the slope at any point P is $\frac{\delta y}{\delta x}$ and the area of the strip under the curve at P is approximately $y \delta x$. To get from y to $\frac{\delta y}{\delta x}$ we *differentiate*: to get from $\frac{\delta y}{\delta x}$ back to y we *integrate*.

This suggests that to get back to the area under a curve we integrate. Figure 9.7 shows the position.

If the curve has as its equation $y = f(x)$ we write:

$$\text{Area (between } x_1 \text{ and } x_2) = \int_{x_1}^{x_2} f(x) \, dx$$

This is known as the *definite integral*. The adjective 'definite' is used because this integral has a definite value associated with it. To find this value we need

Calculus

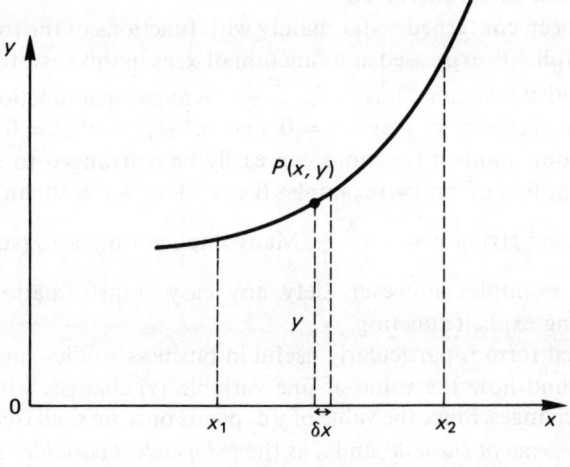

Fig. 9.7

three steps:

1. Find the indefinite integral of the function, using the techniques just discussed.
2. Put in the value $x = x_2$ in the indefinite integral and then the value $x = x_1$.
3. Subtract the value of the integral for $x = x_1$ from that for $x = x_2$. The result is the value for the definite integral. The constant of integration is the same for both x_1 and x_2 and (on subtraction) disappears.

It is difficult to give a proof of either these rules or to explain further the intuitive approach to the meaning of the integral as being the area under the curve, and we do not propose to try to do so. These rules however are widely used. We now show one example of their application.

Past experience has indicated that the pattern of sales (y) in £000 against time (t) in months can be represented by the smooth curve of the function $y = 200 + 6t + 0.4t^3$. It is expected that this pattern will continue into the immediate future. Calculate the total sales which will be made on this basis for the second six months of the coming year.

If $y = 200 + 6t + 0.4t^3$, the sales between months 6 and 12 will be given by:

$$\int_6^{12} (200 + 6t + 0.4t^3)\, dt = (200t + 3t^2 + 0.1t^4)_6^{12}$$

$$= (200 \times 12 + 3 \times 12^2 + 0.1 \times 12^4) - (200 \times 6 + 3 \times 6^2 + 0.1 + 6^4)$$

$$= 4{,}906 - 1{,}438 = \underline{£3{,}468}$$

Hence for the months 6 to 12 estimated sales will be £3,468,000 .

9.11 Partial differentiation

We have been concerned so far mainly with functions of the form $y = f(x)$. When y is explicitly expressed as a function of x, as in this case, the function is called an *explicit function*. Thus $y = \sqrt{x} + x^3$ is an explicit function. Functions of x and y such as $x^2 + xy + y = 0$ and $x^4 + y^4 + x^4 y^5 = 0$ are *implicit* functions. Some implicit functions can easily be rearranged to form explicit functions. The first of our two examples (i.e., $x^2 + xy + y = 0$) can be written as $y(x+1) = -x^2$ giving $y = -\dfrac{x^2}{x+1}$. Many implicit functions (such as that in our second example), however, defy any easy transformation into their corresponding explicit function.

The explicit form is particularly useful in business studies since frequently we wish to find how the value of one variable (y) changes when the other variable (x) changes. Since the value of y depends on x we shall frequently refer to y as the *dependent variable*, and x as the *independent variable*. There are not necessarily any connotations of x being 'freer' than y: merely that we are particularly concerned with the value of y, and are interested in how it changes for values of x.

Frequently the value of the dependent variable will depend on more than one independent variable. Thus the sales demand (D) for a product may vary not only with its price (p_1) but with other independent variables such as the price of other products (p_2), the relevant consumers income (I), and so on. We would write in this case:

$$D = f(p_1, p_2, I, \ldots)$$

Another business example is when we consider the value for the productive output of a factory. If we call this z, then:

$$z = f(x, y)$$

where x might be the labour input and y the capital input. All that we are saying here is that production will vary (and hence be some function of) the basic two inputs into a factory – i.e., the work of people, and of capital (in the form of plant, machinery, buildings, etc. provided). We have not discussed the exact form that the function takes and indeed it will change with differing factories, industries and so on.

In this last example there were three variables – the two independent variables x and y, and the dependent variable. Three-variable functions are quite common in business work, and we propose to discuss them in some detail. A graphical representation in three-dimensional form is possible. Some readers may find such a visual model understandable and hence of some help. Others may not. In any case it is obvious that for functions of more than three variables no visual representation is possible.

In the function $y = f(x)$ which we discussed earlier our concern was

Calculus

principally with the *derivative* – that is, the rate of change of y with respect to x (graphically represented by the slope).

In the three-dimensional function $z = f(x, y)$ we are most interested in the rate of change of z with respect to one of the two independent functions, when the value of the other function is held constant. Graphically, if x, y, z are the three axes at right angles to each other (e.g., this page and for z, a line perpendicular to it), the function $z = f(x, y)$ will be a surface above the plane of the paper, perhaps somewhat like a series of hills and dales on the earth's surface. If we consider a vertical plane parallel to (say) the x axis then for all values on this plane y will be the same. The slope of the surface where it is cut by this plane will then represent the change in z relative to x with y constant. We call this the *partial derivative of z with respect to x*. The term 'partial' indicates that the other variable is to be regarded as constant. We write the partial derivative of z with respect to x as $\frac{\partial z}{\partial x}$. Similarly the partial derivative of z with respect to y is $\frac{\partial z}{\partial y}$. Although in this three-variable case we have only one other independent variable the partial derivative is arrived at, in general, by making *all* other independent variables constant. Reverting to the three-variable case consider the function:

$$z = 2x^2 + 3y^2$$

Here

$$\frac{\partial z}{\partial x} = 4x \quad \text{and} \quad \frac{\partial z}{\partial y} = 6y.$$

in each case the other term disappearing because for partial differentiation purposes it is regarded as a constant.

If the expression involves terms with x and y intermingled the same approach applies. Thus for:

$$z = x^2 + 3xy + 4y^2 + 7$$

$$\frac{\partial z}{\partial x} = 2x + 3y \quad \text{and} \quad \frac{\partial z}{\partial y} = 3x + 8y$$

Second order partial derivatives are derived in a similar way. Thus, continuing with this last example:

$$\frac{\partial^2 z}{\partial x^2} = 2 \qquad \frac{\partial^2 z}{\partial y^2} = 8$$

We can have 'mixed' partial derivatives. Thus:

$$\frac{\partial^2 z}{\partial x \partial y} = \frac{\partial \left(\frac{\partial z}{\partial x}\right)}{\partial y} = \frac{\partial (2x + 3y)}{\partial y} = 3$$

Note that this is the same as:

$$\frac{\partial^2 z}{\partial y \partial x} = \frac{\partial \left(\frac{\partial z}{\partial y}\right)}{\partial x} = \frac{\partial(3x+8y)}{\partial x} = 3$$

and this is an illustration of the general rule that

$$\frac{\partial^2 z}{\partial x \partial y} = \frac{\partial^2 z}{\partial y \partial x}$$

This means that the sequence of differentiation does not matter.

Higher derivatives and derivatives of functions with more variables follow along the same lines and present no special problems. We now give an example of the use of the three-variable function.

THE COBB–DOUGLAS FUNCTION

In econometric analysis a favourite model for the quantity of production (Q) is the Cobb–Douglas function:

$$Q = AL^\alpha K^\beta$$

where A, α and β are suitable constants, and L and K are the labour and capital inputs respectively. Though there have always been problems in the application of this approach, it is still a standard textbook model in this field.

The marginal productivity of labour (that is, the increase in productivity for an increase of one unit of labour input) is obviously of importance. If it is *positive* a greater labour input may be signalled. The marginal product of labour is:

$$\frac{\partial Q}{\partial L} = \alpha A L^{\alpha-1} K^\beta = \frac{\alpha A L^\alpha K^\beta}{L} = \frac{\alpha Q}{L}$$

if capital input is kept the same.

Similarly the marginal productivity of capital is:

$$\frac{\partial Q}{\partial K} = \frac{\beta Q}{K}$$

This indicates that the incremental productivity of each of labour and of capital is a function both of production and of the variable itself. The ways in which these marginal productivities are changing are shown by the second order partial derivatives. These are:

$$\frac{\partial^2 Q}{\partial L^2} = \alpha(\alpha-1)AL^{\alpha-2}K^\beta = \alpha(\alpha-1)\frac{AL^\alpha K^\beta}{L^2} = \frac{\alpha(\alpha-1)Q}{L^2}$$

and

$$\frac{\partial^2 Q}{\partial K^2} = \frac{\beta(\beta-1)Q}{K^2}, \text{ similarly.}$$

Calculus

If α and β each be between 0 and 1 (as would typically be the case) then both $\frac{\partial^2 Q}{\partial L^2}$ and $\frac{\partial^2 Q}{\partial K^2}$ would be < 0.

9.12 Total differentials

We derived the expression $dy = \frac{dy}{dx} dx$ to show the relative rates of change of y and x at a particular point on the curve of the function $y = f(x)$.

For the function $z = f(x, y)$, the corresponding expression is:

$$dz = \frac{\partial z}{\partial x} dx + \frac{\partial z}{\partial y} dy$$

This expression is known as the 'total differential of z'. Its extension into functions of more than three variables follows along identical lines. Thus if:

$$z = f(x_1, x_2, x_3, \ldots x_n)$$

then the total differential of z is:

$$dz = \frac{\partial z}{\partial x_1} dx_1 + \frac{\partial z}{\partial x_2} dx_2 + \frac{\partial z}{\partial x_3} dx_3 + \ldots + \frac{\partial z}{\partial x_n} dx_n$$

As in the two-variable differential the terms dz, dx, dy, etc. merely represent very small increments in the values of z, x and y respectively, and these expressions show their *limiting values*.

The Cobb–Douglas production function provides a good example of the value of this approach.

Suppose the expression $Q = L^{1/2} K^{1/2}$ has been found to be a good model for giving productivity in a factory. The current inputs of labour and capital are (let us say) 900 units and 1,600 units respectively. What will be the rise in productivity if we increase *both* labour and capital by the smallest practical increase, that is 1 unit of each?

We have:

$$dQ = \frac{\partial Q}{\partial L} dL + \frac{\partial Q}{\partial K} dK$$

But

$$\frac{\partial Q}{\partial L} = \tfrac{1}{2} L^{-1/2} K^{1/2}$$

and

$$\frac{\partial Q}{\partial K} = \tfrac{1}{2} L^{1/2} K^{-1/2}$$

giving

$$dQ = \tfrac{1}{2}L^{-1/2}K^{1/2}dL + \tfrac{1}{2}L^{1/2}K^{-1/2}dK$$

Putting $L = 900$, $K = 1,600$ and $dL = dK = 1$, we have as the required marginal increase in productivity:

$$dQ = \tfrac{1}{2}\frac{1}{\sqrt{900}}\sqrt{1600}\times 1 + \tfrac{1}{2}\sqrt{900}\frac{1}{\sqrt{1600}}\times 1$$

$$= \frac{40}{2\times 30} + \frac{30}{2\times 40} = 1.042 \text{ (approx.)}$$

9.13 Three dimensional maxima and minima

In our two-dimensional model $y = f(x)$ we noted that to obtain a stationary point we put the first derivative, $\dfrac{dy}{dx} = 0$, and then used the value of the second derivative to determine whether it was a maximum or minimum, or perhaps a point of inflexion.

In our three-dimensional representation of $z = f(x, y)$ the two-dimensional graph is replaced by a model consisting typically of hills and dales. To determine a proper stationary point (that is one where the increment of z is zero) we need *both* the partial derivatives $\dfrac{\partial z}{\partial x}$ and $\dfrac{\partial z}{\partial y}$ to be equal to zero. This implies that movement parallel to either of the (x or y) axes is level.

Visually, such points correspond to the top of a hill when z is a maximum, or to the bottom of the valley when z is a minimum. There is however another possibility. This occurs when z is at a maximum with respect to variations along one of the x or y axes, and at a minimum for variations along the other axis. This can best be thought of as being a surface shape like a saddle of a horse which has both a raised front and a raised rear. This third case is in fact called a *saddle point*.

The rules for finding maxima, minima or saddle points for the function $z = f(x, y)$ and determining which ones they are of these three alternatives are as follows:

1. Calculate the first order partial derivatives, that is $\dfrac{\partial z}{\partial x}$ and $\dfrac{\partial z}{\partial y}$. Put these equal to zero. A maximum, minimum or saddle point can exist only if both of these are equal to 0.
2. Calculate the value of the second order partial derivatives at the point in question:

   (a) If $\dfrac{\partial^2 z}{\partial x^2} \cdot \dfrac{\partial^2 z}{\partial y^2}$ is greater than $\left(\dfrac{\partial^2 z}{\partial x \partial y}\right)^2$ the function has a maximum or

Calculus

minimum there. If both $\dfrac{\partial^2 z}{\partial x^2}$ and $\dfrac{\partial^2 z}{\partial y^2}$ are negative it is a maximum; if both are positive it is a minimum.

(b) If $\dfrac{\partial^2 z}{\partial x^2} \cdot \dfrac{\partial^2 z}{\partial y^2}$ is less than $\left(\dfrac{\partial^2 z}{\partial x \partial y}\right)^2$ the function has a saddle point if both $\dfrac{\partial^2 z}{\partial x^2}$ and $\dfrac{\partial^2 z}{\partial y^2}$ are equal to zero.

(c) In all other circumstances, and particularly if

$$\dfrac{\partial^2 z}{\partial x^2} \cdot \dfrac{\partial^2 z}{\partial y^2} \text{ equals } \left(\dfrac{\partial^2 z}{\partial x \partial y}\right)^2$$

the function requires further investigation. Putting in values for x and y around the point and finding the corresponding values for z (though tedious) is the best way to determine what sort of 'stationary' value exists there.

These rules appear to be complex. In many practical situations the way they work out is comparatively simple. Consider the function

$$z = x^2 + 2xy + 2y^2 - 5x - 4y + 7$$

Putting the first partial derivatives equal to 0, we have:

$$\dfrac{\partial z}{\partial x} = 2x + 2y - 5 = 0$$

$$\dfrac{\partial z}{\partial y} = 2x + 4y - 4 = 0$$

It is easy to solve these equations for x and y.

Subtracting one from the other, gives $y = -\tfrac{1}{2}$, and putting this value into either of them gives $x = 3$. It follows that at the point on the surface where $x = 3$ and $y = -\tfrac{1}{2}$ the function has a stationary value. To determine which type, we calculate the second partial derivatives. These are:

$$\dfrac{\partial^2 z}{\partial x^2} = 2 \quad \dfrac{\partial^2 z}{\partial x \partial y} = \dfrac{\partial^2 z}{\partial y \partial x} = 2 \quad \dfrac{\partial^2 z}{\partial y^2} = 4$$

$$\dfrac{\partial^2 z}{\partial x^2} \cdot \dfrac{\partial^2 z}{\partial y^2} = 2 \times 4 = 8 \text{ at the stationary point}$$

and

$$\left(\dfrac{\partial^2 z}{\partial x \partial y}\right)^2 = 2^2 = 4$$

Since

$$\dfrac{\partial^2 z}{\partial x^2} \cdot \dfrac{\partial^2 z}{\partial y^2} > \left(\dfrac{\partial^2 z}{\partial x \partial y}\right)^2$$

the stationary point is a maximum or minimum. Since

$$\frac{\partial^2 z}{\partial x^2} = 2 \text{ at the point, and } \frac{\partial^2 z}{\partial y^2} \text{ is } 4$$

and both of these are positive, the stationary point must be a *minimum*.

9.14 Maxima and minima subject to constraint

Almost invariably in the management situation there are constraints in the form of limited capital, labour, machinery or other resources available. The problem of maximising (or minimising) a function subject to limitations or constraints is therefore of importance to us here.

If the function is a production or revenue one the aim will be to *maximise* it. If we are concerned with cost we shall wish to *minimise* it. The term 'optimisation' is useful; it covers both cases.

Constraints usually appear to be in the form of an *inequality*. Thus if we can run a machine for (say) only 40 hours a week, and we have two products competing to use it, then if x and y are the hours programmed for each product $x + y \leqslant 40$. Typically we will want to maximise profit in a situation such as this, and it is obvious that to do this we will use the full number of hours available. In the practical situation this constraint can therefore be expressed as $x + y = 40$, and most constraints can be expressed in this form – that is as an equality rather than an inequality.

We start with the general case for three variables – that is, we wish to optimise $z = f(x, y)$, subject to the constraint $g(x, y) = 0$ and then give one example of this method. There are several ways of proceeding but the most effective is normally by the use of *Lagrange multipliers*. In this approach we form a new expression:

$$L = f(x, y) - \lambda g(x, y)$$

where λ is a variable known as the *Lagrange multiplier*. Since $g(x, y) = 0$, it follows that if we optimise the function L we are, in effect, achieving our objective of optimising $f(x, y)$. We now have an expression L which has three variables x, y, and λ. For optimisation we need to have:

$$\frac{\partial L}{\partial x} = 0 \qquad \frac{\partial L}{\partial y} = 0 \qquad \text{and} \qquad \frac{\partial L}{\partial \lambda} = 0$$

and these three equations need to be solved to give values for x, y and λ. To find whether these are maxima or minima we need to calculate the values of the second order partial derivatives at these points. Sometimes this is unnecessary since whether the values are large or small is fairly obvious.

Suppose we wish to find, using Lagrange multipliers, the critical points for the expression:

$$z = f(x, y) = x^2 - y^2 - y$$

Calculus

and this is subject to the constraint:

$$g(x, y) = x^2 + y^2 - 1 = 0$$

We form a new function:

$$\begin{aligned}L &= f(x,y) - \lambda g(x,y) \\ &= x^2 - y^2 - y - \lambda(x^2 + y^2 - 1)\end{aligned}$$

Putting partial derivatives equal to 0, we have:

$$\frac{\partial L}{\partial x} = 2x - 2\lambda x = 2x(1 - \lambda) = 0$$

$$\frac{\partial L}{\partial y} = -2y - 1 - 2\lambda y = 0$$

$$\frac{\partial L}{\partial \lambda} = -(x^2 + y^2 - 1) = 0$$

The first equation gives directly $x = 0$ or $\lambda = 1$.
If $x = 0$, then either $y = +1$, giving $\lambda = -\frac{3}{2}$
or $y = -1$, giving $\lambda = -\frac{1}{2}$
If $\lambda = 1$, then $y = -\frac{1}{4}$, giving $x = \pm\sqrt{15/16}$

The critical values for $(x$ and $y)$ are therefore:

$$(0, 1)\ (0, -1)\ \left(+\sqrt{\frac{15}{16}}, -\frac{1}{4}\right)\left(-\sqrt{\frac{15}{16}}, -\frac{1}{4}\right)$$

9.15 Multivariate functions and multiple constraints

So far we have discussed the optimisation of a function (z) of two variables (x, y), subject to one constraint. Our approach can be expanded to the case where the value of the function (z) depends on a greater number of independent variables, and also where there are more than one constraint. The general statement is this. To optimise:

$$z = f(x_1, x_2, \ldots, x_n) \qquad (f, \text{ for short})$$

Subject to constraints:

$$g_1(x_1, x_2 \ldots, x_n) = 0 \qquad (g_1 \text{ for short})$$

$$g_2(x_1, x_2 \ldots, x_n) = 0 \qquad (g_2 \text{ for short})$$

$$g_r(x_1, x_2 \ldots, x_n) = 0 \qquad (g_r \text{ for short})$$

We form a new function:

$$L = f - \lambda_1 g_1 - \lambda_2 g_2 \ldots - \lambda_r g_r$$

Where $\lambda_1, \lambda_2, \ldots, \lambda_r$ are Lagrange multipliers. For a critical point:

$$\frac{\partial L}{\partial \lambda_1} = \frac{\partial L}{\partial \lambda_2} = \ldots = \frac{\partial L}{\partial \lambda_r} = 0$$

and we need to solve these equations in addition to:

$$\frac{\partial L}{\partial x_1} = \frac{\partial L}{\partial x_2} = \ldots = \frac{\partial L}{\partial x_n} = 0$$

to get values for $x_1, x_2, \ldots x_n$.

9.16 Elasticity of demand

At the start of this chapter we said we were concerned with the rate at which one co-ordinate (we used y) increases or (more generally) varies relative to the x co-ordinate. There are many application of this which are widely used in economics. Changes in the price (p) of goods will normally affect the quantity (q) of these goods which will be sold. In general, this sensitivity to price change is measured by the ratio of the percentage change in the quantity to the percentage change in price. For very small changes the ratio will be, in the limit,

$$\frac{dq}{q} \times 100 \quad \text{divided by} \quad \frac{dp}{p} \times 100$$

that is:

$$\frac{dq}{dp} \times \frac{p}{q}$$

Assuming that over the relevant range the demand curve (that is, the curve showing the value of q for any value of p) is known, and can be expressed as a function, say, $q = f(p)$, then to find the *elasticity of demand* for any value within this range all we have to do is differentiate.

Thus if the demand curve is, for example:

$$q = f(p) = 1200 - 2p - p^3/3$$

$$\frac{dq}{dp} = -2 - p^2$$

so that the elasticity of demand is:

$$(-2 - p^2) \times \frac{p}{q}$$

If the current price is (say) £9, substituting in this function gives $q = 939$, $dq/dp = -2 - 81 = -83$, so that the elasticity of demand is $-83 \times \dfrac{9}{939} = -0.80$ approx. This means that for a 1% change in price there will be a 0.80% change in quantity. Since this is less than 1%, demand is said to be *inelastic*.

Calculus

If the price is increased to £12, and this is still within the range for which this demand curve applies, then $q = 600$, $\frac{dp}{dq} = -2 - 144 = -146$ and the new elasticity of demand will be $-146 \times \frac{12}{600} = -2.92$ approx. This means that for a 1% change in price there will be a 2.92% change in quantity. Since this is greater than 1%, demand is said to be *elastic*.

9.17 Marginal analysis

Elasticity of demand is just one example of the use of differential calculus in determining the marginal or extra increase of one variable in relation to another related variable at a particular point. We also have *income elasticity of demand* – that is, the ratio of the quantity demanded to change in income.

Finally, consider a total cost curve which shows the relationship between production (x) and total cost (y). The slope, measured by $\frac{dy}{dx}$ is the *marginal cost at a point*. We indeed used the equation (9.3):

$$y = 50 + 2x + \tfrac{1}{2}x^2$$

giving:

$$\frac{dy}{dx} = 2 + x$$

Extensions of this simple two-variable model to three or more variables require the use of *partial derivatives*. If, for example, total cost (y) in a nuclear manufacturing plant depends on two main factors – production (x_1) and the price charged for plutonium (x_2) – then $\frac{\partial y}{\partial x_1}$ and $\frac{\partial y}{\partial x_2}$ will be the marginal cost increments corresponding to small change in x_1 and x_2 respectively.

9.18 The economic order quantity model

There are several practical applications for the economic order quantity (EOQ) model. First, we derive the formula for the model when used in its original form – that is, for the calculation of the quantity of stock that should be ordered to minimise total cost. We then look at the other equally important uses.

Initially, too, we make a number of simplifying assumptions to establish the basic problem. Later some of these assumptions can be relaxed.

Suppose that we are concerned with the stock of only one product in a business. Let us assume that for the coming year S units will have to be carried in inventory to meet sales of that amount during the year. No safety stocks are kept, there is no lead time for ordering (that is, the moment the order is placed the goods will arrive) and sales are at a constant rate throughout the year. The question is this. How often should we order, and what quantity should be ordered?

One way to deal with the situation is to order all S units *ab initio* and run them steadily down throughout the year. (Average quantity in stock would then be $S/2$.) Alternatively, we can order several times during the year. If Q is the quantity ordered, and N the number of times this quantity is ordered:

$NQ = S$ (since in total we must have S units passing through inventory)

Average inventory will be $Q/2$. Figure 9.8 (ignoring the 'new axis') shows the position.

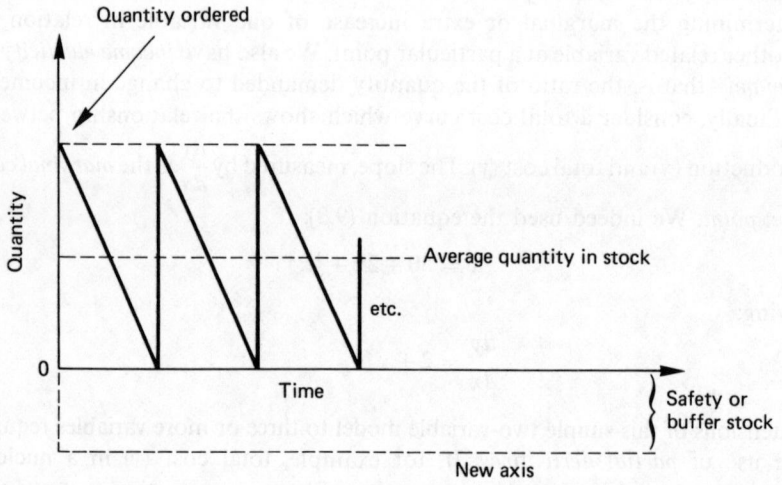

Fig. 9.8

Basically in this situation there will be two types of costs. They are:

(a) *Ordering costs* – If we assume that the cost for an order is fixed (and is O, say), then the amount of these costs will be ON.

(b) *Carrying costs* – The main element of these is likely to be the *interest cost* of the money tied up in stock (though some others, such as rent, insurance, deterioration and supervision can be substantial). Presumably carrying costs will vary with the amount of stock. If the cost per unit per annum is C (say), then since the average stock is $Q/2$, the amount of these costs will be $CQ/2$.

$$\text{Total costs } (TC) \text{ is therefore } ON + CQ/2$$
$$\text{But } NQ = S, \text{ or } N = S/Q$$
$$\text{Hence } TC = OS/Q + CQ/2$$

We wish to *minimise total cost*. Differentiating with respect to Q (and using partial derivatives to emphasise that S, C and O are to be regarded as constant

Calculus

for the present):

$$\frac{\partial(TC)}{\partial Q} = -\frac{OS}{Q^2} + \frac{C}{2}$$

Putting $\frac{\partial(TC)}{\partial Q} = 0$ we have as the value for Q which gives the least cost:

$$Q = \sqrt{\frac{2OS}{C}}$$

This is the so-called economic order quantity (EOQ) formula, or economic lot size formula. Figure 9.9 shows the graphical representation.

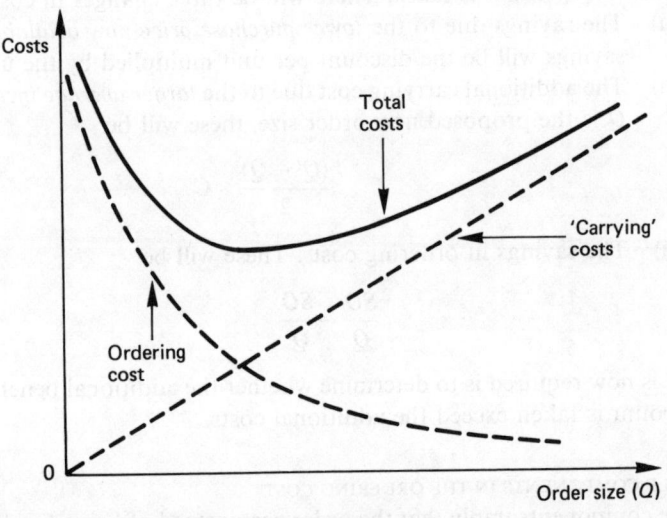

Fig. 9.9

Whilst still remaining with the traditional application of this EOQ model to the product stock control situation, we now consider, briefly, a number of refinements. They are:

1. Safety stocks
2. Lead times
3. Cash discounts
4. Variable components in the ordering cost
5. Stock outs (that is, when stock is zero and orders cannot be met), and non-deterministic situations.

SAFETY STOCKS AND LEAD TIMES

These are both easily dealt with. Safety stocks are *buffer stocks* held against the possibility of stock outs. All that is necessary is to keep this additional stock. Diagramatically the situation is shown by the dotted line in Figure 9.8.

A lead time of (say) L days means raising the re-order point so that the order is 'triggered' L days earlier. Clearly, no additional computation and no change in the basic EOQ model is required for either of these variations.

CASH DISCOUNTS

Cash discount considerations arise if making a larger order will give a reduction in unit cost. To determine whether the discount should be taken we have to work out the total costs in the two situations:

(a) *When the discount is not taken.* The total costs will be derived directly from the basic EOQ model. Let us say the EOQ size is Q units.
(b) *When the discount is taken.* There will be three changes in costs:
   (i) The savings due to the *lower purchase price now available.* These savings will be the discount per unit multiplied by the usage.
   (ii) The additional carrying cost due to the *larger average inventory.* If Q is the proposed new order size, these will be:

$$\frac{(Q'-Q)}{2} \times C$$

   (iii) The savings in ordering costs. These will be:

$$\frac{SO}{Q} - \frac{SO}{Q'}$$

All that is now required is to determine whether the additional benefits when the discount is taken exceed the additional costs.

VARIABLE COMPONENTS IN THE ORDERING COST

These components imply that the order cost instead of being fixed (O), will be of the form

$$O + VQ, \quad \text{where } V \text{ is a constant}$$

This is a particularly important assumption since in all applications of the EOQ model (other than perhaps to stock control) this assumption is very likely to apply. If we replace O by $O + VQ$ in the total cost formula, the new total cost will be:

$$TC = (O + VQ)\frac{S}{Q} + \frac{CQ}{2} = \frac{OS}{Q} + VS + \frac{CQ}{2}$$

Differentiating with respect to Q we have:

$$\frac{\partial(TC)}{\partial Q} = \frac{-OS}{Q^2} + \frac{C}{2}$$

which gives the same formula for the economic order quantity (the constant

Calculus

VS disappears on differentiating). We conclude that the basic EOQ formula is still valid, even if the order cost includes a variable (with quantity) component.

STOCK OUTS AND NON-DETERMINISTIC SITUATIONS

These require the application of *probability analysis*. They are dealt with in Chapter 10.

Production rather than ordering represents the second application of the model. Typically the basic EOQ model applies only to the ordering and control of raw material stock. For work in progress and finished goods the equivalent of 'ordering' will be production. For finished goods stocks those costs equivalent to ordering costs will be the set-up and similar costs associated with scheduling a production run. These costs are likely to be both substantial and truly *additional* costs (it is difficult to see sometimes how raw material ordering costs in the basic EOQ model can be genuinely incremental or even amount to very much). Production scheduling costs are likely to vary with the proposed batch quantity, but as we have seen the EOQ model is unaffected by such an application. This application to finished goods inventory in practice does work, and does enable the business to exercise substantially better control over the size and timing of production runs.

Cash management is the other major application of the EOQ model. In this application the aim is to determine the level of cash which should be held by a business. If substantial sums are available the alternative to keeping them is to put some of them into marketable stocks and securities.

But how much and how often? Since only large businesses usually have sums available this application is limited to those corporations with a substantial treasury function. The particular application that we are considering here also assumes conditions of certainty – for instance, that available market interest rates are known for some time into the future. Typically the period of time involved is one month; for this period it is assumed that the business has a steady demand for cash.

In this cash management application, the carrying cost of cash is the interest foregone on those marketable securities which could have been purchased. The ordering cost is the total cost of transferring marketable securities into cash, or vice-versa. Total cost will include all 'bought and sold' charges and any stamp duties fees, allowances and commissions due.

Balancing cash against marketable securities is a very direct and straightforward application of the basic EOQ. Generally a cushion for cash will be required (i.e., cash will not be allowed to fall to zero) and this corresponds to safety or buffer stock. Generally, too, there are lead times in effecting a financial transaction – the exact amount will depend, in the UK, on the length of time till account settlement day, and this too corresponds to the lead time in ordering raw material stocks.

A very simple illustration of the use of this cash management model might assume that the fixed cost for a typical transfer transaction with cash in a particular business was (say) £50. The business will have estimated steady cash payments during a one-month period of (say) £3.5 million. The interest rate on short-term money market instruments (say the LIBOR rate) is 12%, i.e. 1% per month. No cash cushion is required.

The formula for the optimum transaction size is:

$$Q = \sqrt{\frac{20S}{C}} = \sqrt{\frac{2 \times (50) \times 3{,}500{,}000}{0.01}}$$

$$= \sqrt{35{,}000{,}000{,}000} = \underline{£187{,}083}$$

The average cash balance is $\frac{187{,}083}{2} = 93{,}542$. The business should therefore make $\frac{3{,}500{,}000}{187{,}083} \simeq 19$ transactions in the month converting marketable stock and securities to cash.

9.19 Exercises

Worked answers are provided in Appendix C to questions marked with an asterisk.

1. For the following functions find the value of $\frac{dy}{dx}$:
   (a) $y = 6x - 2$
   (b) $y = 2x^2 - 4x + 1$
   (c) $y = 5x^4 - x^3 + x^2 + 25$
   (d) $y = 6x^3 - x^4/4 + x^3/3 - 1/x^2$
   (e) $y = (2x + 4)(3x + 2/x^2)$
   (f) $y = (2x + 3)^5$
   (g) $y = \log_e (2x + 3)$
   (h) $y = 3e^{2x} + 4e^{-3x} + x^2 e^x$

2. A factory finds that the relationship between its total costs of output (y) and its production (x) is given by the formula $y = 100 + 2x + \frac{x^2}{10}$.
   Find the level of output which minimises average cost.

3. For the following functions integrate with respect to x:
   (a) $3x - 5$
   (b) $2x^2 - 4x + 4$
   (c) $3/(3x - 5)$
   (d) $2e^{3x} + e^{-x}$
   (e) $2e^x - 4e^{3x}$
   (f) $x \log_e x$

4. Sketch the curve $y = x^2 + x + 1$. Calculate the value of the area between the curve and the x axis for values of x between -4 and $+1$.

Calculus

5. For the following functions find the values of $\frac{\partial z}{\partial x}$ and $\frac{\partial z}{\partial y}$:

   (a) $z = 2x + 4y - x^2 - y^2 - 1$
   (b) $z = (x+y)\, e^{(x+y)}$
   (c) $z = \log_e (x^2 + y^2)$
   (d) $z = \dfrac{x+y}{x-y}$

6. Evaluate the following definite integrals:

   (a) $\displaystyle\int_1^2 2x^3\, dx$

   (b) $\displaystyle\int_1^3 3/x^2\, dx$

   (c) $\displaystyle\int_{-1}^3 (2x - x^3)\, dx$

7. Construct a computer program in BASIC (or any other high-level language you like) to find the approximate value of a definite integral $y = f(x)$ using the trapezoidal approach (see Figure 9.10). Note that in

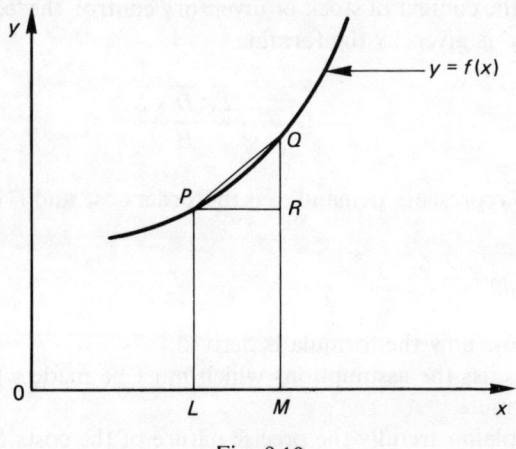

Fig. 9.10

Figure 9.10 the area of the trapezium $PQML$ is equal to:

$$\text{area of triangle } PQR + \text{area of rectangle } PRML$$
$$= \frac{QR \times PR}{2} + LM \times PL$$
$$= \frac{(b-a)}{2} \times h + (a \times h) = \frac{(a+b)h}{2}$$

where $a = LP$, $b = MQ$, $h = LM$

Use this result to split the area under the curve $y = f(x)$ into a series of trapeziums (trapezia for the pedantic!). By reducing the distance h, and hence finding a large number of values for y (evaluated by the computer), the area under the curve can usually be found pretty accurately.

8. Construct a computer program in BASIC (or any other high-level language you like) to find the break even point for a manufacturing plant producing x units of one type of product only. Assume linear relationships with production for both total costs and revenue. In this form, the program is very simple (merely incorporating the calculation of the solution to the two linear equations).

9. Incorporate in your program in question 8 an output giving the profit at any level of production.

10. Rework the programs in questions 8 and 9 for the situations:

    (a) When total costs are of the form $y = a + bx + cx^2$
    (b) When total revenue is of the form $y = px + qx^2$.

    Under what circumstances, if any, do expressions for maxima and minima profits exist?

11. Within the context of stock or inventory control, the 'economic order quantity' is given by the formula

$$Q = \sqrt{\frac{2 \times D \times S}{H}}$$

where D represents 'demand', S is the 'order cost' and H is the 'carrying cost'.

Required:

(a) Show how the formula is derived.
(b) Discuss the assumptions which must be made when using this formula.
(c) Explain carefully the precise nature of the costs S and H.
(d) Describe briefly how the 're-order level' (R) would be determined in the case of variable demand and lead time. Explain how R and Q are used within the usual 'two-bin' system of stock-control.

(*The Association of Certified Accountants*, December 1984)

12. (a) The quantity of a commodity demanded or supplied is represented by q tonnes and £p represents its price. The demand equation is $2p - q = 15$ and the supply equation is $-p + 3q = 5$.

Calculus

Required:

(i) Express these equations in matrix form.

(ii) Using matrix methods establish the values of p and q which satisfy the demand and supply equations simultaneously.

(b) Another demand equation is given by $q = 2p^3 - 21p^2 + 36p + 9$.

Required:

Using the method of differential calculus establish the maximum demand and its corresponding price.

(c) Elasticity of demand is given by the expression

$$\left(\frac{dq}{dp}\right) \cdot \frac{p}{q}$$

Required:

If the demand function is given by $q = \frac{90}{p^3}$, show that the elasticity of demand is constant.

(*The Association of Certified Accountants*, December 1984)

13.* A ship travels between two ports. The cost of fuel is

$$£100\left(aX + \frac{b}{X} + 10\right),$$

where X is the average speed of the ship in knots ($X > 0$), and a and b are constants. If the ship travels at 4 knots the cost of fuel is £9,000, but at 6 knots it is £7,000. [1 knot = 1 nautical mile per hour.)

(a) Find the values of a and b.

(b) By a graphical method, or otherwise, find the speed giving optimum fuel economy and the cost of fuel at that speed.

(*Institute of Cost and Management Accountants*, May 1984)

14.* Last year a company began marketing a new (copyrighted) board game, Zqkopoly. Experience to date has shown that about 300 games per week can be sold at a selling price of £20 each, but if the price is reduced to £13, about 440 can be sold.

You are required:

(a) assuming a linear relationship, to show that the approximate demand equation is:

$$P = 35 - \frac{1}{20}X$$

where P equals price per unit ($P > 0$),

X equals quantity of units sold each week ($X > 0$);

(b) to write down the company's revenue (R) equation in terms of X;
(c) the company having fixed costs of £2,000 per week and variable costs of £10 X per week, to write down the total cost (C) equation in terms of X;
(d) to determine at what point(s) the company breaks even;
(e) by drawing a graph or examining the break even position or otherwise, to advise the company on production and pricing policy for Zqkopoly if it wishes:
  (i) to maximise profit;
  (ii) to maximise revenue.
(*Institute of Cost and Management Accountants*, May 1985)

15.* Raw material Z is used by your organisation at the rate of 200 kg per week in a 50-week year. The cost of the purchasing office for the forthcoming year has been budgeted using the formula:

Total cost (£) = 10,000 + 15 × number of orders

when the number of orders is between 1,000 and 1,500 per annum. The cost of Z is expected to be stable throughout the year at £30 per kg.

There are no variable costs associated with receipt of the material as it is delivered by the suppliers directly into storage.

Stockholding costs have been estimated at 25% per annum.

You are required to:

(a) calculate the economic order quantity using the basic model

$$q^* = \sqrt{\frac{2 \times \text{annual demand} \times \text{cost of order}}{\text{cost of capital} \times \text{unit price}}}$$

(b) describe briefly and calculate the effect on the economic order quantity of the following independent events:
  (i) an increase in stockholding costs to 35%;
  (ii) a reduction in the price of Z to £20 per kg;
  (iii) an increase to 250 kg in the quantity used per week.

(c) comment on the sensitivity of the economic order quantity to changes in the variables as evaluated in (b).

(*Institute of Cost and Management Accountants*, November 1985)

16. Your organisation uses 1,000 packets of paper each year of 48 working weeks. The variable costs of placing an order, progressing delivery and payment have been estimated at £12 per order.

Storage and interest costs have been estimated at £0.50 per packet per annum based on the average annual stock.

The price from the usual supplier is £7.50 per packet for any quantity. The usual supplier requires four weeks between order and delivery.

A potential supplier has offered the following schedule of prices and quantities:

£7.25 per packet for a minimum quantity of 500 at any one time
£7.00 per packet for a minimum quantity of 750 at any one time
£6.94 per packet for a minimum quantity of 1,000 at any one time

If more than 450 packets are received at the same time an additional fixed storage cost of £250 will be payable for the use of additional space for the year.

Assume certainty of demand, lead time and costs.

You are required to:
(a) calculate and state the EOQ from the existing supplier;
(b) calculate and state the stock level at which the orders will be placed;
(c) calculate and state the total minimum cost for the year from the existing supplier;
(d) calculate and state the total minimum cost if you change to the new supplier.

(*Institute of Cost and Management Accountants*, May 1986)

17. The entire share capital of Sword Ltd has just been purchased by an overseas company which intends to continue with Sword Ltd's sole business of manufacturing and selling washing machines. At present the factory capacity is 175,000 per annum. The purchasing company, Damocles Inc., already makes and sells washing machines in Spain. Damocles Inc. received three market surveys, one each for Great Britain, Holland and Denmark. The demand functions for the luxury washing machine Sword Ltd manufactures for each of the three countries are as follows:

Britain: Price (£) $= 375 - 0.0012Q$
Holland: Price (£) $= 435 - 0.0036Q$
Denmark: Price (£) $= \dfrac{5}{Q} + 141 - 0.00055Q$

where Q represents annual sales.

There will be a small transportation cost from the manufacturing plant in Britain to Holland and to Denmark, but that is incorporated in the demand functions. All other costs are to be treated as common to all the products no matter in which market the product is sold. The total cost function for Sword Ltd is estimated as:

Total Cost (£) $= 20{,}000{,}000 + 50Q + 0.0001Q^2$

where Q is the annual production.

The prices quoted are in £ and a fixed exchange rate between the three countries is assumed. The Sword Ltd washing machine is a luxury model the pricing of which is in keeping with the corporate objective of profit maximisation.

Some of the parts assembled by Sword Ltd are bought in and include electrical parts which Damocles Inc. is considering manufacturing in the Sword Ltd factory. This decision would produce a different form of total cost curve of the form given below:

$$\text{Total cost } (£) = 23{,}000{,}000 + 2a\left(\frac{Q}{100}\right)^{(1-b)} + 40Q + 0.0001Q^2$$

where Q is the annual production,
a is the time taken to produce the first batch of 100 electrical parts
b is the learning effect

Requirements

(a) Calculate separately the optimal annual sales volume for each of the three markets under review, assuming that the company sells solely in that market and does not manufacture the electrical parts.
(b) Prepare an optimal sales and profit plan for each of two pairs of countries, Britain and Holland, and Britain and Denmark, assuming Sword Ltd does not manufacture the electrical parts.
(c) Interpret the difference between the sales plans prepared in (a) and (b) above and then comment on the effect the revised total cost curve, which assumes manufacture of electrical parts, might have on sales volumes and profits.
(d) Justify, briefly, how a firm can have a differential pricing policy for the same product for different European countries.

(*Institute of Chartered Accountants in England and Wales*, July 1985)

18. If $y = x^3$, show that $\dfrac{dy}{dx} = 3x^2$.

Use the formula (easily shown to be correct by multiplying out):

$$(a+b)^3 = a^3 + 3a^2b + 3ab^2 + b^3$$

Argue that if δx and δy are the appropriate small increments in x and y if the point $(x+\delta x, y+\delta y)$ lies on the curve $y = x^3$, then

$$y + \delta y = (x + \delta x)^3$$

Expand $(x+\delta x)^3$ by the above formula, subtract from this equation $y = x^3$, divide through both sides by δx and then consider the limit as $\delta x \to 0$.

Calculus

19. If $y = x^n$ show that $\dfrac{dy}{dx} = nx^{n-1}$ for positive integer values for n.

    Use the binomial theorem, given at the end of Chapter 10, and proceed along the same lines as in Question 18 above. The formula in Question 18 is, of course, only the special case of the binomial theorem with $n = 3$.

20. Using Lagrangian multipliers determine for what values of x and y the function:
    $$z = 10x - x^2 + 14y - 2y^2 + xy$$
    will have its maximum, subject to:
    $$2x + 3y = 100$$

21. An investor starts with an initial sum now (time 0) of £16,000. The appropriate project investment opportunities curve is:
    $$C_1 = 240\sqrt{(16,000 - C_0)}$$
    The interest rate (r) for lending and borrowing is 20%. His appropriate utility function is the curve:
    $$u = C_0 C_1$$
    where C_0 is the consumption in the first current period
    C_1 is the future cash flow, resulting from investment in the current period.
    Obtain the optimal investment decision (I).
    This is an example of the economic approach to investment decision making outlined in Chapter 8.
    Differentiate the project investment opportunity curve, and set the slope equal to that of the money market line. This will give an optimal investment strategy for all investors of $I = 10,000$ with next period return of $C_1 = 24,000$.
    To determine the investor's optimal financial decision find the tangency point between the market line through the above optimal point and the investor's indifference function. Take the total differential of the utility curve $u = C_0 C_1$ and equate to the slope of the money market line. (This is based on an example by Levy and Sarnat.)

Chapter 10
Probability

10.1 Probability theory

Probability theory has long been part of most mathematical texts. The way it has been approached has varied very much with the prevailing attitude. At one time a very traditional approach was used. Recently with the increased use of modern maths, the approach has been through the use of sets. In line with our general views we use set theory, supported by traditional mathematics.

Probability theory deals with experiments, events, occurrences which are not deterministic in nature. For instance if we decide to throw a dice (strictly, we should use 'die' for the singular, but we follow modern usage) in the air it is certain (determined) that it will come down. What number will come up on top is not determined in advance. Commonsense might lead us to believe that as there are six possible numbers and each is presumably equally likely to appear – the chance that any particular number (say, a 4) will be thrown is $\frac{1}{6}$th. But for any given throw nothing is certain. For a large number (n) of throws the ratio that the number (s) of times a 4 will appear to the number of throws (i.e., s/n), will, we suppose be close to $\frac{1}{6}$th. But how close? Some courses on statistics start (very boringly!) with students throwing dice repeatedly and observing the results. These experiments are designed to show that as the total number of throws increases then the ratio s/n tends to $\frac{1}{6}$. This is empirical evidence of the 'law' that the relative frequency becomes stable in the long run. (It is also sometimes referred to as 'the law of large numbers'.) These limiting values for the ratios are what we, in a commonsense way, term *probabilities*.

Let us continue with our example of the dice and relate it both to the theory of sets and also, where appropriate, include traditional probability theory and terminology. For any one throw of the dice there are six possible outcomes. In the set theory approach the set or group of all possible outcomes is called the *sample space*. It is best seen as the relevant universal set. This universal set is always the set of all possible outcomes of the particular experiment. This set is usually denoted by S. In our case, we have:

$$S = \{1, 2, 3, 4, 5, 6\}$$

A particular outcome (e.g., throwing a 4) is called a *sample point*. A single sample such as this is also called an elementary event, since it is one of the basic members of the set. In traditional probability analysis the term outcome is

Probability

always used and outcomes which cannot occur together are termed 'mutually exclusive'. In traditional probability a *combination* or group of outcomes is called an event; fortunately in the set theory approach the same term is used. For our experiment we could define a set $A = \{2, 4, 6\}$, that is, the set of all relevant even numbers. This is an *event*. Another possible event would be the set $B = \{1, 3, 5,\}$, that is the set of all the possible odd numbers in our experiment.

All these outcomes (or sample points, elementary events, basic members) and events (groups of outcomes) are *sets*. We now look at the meaning of the three set operations – union, intersection and complement – that we gave in Chapter 4, and see how these work out when related to outcomes and events, and to our dice example in particular.

The union of two sets A and B (that is, $A \cup B$) is the event that occurs whenever either A or B occurs, or both. In probability analysis this will be true whether A and B are outcomes (sample points) or events. If we continue with our dice example, where A and B were both events (each being a group of three outcomes), then $A \cup B = S$, since if A or B occurs we cover all the possible outcomes. In our case, it so happens that both A and B cannot occur together, but, in general, this is not necessarily true.

The *intersection* of A and B (i.e., $A \cap B$) is the event that occurs when *both* A and B occur. In traditional probability analysis it is referred to as *joint probability*. In our simple case A and B, as we noted, cannot occur together. In set theory terminology we say they are disjoint (i.e., $A \cap B = \emptyset$, where $\emptyset$ is the empty set). In probability theory this event which cannot occur is called the *impossible event*. In traditional probability terms A and B would be mutually exclusive.

The *complement* of A (i.e., A^c, or A' or $\bar{A}$)) is the event that occurs whenever A does not occur. In our case $A^c = B$.

10.2 Finite probability spaces

So far we have tried to give some intuitive and also some experimental indication of what we understood by probability. We have also discussed possible outcomes (sample points) and combinations of outcomes (events) using basically a set theory approach. We now need to associate *probability* to sample points. We give the accepted mathematical definitions first and then consider whether these reflect in a satisfactory manner the concepts and understanding that we have arrived at so far.

Consider a finite sample space $S = \{a_1, a_2, a_3, \ldots, a_n\}$, that is, the set of the sample points, $a_1, a_2, \ldots, a_n$. The corresponding finite probability space with which we are concerned is obtained by assigning to each of the sample points a_1, a_2, etc. a real number p_1, p_2, etc. called the probability of a_1, the probability of a_2, etc. The following conditions must apply:

1. Each p_1, p_2, etc. is non-negative – i.e., $p_1 \geq 0, p_2 \geq 0$ and so on.
2. The sum of p_1, p_2, etc. is 1 – i.e., $p_1 + p_2 + \ldots + p_n = 1$.

The probability that a_1 occurs is written as $P(a_1)$. If we have an event A consisting of several points or outcomes we define the probability of A, written $P(A)$ as being the sum of the probabilities of all the sample points that make up the event A.

Do these two *conditions* and our *definition* of the probability of an event make sense? In the two *conditions* we are saying only that the probabilities must be positive and that the probability of all possible, mutually exclusive states must add up to 1. We have already used this latter several times by implication; for instance, when we said that intuitively the chance that a '4' was thrown on a dice was $\frac{1}{6}$th. Each of the 6 possible throws was, we said, equally probable, so we divided the total of *one* by 6 to give $\frac{1}{6}$.

This *definition*, too, does tie up with our normal commonsense approach. The probability that either one or the other of two or more mutually exclusive outcomes will occur is obviously the sum of their individual probabilities. Taking the dice example the probability of event A (one of 2, 4, 6 occurring) is $\frac{1}{6} + \frac{1}{6} + \frac{1}{6} = \frac{1}{2}$.

10.3 Equiprobable spaces

Frequently, as in our basic dice example, the physical characteristics of the situation or the experiment suggest that the various outcomes be assigned *equal* probability values. A finite space S which has for each sample point (or outcome) the *same* probability is called an *equiprobable space*. If S contains n points it follows that the probability of each point or outcome is $\dfrac{1}{n}$. For an event or set A which contains r of these points the probability that A occurs is $r \times \dfrac{1}{n} = \dfrac{r}{n}$.

Using probability notation, we say:

$$P(A) = \frac{\text{number of elements in } A}{\text{number of elements in } S} = \frac{r}{n}$$

but only provided the space is equiprobable.

10.4 Conditional probability

Sometimes, in life, probability estimates will depend on (and hence will vary with) the occurrence (or otherwise) of certain other events or conditions. In the business situation most probability work is in connection with *sales* estimates. This is understandable since business projections – whether for profit or for cash requirements – depend on the level of sales achieved. (In particular, many expenses are directly proportional to sales or production throughput.) The dependence of *sales* on certain events is obvious. If our main rival for a large contract does not submit a bid then our sales are likely to be higher. If supplies of sub-contracted parts or key materials are restricted, production and sales

Probability

must be lower. The level of the sales we achieve is *conditional* on these factors, as indeed on many other factors and events.

Conditional probability theory is therefore important to us. We give the mathematical theory first and then consider how this works out in a simple example.

Let A be an arbitrary event in a sample space S. The probability that B will occur, provided that A has occurred – that is the *conditional probability* of B given A – is defined as:

$$P(B|A) = \frac{P(B \cap A)}{P(A)}$$

provided $P(A) > 0$.

That is, in words

$$\frac{\text{The joint probability of } B \text{ and } A}{\text{probability of } A}$$

We have already found that Venn diagrams are helpful to depict sets. Both B and A are sets. S is defined as the appropriate universal set (i.e., the set of all possible outcomes of this particular situation or experiment). Figure 10.1 shows the position:

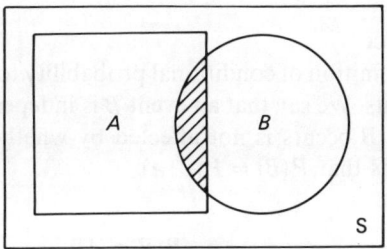

Fig. 10.1

The event B is represented by the disc shape in Figure 10.1. We can see all the possible outcomes which we call the event B as being made up of two sections. First, there are those outcomes which occurred when A *did not occur* (represented by the unshaded part); second there are those outcomes which occurred when A *did occur* (represented by the shaded part).

If the space is equiprobable, the probability of an event, such as A, occurring is proportional to the number of sample points or elementary events which it contains. Hence in an equiprobable space,

$$P(B|A) = \frac{\text{number of elements in } B \cap A}{\text{number of elements in } A}$$

$$= \frac{\text{number of ways both } B \text{ and } A \text{ can occur}}{\text{number of ways } A \text{ can occur}}$$

We can use our dice experiment (to which, it may be felt, we are inordinately attached!) to give a simple example of conditional probability.

Suppose we throw two dice. In this example we wish to find the probability that a 2 occurs, but conditional on the sum of the two dice thrown being 6. Using the same notation we have to find $P(B|A)$ where

$$A = \{\text{sum is } 6\} = \{(1, 5), (2, 4), (3, 3), (4, 2), (5, 1)\}$$

$$B = \{\text{a 2 appears uppermost on the dice}\}$$

It is evident that the space is equiprobable, hence we can use the formula

$$P(B|A) = \frac{\text{number of ways both } A \text{ and } B \text{ can occur}}{\text{number of ways } A \text{ can occur}}$$

Looking at the elements that make up the set A we can see that there are five of them in total, and in only two of them does the 2 appear. Hence

$$P(B|A) = \tfrac{2}{5}$$

10.5 Independence

We can use our definition of conditional probability to define what we mean by independent events. We say that an event B is independent of an event A if the probability that B occurs is not affected by whether A has, or has not, occurred. This means that $P(B) = P(B|A)$.

But we know that

$$P(B|A) = \frac{P(B \cap A)}{P(A)}$$

that is,

$$P(B) = \frac{P(B \cap A)}{P(A)}$$

Hence

$$P(B \cap A) = P(A).P(B)$$

In Chapter 4 we showed that $(B \cap A) = (A \cap B)$. Therefore, for independent events:

$$P(A \cap B) = P(B \cap A) = P(A).P(B).$$

This is our mathematical definition of independent events. It is obvious that if B is independent of A, then A must be independent of B. If this equation above does not apply we say that the events A and B are *dependent*.

10.6 Probability and combinatorial analysis

This completes our work on probability theory. In the next sections we give a number of applications of this probability approach. We conclude this chapter with a section on *combinatorial analysis* (that is, the analysis of the number of ways an event can take place). Combinatorial analysis is concerned with – and traditionally has often been entitled – 'permutations and combinations'.

10.7 Inventory stock out costs

In Chapter 9 we discussed the economic order quantity (EOQ) model and a number of its applications. Here we are concerned solely with inventory and the probability of a *stock out*. Unless large buffer stocks are kept stock outs will occur. When calculating the best level of buffer or safety stock to hold we have to weigh the cost of carrying the additional inventory against some calculation of the cost to the business of a stock out.

It is not easy to assess this stock out cost. If the stock is of raw materials required for the production process the cost may be very substantial. If the stock is of finished goods waiting to be sold, the cost will depend on what effect the lack of stock has on the customer. If he goes elsewhere for this one order, and the trading relationship is unharmed, the only loss will be the contribution from the sale of that particular item of stock. If there is a loss of goodwill – if, for instance, the customer takes his business elsewhere permanently – the effect will be much more substantial. Here we assume that by using past records intelligently we can arrive at some good indication of the likely cost of a stock out. We also assume in this example, which shows this type of calculation for a particular item of finished goods inventory, that past records show that its usage is likely to follow this same pattern over the next month.

Finished goods item of stock

Usage (in units)	150	200	250	300	350
Probability	0.20	0.35	0.30	0.10	0.05

We suppose that we schedule a typical production batch of 200 units, and that the stock out cost has been estimated at £12 per unit and the carrying cost for the month is £1.50 per unit.

Table 10.1 shows the calculation of the expected stock out cost and the carrying cost (which together make up the total cost) for these various additional levels of safety stock. If, for example, 100 units of safety stock are kept, a stock out of 50 units may occur. The stock out cost of this is (50 × £12), i.e. £600. Since the probability of this occurring is 0.05, the expected value of the stock out cost is 0.05 × £600 = £30. Together with the carrying cost, this

TABLE 10.1

Safety stock (units)	Stock out (units)	Stock out cost (£12 per unit)	Probability	Expected stock out cost	Carrying cost	Total cost
150	0	0	0	0	225	225
100	50	600	0.05	30	150	180
50	100	1200	0.05	60		
	50	600	0.10	60		
				120	75	195
0	150	1800	0.05	90		
	100	1200	0.10	120		
	50	600	0.30	180		
				390	0	390

gives a total cost of £180. The other calculations follow along the same lines. We are concerned with the *total cost*, and the last column shows that the minimum value is £180, and that this occurs when we have the safety stock of 50 units.

10.8 Risk and uncertainty in business forecasting and decision making

'What's to come is still unsure'. From Shakespeare through Omar Khayyam: 'the Worldly Hope men set their hearts upon turns ashes – or it prospers'; we all have to accept that the future is not certain. For a business planning ahead this presents problems. How can it cope with them?

There are a number of ways.

(1) SEVERAL SINGLE-FIGURE ESTIMATES

Typically a high (*HI*) and a low (*LO*) as well as the best single-figure estimate will be made. These high and low figures should not be the most extreme figures – the *LO* should not be a real Doomsday estimate – but (in reasonable terms) the likely highest and the likely least figure.

(2) SENSITIVITY ANALYSIS

The first approach is really a subset of the more general SENSITIVITY ANALYSIS method in which projections are based on a variety of figures for the important variables, such as sales. Estimated Profit and Loss Accounts, and Cash Flow Forecasts, are then prepared for each value of these variables. These projections are examined to see how sensitive *key factors* – typically profit and cash – are to changes in the original input figures.

This 'what if' analysis, works out the new scenarios for the variations. Thus: 'what if sales drop by 10%, what will my profit figure be, will our cash reserves be adequate?'. The alternative approach – really only the other side of the same coin – is to start at the required levels for key result factors and then see by how

Probability

much the original input can change *within this restraint*. Thus, 'I do not want reported profit to fall below £X, what is the minimum sales figure I must achieve?'

Sensitivity or 'what if' analysis is a most important tool in forward planning. The widespread use now of micros in business of all sizes makes it a very practical and simple technique. All spreadsheets (Beebcalc, Visicalc, Viewsheet, Lotus) are ideally suited to sensitivity analysis. If we change any input figure (such as sales or some significant expense item), the program will automatically recalculate all the figures. The values for key factors can be read directly and immediately from the screen.

(3) PROBABILISTIC ANALYSIS

A prerequisite of this technique is that estimates of the probability of occurrence for each of the factors involved can be made. Some techniques it is true do exist (see for example, Dewhurst and Burns, 1983) but it is not easy to see how this can be done at all accurately. We, however, are not concerned with these matters but with the mathematical theory and manipulation of these basic probability figures.

The probabilistic approach involves the idea of *expected value*. This is a straightforward commonsense concept, best shown by a simple example. Suppose you are told that a normal, fair coin will be tossed, that if it comes down heads you will receive £1, and if tails, nothing. What is the value of this to you? Presumably the answer is 50 p, that is as an equation.

$$\tfrac{1}{2} \times £1 = 50\text{p}$$

or, as a more general statement:

$$\text{Probability} \times \text{Sum} = \text{Expected value } (EV)$$

If the offer were changed to include an offer of 60p if tails had come down, we can calculate the total expected value as:

$$\tfrac{1}{2} \times £1 = 50\text{p}$$
$$\tfrac{1}{2} \times 60\text{p} = \underline{30\text{p}}$$
$$\text{Total expected value } \underline{80\text{p}}$$

Bayes decision rule states that when alternative courses of action are possible the correct decision is to make that choice which has the *highest total expected value*. If the alternative to this coin choice above is to throw a dice and the offer is that, if a 1, 2, 3 are thrown 10p will be received, if a 4 or a 5, 60p and if a 6, £3, the total expected value of this choice would be:

$$\tfrac{1}{2} \times 10\text{p} = 5\text{p}$$
$$\tfrac{1}{3} \times 60\text{p} = 20\text{p}$$
$$\tfrac{1}{6} \times £3 = 50\text{p}$$

$$\text{Total expected value } \overline{75\text{p}}$$

Decision trees are widely used for presenting alternative choices. The above coin and dice alternatives, shown in this form would be as in Figure 10.2.

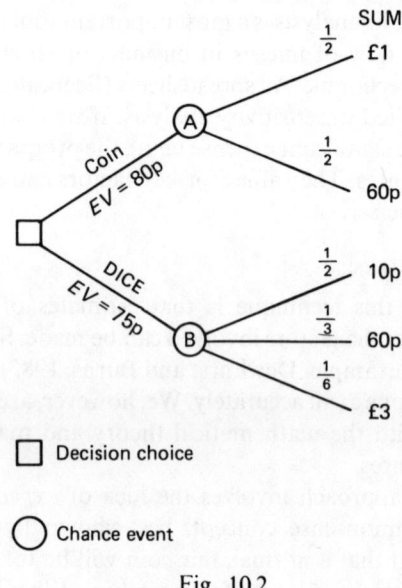

Fig. 10.2

At chance event *A* in Figure 10.2 the total expected value is 80p, at *B* it is 75p. The correct decision is to choose *A* as this has the higher total expected value. The *rules for decision tree analysis* can be summarised as:

1. Proceed from right to left
2. At each chance event find the total expected value
3. At each decision point choose that course of action with the highest total expected value.

In the business situation the alternatives are likely to include cash projections forward over several years as well as joint probabilities (since the projection for any particular year is likely to be dependent on the figures achieved in the preceding years) and a sequence of abandonment possibilities. The latter two – conditional probability and abandonment values – we will consider briefly later. First, we give a simple business example involving sales forecasts and other associated factors several years into the future.

A MANUFACTURING BUSINESS

For a business estimated sales figures are likely to be around £1,000,000 per year, with possibilities of up to around £1,100,000 or a drop to around £800,000. Since it is quite unrealistic to imagine that sales will be exactly any of these figures we need to regard these as *mid-points of ranges*.

Probability

As we noted, it is very difficult to arrive in practice at probabilities for figures in these ranges. Nevertheless we assume that these probabilities have been estimated, and are 0.5 for the £1m figure, and 0.3 and 0.2 respectively for sales around the £800,000 and £1,100,000 figures. Variable costs, too, may alter with the sales level achieved. The estimated probabilities are given in Figure 10.3. The expected cash flow arrived at is £344,700. If we take it that our estimated figures will remain the same for a six-year period and that the criterion rate to apply for discounting is 22%, the expected net present value is

$$3.167 \times 344,700 = £1,091,664.9$$

Here we have assumed, for simplicity, that at the end of the six-year period the plant and equipment would have no material scrap value. If this is not the case, the *estimated net disposal proceeds* (discounted by the appropriate factor for the year in which the sale took place) will be added to this estimated net present value.

It may be that another sensible alternative is to abandon the business and sell the factory, plant and machinery now. If the estimated net proceeds would be £1,000,000 a summarised version of the decision tree diagram would be as in Figure 10.4.

Where cash flows into the future have to be considered, discounting is the appropriate technique to use and the third rule for decision tree analysis needs to be re-written to say:

3. At each decision point choose that course of action with the highest total expected *present* value.

CONDITIONAL PROBABILITY

Already we have used the term 'conditional' when considering the probability figures from one year to another for sales. For if 'high' sales are achieved in the first year it is more probable that sales in the second year will be high, too. The second year's probability will be conditional on the first year's results, and we shall need a new branch on the decision tree. Our naive assumption (in our business example) that each year's probabilities will be unchanged was unrealistic. We could use conditional probability notation for the new probabilities (but it would be simpler just to put in the new probability figures). This use of conditional probability is purely notational.

We can use conditional probability theory much more directly, however, in the calculation of *prior* and *posterior probabilities*, as in the following example.

Suppose a pharmaceutical company has developed a new drug to eliminate sexual deviationist tendencies. Based on laboratory work it is estimated that:

1. the probability that it will be effective, i.e. $P(E) = 0.8$
2. the probability that it will not be effective, i.e. $P(N) = 0.2$

These are called *prior probabilities*.

Sales	Variable manufacturing cost		Contribution	Joint probability (JP)	JP × Contribution
(1) (10 per unit)	(2) COST PER UNIT £	(3) COST £	(4) £	(5)	(6) £
£800,000 (0.3)	5.60 (0.2)	448,000	352,000	(0.06)	21,120
	5.50 (0.6)	440,000	360,000	(0.18)	64,800
	5.30 (0.2)	424,000	376,000	(0.06)	22,560
£1,000,000 (0.5)	5.60 (0.2)	560,000	440,000	(0.10)	44,000
	5.50 (0.6)	550,000	450,000	(0.30)	135,000
	5.30 (0.2)	530,000	470,000	(0.10)	47,000
£1,100,000 (0.2)	5.60 (0.1)	616,000	484,000	(0.02)	9,680
	5.50 (0.8)	605,000	495,000	(0.16)	79,200
	5.30 (0.1)	583,000	517,000	(0.02)	10,340

Expected value of contribution   433,700
Less fixed costs   131,000

                                 302,700
Add depreciation   42,000

Expected cash flow from profits £344,700

Fig. 10.3

1. Probability figures are in brackets.
2. Fixed costs of £131,000 are the same at all levels.
3. The variable manufacturing cost (3) is the cost per unit (2) × the no. of units ((1) ÷ 10) – e.g., for the first row 448,000 = 5.6 × 80,000.
4. Depreciation is a non-cash expense which has been deducted, and therefore needs to be added back to arrive at the cash flow figure.

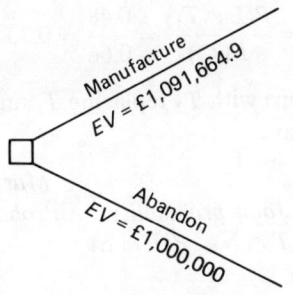

Fig. 10.4

It is possible to put the drug to a test. Such a test, however, costs a lot of money and in the past has not always been a reliable predictor. The test's historical record, and hence the best guide to its likely success or otherwise, may be summed up as in this table of probabilities:

	Actual results	
	Not effective (N)	Effective (E)
Test indicates it will not be effective (T_1)	0.9	0.6
Test indicates it will be effective (T_2)	0.1	0.4

In conditional probability terms, we write:

$$P(T_1|N) = 0.9 \qquad P(T_1|E) = 0.6$$
$$P(T_2|N) = 0.1 \qquad P(T_2|E) = 0.4$$

We are concerned to find new posterior probabilities that the drug will or will not be effective, given that our test implies either that it will not be effective, or that it will. We wish to find:

$P(N|T_1)$, $P(E|T_1)$, the new probabilities if the test indicates it will be not effective

$P(N|T_2)$, $P(E|T_2)$, the new probabilities if the test indicates it will be effective

First we calculate the *joint probabilities*:

$$P(T_1 \cap N) = P(T_1|N) \times P(N) = 0.9 \times 0.2 = 0.18$$
$$P(T_1 \cap E) = P(T_1|E) \times P(E) = 0.6 \times 0.8 = 0.48$$

Adding
$$P(T_1 \cap N) + P(T_1 \cap E) = P(T_1) = 0.66$$

(since N and E are the only states T_1 can be associated with)

But
$$P(N|T_1) = \frac{P(N \cap T_1)}{P(T_1)} = \frac{0.18}{0.66} = 0.27 \text{ approx.}$$

and $$P(E|T_1) = \frac{P(E \cap T_1)}{P(T_1)} = \frac{0.48}{0.66} = 0.73 \text{ approx.}$$

Repeating these calculations with T_2 replacing T_1 and setting them all out in the form of a table, we have:

	Joint probability $P(T \cap N)$ $P(T \cap E)$		Marginal Probability	Posterior Probability $P(N\|T)$ $P(E\|T)$	
Test indicates:					
T_1 (not effective)	0.18	0.48	0.66	0.27	0.73
T_2 (effective)	0.02	0.32	0.34	0.06	0.94
$P(N)$	= 0.20	$P(E) = 0.80$	1.00		

The *posterior probabilities* are:

$$P(N|T_1) = 0.27 \quad P(E|T_1) = 0.73$$
$$P(N|T_2) = 0.06 \quad P(E|T_2) = 0.94$$

It is interesting to see how the probabilities have changed. If we suppose that the test implies that the drug will be effective our initial 0.8 probability will increase to 0.94. If the test is negative the probability falls to 0.73. Neither are very dramatic changes but then this particular test does not have too good a track record!

(4) RISK ANALYSIS

When probabilistic figures can be estimated for key variables there is another approach. We construct an *uncertainty profile* for *each* key factor. A project appraisal example will explain this approach.

Suppose that estimates of sales from a project indicate that there is a two-thirds chance of their falling within £40,000 of an 'expected' – or average over the long run – figure of £250,000 a year; that there is only a 0.1 chance of their falling below £180,000, and that there is also a 0.1 chance that they will exceed £350,000.

From this and similar information for other key factors the uncertainty profile for each key factor (sales, variable costs, etc.) can be drawn up. Only a few points usually need to be entered. Each curve is generally drawn through these points on the assumption that the spread follows a normal probability distribution pattern (Figure 10.5).

Once an uncertainty profile has been established for each key factor, in this ('Monte Carlo') technique we sample, repeatedly, at random from the distribution of all of these variables. Each combination will give a different rate of return on investment. Each calculation would be long and tedious. Fortunately, computer programs easily cope with this situation.

Probability

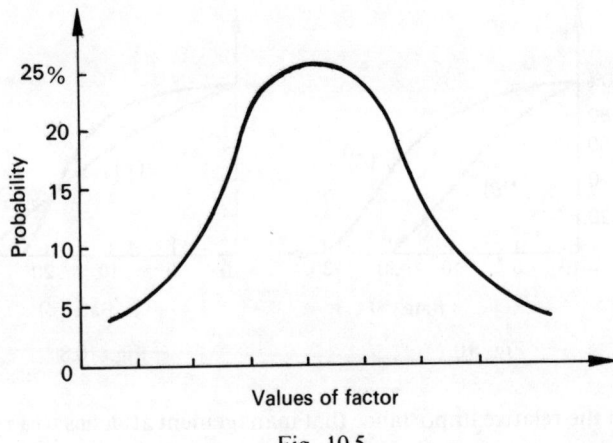

Fig. 10.5

Some 1,000 or so different combinations will usually suffice, and from the results a *probability distribution*, or 'risk profile' is built up.

Figure 10.6 shows the results of a typical simulation. It is easy to read off from this figure the probability that any required rate of return will be achieved. In this case, there is a 90 per cent probability that at least a 12% return will be obtained.

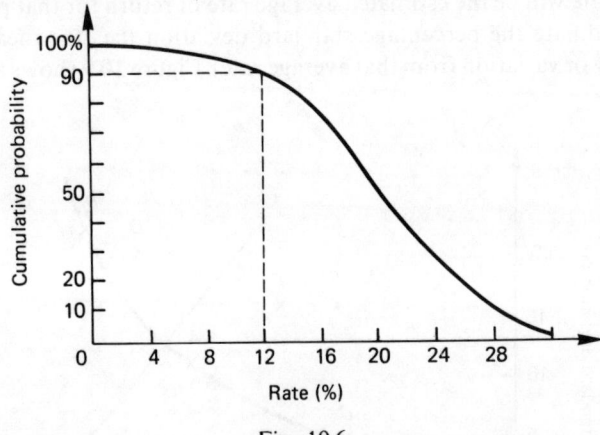

Fig. 10.6

It is now possible to compare accurately two or more investment alternatives. If the risk profiles for two possible investments are drawn, they may either not overlap, as in Figure 10.7, or they may cross as in Figure 10.8.

If the risk profiles are as in Figure 10.7 then (*A*) is absolutely better than (*B*), as it offers a greater probability of achieving each and every given rate of return. As between (*X*) and (*Y*) the position is not so clear. It depends, in the last

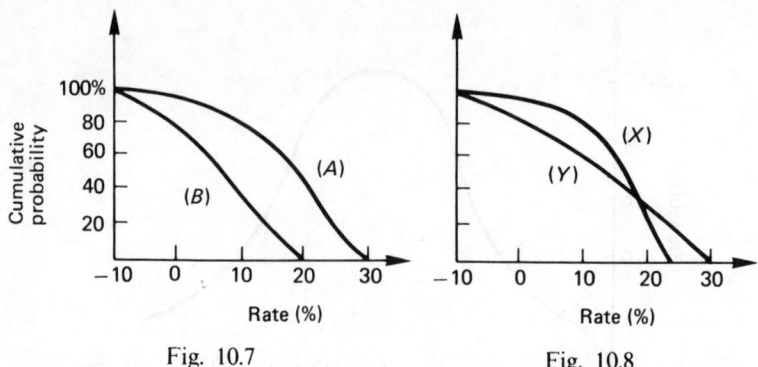

Fig. 10.7 Fig. 10.8

analysis, on the relative importance that management attaches to a reasonable chance of a high rate of return (Y), or being fairly sure that the rate will not fall to a very low figure (X).

The problem, then, is a *risk–return trade off* – that is, weighing the relative advantages of a high rate of return against a high variability. The measure of the spread of a distribution that statisticians use is the 'standard deviation'. The approach, therefore, is to plot a graph of the rate of return against the standard deviation.

Each point on the graph will represent an individual investment project. Its x co-ordinate will be the estimated average rate of return for that project, and its y co-ordinate the percentage standard deviation (i.e., the measure of its uncertainty or variation from that average rate). Figure 10.9 shows a typical set of results.

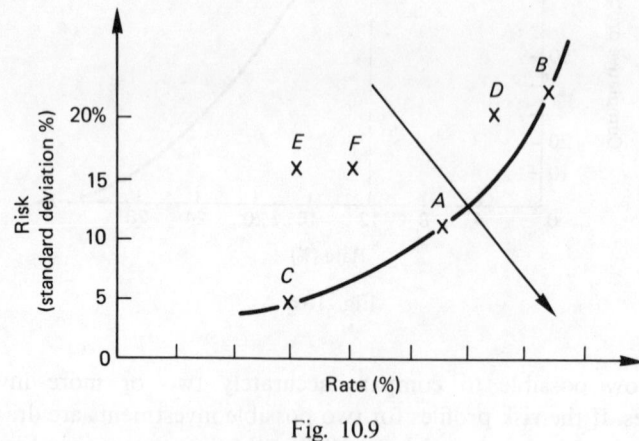

Fig. 10.9

Clearly, project C is better than E, and F than E. What is required is a combination of a small risk and a high rate of return, so the results should be in the general direction indicated by the arrow. The points A, B, C (each

Probability

representing an investment project) lie on the 'efficiency frontier'. Once the efficiency frontier for a company has been determined, the objective must be to improve on this position – that is, to take on projects *beyond this frontier*.

(5) STATES OF NATURE

Frequently outside factors need to be considered in probability and risk analysis. All businesses are affected by the general economic climate, and some businesses by much more local, temporary, factors such as the weather. We need to consider the appropriate techniques to use which take into account the probability that these 'states of nature' will occur. Arriving at these probability figures, is, we say again, not easy, but is not within the domain of this book. In fact the independent Paris-based OECD gives fairly reliable economic forecasts for its member countries, and – to take an example at the other end of the spectrum – an ice cream supplier for a festival may well feel he can place some reliance these days on local weather forecasts.

However, let us concentrate on the more general business situation where a particular company is considering three alternative proposals E, C and P. It obtains the best available information as to the likely economic conditions, and based on this puts probabilities of 0.6 on 'boom', 0.3 on 'normal' and 0.1 on 'depression'. The three alternatives will each have different estimated returns under the three different states of nature. These are:

	Boom (0.6)	Normal (0.3)	Depression (0.1)
E	+9	0	−9
C	−1	+5	+10
P	+15	0	−30

These figures will be the *net cash flow* over the next year, if it is appropriate to restrict the horizon to such a short period. For capital investment projects whose estimated cash flows will continue for a number of years, these figures would be the *net present values*.

To obtain the expected values we multiply each figure by its associated probability. In matrix form, we have:

Choices $\qquad\qquad\qquad\qquad\qquad\qquad$ Expected value

$$\begin{matrix}E\\C\\P\end{matrix}\begin{pmatrix}+9 & 0 & -9\\-1 & +5 & +10\\+15 & 0 & -30\end{pmatrix}\begin{pmatrix}0.6\\0.3\\0.1\end{pmatrix} = \begin{pmatrix}0.6\times 9+0.3\times 0+0.1\times -9\\0.6\times -1+0.3\times 5+0.1\times 10\\0.6\times 15+0.3\times 0+0.1\times -30\end{pmatrix} = \begin{pmatrix}4.5\\1.9\\6.0\end{pmatrix}$$

This gives the expected value of P, that is $EV(P)$, as the best at 6.0. The values, and their rankings are:

		Rank
$EV(E) = 4.5$		2
$EV(C) = 1.9$		3
$EV(P) = 6.0$		1

To take account of the riskiness of each choice, we can calculate the standard deviations and the coefficients of variation as follows:

STANDARD DEVIATIONS (SD)

Deviations of P from $EV(P)$ are $(15-6=)9$, and similarly 6, 36
Hence SD of P is $\sqrt{(0.6 \times 9^2 + 0.3 \times 6^2 + 0.1 \times 36^2)} = \sqrt{189} = 13.75$
Similarly SD of C is 3.81 and SD of E is 6.04

COEFFICIENT OF VARIATION (SD/EV)

		Ranking
Coefficient of Var (E) = 6.04/4.5 = 1.3		1
Coefficient of Var (C) = 3.81/1.9 = 2.0		2
Coefficient of Var (P) = 13.75/6 = 2.3		3

and we can combine the expected value and the risk (measured here by coefficients of variation to take account of the differing sizes of the expected values) in a similar fashion to that shown in Figure 10.9.

We can also consider whether it is worth the cost of obtaining (assuming it *can* be got) perfect information (*PI*). Let us see the possible value of such information in our business example. Suppose, first, that the perfect information tells us that boom will occur. What choice will we make? The answer must be P, since this gives us the greatest return. If either normal or depression are confidently forecast, our choice in either case must be C. However, it is not equally likely that each of these states will be indicated. Based on the best information available to us at the time of the decision, the probabilities are 0.6, 0.3 and 0.1 respectively. Hence:

$$EV(PI) = 0.6 \times 15 + 0.3 \times 5 + 0.1 \times 10 = \underline{11.5}$$

The gain – and hence the amount we should consider paying for perfect information – is $11.5 - 6.0 = 5.5$.

(6) TOTAL UNCERTAINTY

More for completeness of treatment than for the mathematical content (which is minimal) we note that where total uncertainty* occurs (that is, no probability figures at all can be put on each of the likely alternatives) two approaches are possible. In the first approach we run through the range of probability figures and observe how the expected values of the choices alter. Some pattern may emerge from this 'sensitivity analysis' method which will be of help in decision making.

The second approach is to consider the alternatives from the point of view of our emotional attitude in the event that we are successful or not. There are three basic ways to look at this.

*(This will be the case if unknowns enter into the calculations. The worst type of decisions made under uncertainty (DMU, in Student jargon) are where unknown unknowns (unk unks or DMU^2) occur. These are not totally unknown – (if the adjective be allowed) in real life!)

Probability

1. *Maximin* ('choosing the best of the worst') will indicate, for our business example, that C is best, since this means that the maximum possible loss is only -1.
2. *Maximax* (choosing the best of the best) indicates P, since this means that the maximum gain of 15 is possible. These two approaches respectively represent pessimistic and optimistic attitudes.
3. *Minimax regret* (minimising the regret for having made the wrong choice) indicates C, since of all three this has the minimum regret of $15-(-1) = 16$.

10.9 Markov chains

Many mathematical models consider a *transition* from one state to another. If the change is dependent solely on the preceding event or state then we may have a Markov Chain – that is, a sequence of Markov processes. Strictly for this to occur we need three conditions to apply:

1. the states and transitions are discrete and (for an initial analysis) finite in number;
2. transitions depend only on the immediately preceding event and not on events or states before that;
3. transitions can be expressed as probabilities which remain constant over time.

Condition 2, which eliminates the influence of events other than the immediate past event, restricts the process to a 'first order' Markov Chain.

Examples of Markov processes abound. They are used widely in the behavioural sciences, and they have applications in reliability analysis where the performance of equipment at various levels of completeness is being assessed. Later we consider a special type of application suitable for the ageing of debtors, but the most common use of this process in the business situation is in brand loyalty, and we start off with a typical example of this.

In that wet, wild country known as Ruritania there are two suppliers of rainwear who dominate the market. Other manufacturers are so small in number and market power that they can be ignored. The two main suppliers are Aquatex (X) and Aquadry (Y).

Consumers do switch from one of the two main suppliers to the other for various reasons. The probabilities of a switch or a retention over a three-month period can be estimated fairly accurately from past data. In matrix form, they are shown as:

		To	
		X	Y
From	X	0.90	0.10
	Y	0.05	0.95

Clearly users of rainwear Y have a high brand loyalty. The chance that a customer of Y will remain a customer at the end of three months is 0.95. The present market share expressed as a row vector is

$$\begin{matrix} X & Y \\ (0.6 & 0.4) \end{matrix}$$

so that the product

$$(0.6 \quad 0.4) \begin{pmatrix} 0.90 & 0.10 \\ 0.05 & 0.95 \end{pmatrix}$$

$$= (0.56 \quad 0.44)$$

gives the *market share after one quarter*.

X's market share, for instance, is made up of these who start with X and have not changed (0.6×0.9) and those who started with Y and changed to X (0.4×0.05). Similarly for Y.

If the transition matrix:

$$\begin{pmatrix} 0.90 & 0.10 \\ 0.05 & 0.95 \end{pmatrix}$$

is denoted by A for short, the market share after one quarter can be written as:

$$(0.6 \quad 0.4) A$$

Assuming the same probabilities apply to the second quarter, the market share at the end of six months will be

$$(0.6 \quad 0.4) AA = (0.6 \quad 0.4) A^2$$

and the market share at the end of three quarters or (in general) n quarters will be, by analogous reasoning:

$$(0.6 \quad 0.4) A^3$$

and

$$(0.6 \quad 0.4) A^n$$

Working out the market shares for the third and fourth quarters and putting them down together with the current (Quarter 0) market share, we have the results shown in Table 10.2

TABLE 10.2

Quarter	X	Y
0	0.600	0.400
1	0.560	0.440
2	0.526	0.474
3	0.497	0.503
4	0.473	0.527

Probability

From this table, as it stands, it is not too easy to see to what, if any, final limiting values X and Y will tend. Plotting the results on graph paper (Figure 10.10), and carrying on the curves by eye, seems to indicate that the figures for the market share of both X and Y are becoming more fixed and stable. Y is obviously tending to settle down in some position well into the top half of the market share, with X in the lower half. If we suppose that these final fixed shares are p and q, it follows that:

$$(0.6 \quad 0.4) A^n \to (p \quad q)$$

when n tends to infinity, and also

$$(p \quad q) A = (p \quad q)$$

since, once we have arrived at the final steady state, a further transition must, by definition, leave the market share unchanged.

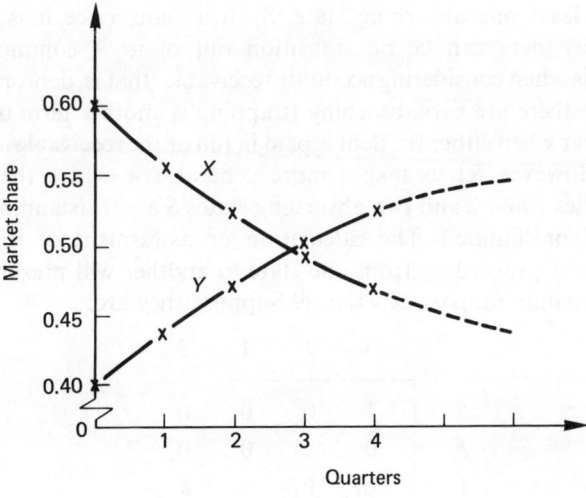

Fig. 10.10

The second equation, in full, is

$$(p, q) \begin{pmatrix} 0.9 & 0.1 \\ 0.05 & 0.95 \end{pmatrix} = (p \quad q)$$

Multiplying out gives:

$$0.9p + 0.05q = p$$
$$0.1p + 0.95q = q$$

We cannot solve these equations for p and q as they stand. All we can get is the ratio p/q. However since p and q together take up the full market share, we also have $p+q=1$. Using this to solve for p and q, we get:

$$p = \tfrac{1}{3}$$
$$q = \tfrac{2}{3}$$

Steady state market shares are therefore $\tfrac{1}{3}$ for X and $\tfrac{2}{3}$ for Y.

It will be noted that in the calculation of these market shares the original market share figures of 0.6 and 0.4 did not enter into it at all. It follows that final market shares depend only on the *transition probabilities*. The original market shares are irrelevant, and whatever they may have been, the final market shares reached will be the same. Naturally, it will take longer to get near the steady state market shares if the initial shares are way out of line with these.

10.10 Absorbing states

A type of Markov Chain of particular interest occurs when the model contains at least one absorbing state. Such a state, once it is reached, is final – that is, there can be no transition out of it. A common business application is when considering accounts receivable (that is, debtors), and their ageing. Here there are two absorbing (trapping is another term used) states, and they occur when either the debt is paid in full or the receivable is declared a bad debt. However, let us take a more general case where there are two transient states 1 and 2 and two absorbing states S and F (standing, perhaps, for 'success' or 'failure'). The calculation or assessment of the expected probabilities of proceeding from one state to another will presumably have been based mainly on past experience. Suppose they are:

	S	F	1	2
S	1	0	0	0
F	0	1	0	0
1	b_{11}	b_{12}	a_{11}	a_{12}
2	b_{21}	b_{22}	a_{21}	a_{22}

Since S and F are absorbing states the probability of proceeding from either of them to any other state is nil. Conversely, the probability of remaining in the same absorbing state is necessarily 1.

Let us concentrate on the probability of ending up in state S. The simplest approach is to list the ways that this can be done, and then add up all their probabilities. If the initial shares or quantities in states 1 and 2 are, say, x and y respectively, then the likelihood of them finishing up in S after one step is:

$$(x \ \ y) \begin{pmatrix} b_{11} \\ b_{21} \end{pmatrix}$$

This is because the chance of going straight from 1 to S is b_{11}, so that the share

Probability

or quantity will be xb_{11}. From 2 to S it will be yb_{21}, and the total is therefore $xb_{11} + yb_{21}$.

The likelihood of going to S in two steps is more complicated. The possible ways are:

$$1 \to 1 \to S, \text{ i.e. } xa_{11}b_{11}$$
$$1 \to 2 \to S, \text{ i.e. } xa_{12}b_{21}$$
$$2 \to 1 \to S, \text{ i.e. } ya_{21}b_{11}$$
$$2 \to 2 \to S, \text{ i.e. } ya_{22}b_{21}$$

The total of these is:

$$(x \ \ y)\begin{pmatrix} a_{11} & a_{12} \\ a_{21} & a_{22} \end{pmatrix}\begin{pmatrix} b_{11} \\ b_{21} \end{pmatrix}$$

The likelihood of going in three or more steps follows the same pattern, so that the likelihood of going to S in n steps is:

$$(x \ \ y)A^{n-1}\begin{pmatrix} b_{11} \\ b_{21} \end{pmatrix}$$

where

$$A = \begin{pmatrix} a_{11} & a_{12} \\ a_{21} & a_{22} \end{pmatrix}$$

Similarly, the likelihood of going to F after exactly n steps is:

$$(x \ \ y)A^{n-1}\begin{pmatrix} b_{12} \\ b_{22} \end{pmatrix}$$

As S and F are absorbing states, once one has arrived at either of these states there is no coming out again. Hence the likelihood of being in S after N steps is the sum of the likelihood of getting there in one step, or in two steps, or in any number of steps up to n steps.

Hence, the likelihood of being at S after n steps is:

$$(x \ \ y)\begin{pmatrix} b_{11} \\ b_{21} \end{pmatrix} + (x \ \ y)A\begin{pmatrix} b_{11} \\ b_{21} \end{pmatrix} +$$
$$+ (x \ \ y)A^2\begin{pmatrix} b_{11} \\ b_{21} \end{pmatrix} + \ldots + (x \ \ y)A^{n-1}\begin{pmatrix} b_{11} \\ b_{21} \end{pmatrix}, \text{ i.e.}$$

$$(x \ \ y)(I + A + A^2 \ldots A^{n-1})\begin{pmatrix} b_{11} \\ b_{21} \end{pmatrix}$$

where I, the identity matrix, has been introduced to preserve the general layout.

The sum to infinity of a geometric series $1 + x + x^2 + \ldots$ is $(1-x)^{-1}$, provided $x < 1$ and by analogous reasoning we may infer that the sum to infinity of the above series is:

$$(x \quad y)(I - A)^{-1} \begin{pmatrix} b_{11} \\ b_{21} \end{pmatrix}$$

Hence, the final probability of ending up in the absorbing state S is:

$$(x \quad y)(I - A)^{-1} \begin{pmatrix} b_{11} \\ b_{21} \end{pmatrix}$$

and of ending up in the absorbing state F is:

$$(x \quad y)(I - A)^{-1} \begin{pmatrix} b_{12} \\ b_{22} \end{pmatrix}$$

The matrix $(I - A)^{-1}$ is referred to as the fundamental matrix of an absorbing Markov chain.

We have now arrived at our answer. The product matrices above show the amounts of x and y (i.e., the amounts currently in the transient states 1 and 2) which will eventually finish up in the absorbing states S and F.

The approach which we have described is quite general. Although we used only two transient states we could have used the same format and extended the transient states to any number.

The layout will still be of the form (referred to as *canonical*)

$$\begin{pmatrix} I & | & O \\ --- & | & --- \\ B & | & A \end{pmatrix}$$

where I and O are the identity and null matrices.

10.11 Combinatorial analysis

In this section we consider how to determine the number of times (and hence the *probability*) that some discrete event takes place without necessarily having to list or lay out each possible event. Typical events that are customarily used as examples are whether a coin comes down heads or tails, which side of a dice comes up on top (these are the two examples we use in the main section on probability), which card will be drawn from a pack of 52 playing cards, and which letter will be drawn from a set of all the letters of the alphabet. When the events or objects have to take place in a given order this is called a *permutation*. When the order does not count this is called a *combination*. For example, if we wish to consider the number of ways that we can lay down two cards from three specific cards – and let us say these are an Ace, a King, and a Queen (AKQ) – these will be:

AK, KA, AQ, QA, KQ, QA – if in a given order, or

$AK, AQ, KQ,$ – if the order does not count.

Probability 209

The first is a permutation, since the Ace and King in AK and KA are in a different order. The second is a combination since AK and KA are the same if order does not count. The same applies to the other permutations and combinations in which the Queen is included. The probability of permuting an Ace and a King (in that order) from the three cards Ace, King, Queen is therefore 1/6. The probability of a combination of an Ace and a King being chosen (that is, when the order as between these two does not matter) is therefore 1/3.

Combinatorial analysis relies on one fundamental theorem. This can be seen – if so desired – as being so obvious that it does not need proving! We merely state it. It is analogous to the statement (again in the main section on probability) that the probability of both of two independent events A and B occurring is the product of their individual probabilities – i.e., $P(A \cap B) = P(A) \times P(B)$ and, by an obvious extension, if we have n independent events $A_1 A_2 \ldots A_n$ the probability that they all occur is $P(A_1 \cap A_2 \cap \ldots \cap A_n) = P(A_1) \times P(A_2) \times \ldots \times P(A_n)$.

In an analogous way, too, we say here that if an event can occur in A_1 different ways, another event in A_2 ways, a third event in A_3 different ways and so on up to and including the last event in A_n ways, then the number of ways the events can occur in the given order is $A_1 \times A_2 \times A_3 \times \ldots \times A_n$.

One example of this long and diffuse (but really almost tautological statement) may help. Let us take just two events. The first is choosing one out of the first two letters of the alphabet (i.e., *a* and *b*) the second is choosing one out of the last three letters of the alphabet (*x*, *y* and *z*). The first can be done in two ways; the second in three. The total number is therefore $2 \times 3 = 6$ ways of choosing two letters.

10.12 Permutations

Any arrangement, then, of a set of *n* objects or events in a given order is called a *permutation*. If the number of these objects or events (we use these two terms interchangeably in this section on combinatorial analysis) is *r*, we call the permutation an '*r*-permutation' or a 'permutation of the *n* objects taken *r* at a time'. If all the *n* objects are to be taken, we call it a 'permutation of the objects (taken all at a time)'.

Reverting to our AKQ example it follows that:

1. $AKQ, AQK, KAQ, KQA, QKA, QAK$ are the permutations (taken all at a time).
2. AK, KA, AQ, QA, KQ, QK are the permutations taken two a time.
3. $A\ K\ Q$ are the permutations taken one a time.

In general the number of permutations in two groups are *not* the same (as they happen to be for this simple case in 1 and 2 above).

We need a formula to tell us how to calculate an *r*-permutation from *n* objects – that is, in how many ways we can list *r* objects from *n* when the order does matter. Let us attempt to list all the possible ways in a sequence from left to right (as we have done in our example above with AKQ). Since there are initially n objects to choose from, the first object can be put down in n ways. For the next position there will be only $n-1$ objects available since one object will have been already put into the first (left-hand) slot. There are $n-1$ ways in which we can put this second object into position. For the third slot there will be $n-2$ objects available, and hence $n-2$ ways of writing it down. This process will continue till for the last slot (the rth), there will be only $n-r+1$ objects left and hence only $n-r+1$ ways to write it down. The numbers of ways each can be done are respectively n, $(n-1)$, $(n-2)$, ... $(n-r+1)$.

Now we know that if one event can occur in n ways, another independent event in $n-1$ ways and so on, the number of ways, in total, that the event can occur in a given order is

$$n \times (n-1) \times (n-2) \times \ldots \times (n-r+1)$$

There is a standard term and notation which will help us here. It is called *factorial n*, or *n* factorial, and is written $n!$.

Hence:

$$n! = n(n-1)(n-2)\ldots 1$$

Obviously $1! = 1$, and conventionally we say that $0! = 1$ (but see below).

The r permutation above, that is:

$$n(n-1)(n-2)\ldots(n-r+1)$$

can be written as:

$$\frac{n(n-1)(n-2)\ldots(n-r+1)(n-r)(n-r-1)(n-r-2)\ldots 1}{(n-r)(n-r-1)(n-r-2)\ldots 1}$$

i.e., as $\dfrac{n!}{(n-r)!}$.

Hence the permutation of n objects taken r at a time is:

$$\frac{n!}{(n-r)!} \quad \text{or} \quad n(n-1)(n-2)\ldots(n-r+1)$$

There are a number of alternative notations for this, too. The two most common are:

$$P(n,r) \quad \text{and} \quad {}_nP_r$$

Hence the r-permutation is:

$$P(n,r) = {}_nP_r = \frac{n!}{(n-r)!}$$

In the special case of a permutation of all the n objects (i.e., $r = n$), we have:

$$P(n,n) = n!$$

Probability

(which on other, commonsense, grounds we know to be true, and hence can be seen as some sort of justification for the assumption we made a few lines back that $0! = 1$).

Occasionally we may want to find the number of permutations when some are identical (or are so alike that they can be considered to be identical in the given situation). Suppose, for example, that we wish to find all possible five letter permutations of the word *LOLLY*. There are five letters in this word, and hence a total of 5! possible permutations. However, three of the letters are identical. Clearly transposing one L with another leaves the 'word' unaltered. In how many ways can this be done? The number of permutations of the three Ls is $3! = 6$. Hence there are $5!/6$ different ways of arranging the letters of the word *LOLLY*. We could argue that there are $1! = 1$ ways of arranging each of the other two letters (O and Y). To be pedantic, we could therefore say that the number of permutations is:

$$5!/1!\ 1!\ 3! = 20$$

Again by an obvious extension this can be seen as just one example of the general proposition that if we have n objects of which a_1 are identical, a_2 are also identical and so on up to a_r, then the number of different permutations is

$$\frac{n!}{a_1!\ a_2!\ \ldots\ a_r!}$$

where $a_1 + a_2 + \ldots + a_r = n$ and all are positive integers.

10.13 Combinations

Any arrangement of a set of n objects or events, where order does not count, is called a combination. If the number of events or objects is r, we call this a 'r-combination of a set of n objects'. Since the number of permutations of the r objects taken r at a time is $r!$, the number of combinations of n objects taken r at a time is

$$\frac{\text{number of permutations}}{r!}$$

$$= \frac{n!}{(n-r)!} \div r! = \frac{n!}{(n-r)!\,r!}$$

This argument, too, is analogous to that in the preceding section where we used the word *LOLLY* as an example.

There are a number of alternative notations for a r-combination of n objects. Two of the most frequently used are $C(n, r)$ and $_nC_r$. However perhaps the most common is $\binom{n}{r}$.

Hence we have:

$$\binom{n}{r} = C(n, r) = {}_nC_r = \frac{n!}{r!(n-r)!}$$

It is important to realise that the first three are just different notations for the same thing.

10.14 Binomial theorem

Consider the expression $(a+b)^n$. Later (in question 16) we shall indicate how to show that where n is a positive integer:

$$(a+b)^n = \sum_{k=0}^{n} \frac{n!}{k!(n-k)!} a^{n-k} b^k$$

or, using the notation from the preceding section on combinations:

$$(a+b)^n = \sum_{k=0}^{n} \binom{n}{k} a^{n-k} b^k$$

This is called the *bionomial theorem* and the coefficients are called the *binomial coefficients*. These coefficients for successive powers of n can be arranged in a symmetrical triangular form. The coefficients have the interesting property that any number in one row can be obtained by adding together the two numbers in the row directly above it. This display form is called *Pascal's triangle*. The first few rows are as in Figure 10.11.

```
(a + b)⁰ = 1                                                              1
(a + b)¹ = a + b                                                        1   1
(a + b)² = a² + 2ab + b²                                              1   2   1
(a + b)³ = a³ + 3a²b + 3ab² + b³                                    1   3   3   1
(a + b)⁴ = a⁴ + 4a³b + 6a²b² + 4ab³ + b⁴                          1   4   6   4   1
(a + b)⁵ = a⁵ + 5a⁴b + 10a³b² + 10a²b³ + 5ab⁴ + b⁵            1   5  10  10   5   1
(a + b)⁶ = a⁶ + 6a⁵b + 15a⁴b² + 20a³b³ + 15a²b⁴ + 6ab⁵ + b⁶  1   6  15  20  15   6   1
```

Fig. 10.11

There are many uses of the binomial theorem: some of these are given in the following questions.

10.15 Exercises

Worked answers are provided in Appendix C to questions marked with an asterisk.

1. An ordinary six-sided dice has four of its faces coloured red and the remaining two coloured black. What is the probability that when the dice is thrown the top face will be:
    (a) red?
    (b) black?

Probability

    (c)   red or black?
    (d)   neither red nor black?

2. Two dice are thrown simultaneously. What is the probability that both will be 3s?

3. Two dice are thrown simultaneously. What is the probability that the total score will be:
                (a) 5,   (b) 6?

4. In how many ways is it possible to arrange in an ordered pattern the letters in the words:

    (a)   AT?
    (b)   RED?
    (c)   LUCK?

5. The committee of a racket club is unable to come to a decision as to how to choose the best mixed doubles pair. They are considering 8 lady players and 10 men players. Finally, in complete disarray they decide to draw lots. Muriel is one of the ladies and Jim is one of the men being considered.
Calculate the probability that:

    (a)   Muriel is in the pair, but Jim is not;
    (b)   At least one of these two is chosen.

6. Two events E and F are independent of each other. The respective probabilities that they will occur are $1/7$ and $2/9$.
Calculate the probability that:

    (a)   both E and F occur;
    (b)   neither of these events occurs;
    (c)   at least one of them occurs;
    (d)   exactly one occurs.

7. In a survey of 200 owners of electronic appliances made by a certain very well-known manufacturer, it was found that:

                110 had their basic micro (BM)
                50 had their word processor (WP)
                60 had their audit synchroniser (AS)

and that

                40 had both their BM and WP
                25 had both their AS and WP
                35 had both their BM and AS
                20 had all three

Calculate the probability that a particular owner has:

(a) a *BM* and/or a *WP* and/or a *AS*;
(b) a *BM* only.

8. Produce a program in BASIC or any other high-level language you like to:

   (a) *Calculate the factorial of a number*
   The best definition to use for this purpose is:

   $$n! = n \times (n-1)! \quad \text{for } n > 1$$
   $$n! = 1 \quad \text{for } n = 1$$

   (b) *Find the number of permutations* of n objects taken r at a time.
   (c) *Find the number of combinations* of n objects taken r at a time.
   The programs in (b) and (c) can be quite simple, merely incorporating the formulae given in the text for permutations and combinations.

9. The credit manager of a company knows from past experience that if an applicant for a £60,000 mortgage is a 'good risk', and the company accepts the applicant, the company's profit will be £15,000. On the other hand, if it accepts an applicant who is a 'bad risk' the company will lose £6,000. Rejecting a 'bad risk' applicant costs and gains the company nothing, but it values the goodwill lost by rejecting an applicant who is a 'good risk' as equivalent to a loss of £3,000.

   (a) Complete the following profit and loss table for this situation:

Decision	Class of Risk	
	Good	Bad
Accept		
Reject		

   (b) The credit manager assesses a certain applicant to be twice as likely to be a bad risk as a good one. What should be the decision so as to maximise the company's expected profit?
   (c) Another manager independently assesses the same applicant to be four times as likely to be a bad risk as a good one. What should be this manager's decision?
   (d) What estimate of the probability of being a 'good risk' would make the company indifferent between accepting or rejecting applications for such a mortgage?

   (*Chartered Insurance Institute*, April 1983)

Probability

10. A specialist car manufacturer uses 2,000 gearboxes a year and can either buy them locally at a cost of £125 per unit or import them for £120 per unit. The annual holding cost is £12 per unit. The cost of placing an order locally is £5; for imported gearboxes it is £15. The handling costs per order are £10 for the local product and £100 for the imported units.

You are required to:

(a) calculate the minimum average annual cost for each supplier and determine which is the more economic source of supply;
(b) assuming that the imported gearboxes come by sea and the probability of a dock strike lasting more than 10 days is 0.1 and the consequent stock out costs £80,000, discuss whether the decision based on the results in (a) needs to be changed; and
(c) discuss the usefulness of the calculations in (a) and (b) in making management decisions.

(*Institute of Chartered Accountants in England and Wales*, November 1984)

11. Cadmus, Carna and Clytie, a firm of investment consultants, have been approached by one of their clients with regard to the investment of a sum of £100,000 over a period of two years. After a thorough survey of the available opportunities, two alternatives (*A* and *B*) are proposed, one involving a small amount of risk, the other being risk-free.

Investment *A* will lead to a return of either 8%, 10% or 12% in each year but, due to the nature of the investment, there will be some correlation between year 1 and year 2 returns. This is shown by the following table which gives the probability of various returns in year 2 given the returns in year 1.

Year 1	Year 2 8%	10%	12%
8%	0.6	0.3	0.1
10%	0.2	0.5	0.3
12%	0.1	0.2	0.7

At this stage, the three different returns in year 1 are considered to be equally likely.

Investment *B* will produce a certain return of 9.5% per year.

You may ignore the effects of taxation, and you may assume that interest earned in year 1 is re-invested for the second year.

Required:
(a) Assuming that whichever alternative is chosen, the investment will be made for the full two-year period:
  (i) draw a decision tree to represent the alternative courses of action and outcomes;
  (ii) on the basis of the expected value of returns, which investment would you recommend?
  (iii) What is the probability that investment B produces a greater return than investment A?
(b) Indicate how your decision tree should be modified if it is possible to switch from investment A to investment B at the end of year 1. In this case, what investment strategy will produce maximum expected return?
(*The Association of Certified Accountants*, December 1984)

12.* The distribution of cash flows for two investment projects is given below:

Project A		Project B	
Net cash flow (£)	Probability	Net cash flow (£)	Probability
1,000	1/3	0	1/3
3,000	1/3	3,000	1/3
5,000	1/3	6,000	1/3

Either project A or B but not both may be undertaken.

You are required to:

(a) calculate the expected values of projects A and B and their respective variances;
(b) discuss general criteria for making a choice between projects such as A and B, and to determine which project should be selected, giving reasons for your conclusion;
(c) discuss briefly the use which may be made of the coefficient of variation in helping to decide between projects such as A and B, illustrating your answer with regard to these two potential investments.

(*Institute of Cost and Management Accountants*, November 1985)

13.* A centralised kitchen provides food for various canteens throughout an organisation. Any food prepared but not required is used for pig food at a net value of 1p per portion.

Probability

A particular dish, *D*, is sold to the canteens for £1.00 and costs 20p to prepare.

Based on past demand, it is expected that during the 250-day working year the canteens will require the following quantities:

On	100 days	40
On	75 days	50
On	50 days	60
On	25 days	70

The kitchen prepares the dish in batches of 10 and has to decide how many it will make during the next year.

You are required to:

(a) calculate the highest expected profit for the year that would be earned if the same quantity of *D* were made each day;

(b) calculate the maximum amount that could be paid for perfect information of the demand (either 40, 50, 60 or 70) if this meant that the exact quantity could be made each day.

(*Institute of Cost and Management Accountants*, November 1985)

14. Hyper Furniture Store plc is considering a major extension to one of its stores. Its assessment of the current economic situation leaves it with the option of starting work on the expansion scheme now, delaying the start for one year or deciding after a year not to expand at all. The capital cost of the scheme will be £1 million if it takes place now or £1.1 million if it is delayed by one year.

The potential future market has been estimated at three levels and certain likelihoods of achievement have been assigned:

Market	Probabilities
Weak	0.1
Average	0.6
Strong	0.3

The present values of the net revenues as at the end of the first year of trading if the extension goes ahead straight away are given as:

Market	Present value of net revenues in one year's time £m
Weak	1.45
Average	1.65
Strong	1.85

However, if the start of the scheme is delayed a year or alternatively it is not proceeded with at all, then the following outcomes have been estimated:

Market	Present value of net revenues in one year's time	
	Extension undertaken £m	Extension not undertaken £m
Weak	1.35	0.45
Average	1.45	0.55
Strong	1.85	0.65

Assume that the respective capital outlays take place either immediately or in one year's time. Cost of capital is 10%.

You are required to:

(a) prepare a decision tree diagram of the situation to be evaluated by Hyper Furniture Store plc's management;
(b) make an evaluation of the tree on the basis of the expected net present value;
(c) describe briefly how you would interpret the results from (b) above.

(*Institute of Cost and Manangement Accountants*, May 1986)

15. Prove that

$$\binom{n+1}{k} = \binom{n}{k-1} + \binom{n}{k}$$

This is the general statement of the property, already noted in the text, for Pascal's triangle – that is, a number in any row can be obtained by adding together the numbers in the row directly above.

The first step in your proof should be to arrange for the two factors on the right to have a common denominator.

16. Show, by multiplying through both sides by $(a+b)$ that if the binomial theorem:

$$(a+b)^n = \sum_{k=0}^{k=n} \binom{n}{k} a^{n-k} b^k$$

is true for any positive integer value for n then (since it will still be in the same form, with $n+1$ replacing n), it is also true for $n+1$. Hence argue that since it is true for $n = 1$, it must be true for $n = 2$ and therefore for $n = 3$, and so on.

It must therefore be true for *all* positive integer values of n.

This method of proof is called *proof by induction*. It is a rather specialised (but powerful) method. Use the result from Question 15 above in your proof.

17. Write a program in BASIC (or any other suitable high-level language if you prefer) with inputs a, b and N to calculate the value of the right-hand terms of the binomial expansion, that is:

$$\sum_{k=0}^{k=n} \binom{n}{k} a^{n-k} b^k$$

18. Use the program in Question 17 to find the value of the binomial expansion when $a = 1$, $b = 1/n$ for the following cases:
    1. $n = 5$
    2. $n = 6$
    3. $n = 7$
    4. $n = 8$
    5. $n = 9$

    It is possible to use the binomial theorem to arrive at an approximate value for e. This follows because (though we do not prove it), the value of e is given by:

    $$e = \lim_{n \to \infty} (1 + \tfrac{1}{n})^n$$

    Compare your results above with the value (to 4 decimal places) for e of 2.7183.

19. Items come off the end of a production line. Some items will be defective, most will not. Samples are taken from the production line to monitor for quality.

    (a) For a sample of two show that:

        (i) if the proportion of defective items is 10% of the whole then the probabilities of two defective items, one defective item or no defective items being found in the sample are respectively 0.01, 0.18 and 0.81.

        (ii) if the proportion of defective items is p (so that the probability of a non defective $q = 1 - p$) show that the corresponding probabilities are now p^2, $2pq$, q^2.

    (b) For a sample of three show that the corresponding probabilities (with an obvious notation) are:

P (3 defectives)	0.001	p^3
P (2 defectives)	0.027	$3p^2q$
P (1 defective)	0.243	$3pq^2$
P (no defectives)	0.729	q^3

    (c) Show that these are just special cases of the binomial distribution (which applies to independent trials with two outcomes). (The binomial distribution is also called the *Bernouilli distribution*).

Chapter 11
Regression Analysis

11.1 Introduction

In all life we see examples of how one thing depends on another. In a factory direct material costs go up with throughput, at home, electricity consumption with the number of lights an hour turned on. Many examples have been already met with in this book.

Sometimes the relationship is not too clear. For instance a factory worker's performance on a complicated job will vary with the level of his tiredness. In what precise way, however, is not intuitively obvious. Sometimes, as in the example of factory direct material, the connection is obvious and direct. If the cost of the materials going into the production of one unit is (say) £2, then ignoring such complications as wastage and so on, we can make the statement 'Materials cost is twice the throughput' (i.e., the number of units produced). Alternatively we can use symbols. If Y is the direct materials cost in money terms, and X is the number of units produced we say:

$$Y = 2X$$

Usually, however, we can only *infer* a relationship of this nature. Suppose that in the last four months the figures for direct labour costs and throughput had been as in Table 11.1. Whether we had guessed from our knowledge of the situation or not that these costs should have been twice throughput, we could have inferred this by looking at the figures. In each month, cost is almost exactly twice throughput.

Frequently, a relationship will not be easy to determine simply by inspection of the sets of data. Suppose the figures for some overhead cost (e.g.,

TABLE 11.1

Monthly	Cost	Throughput
1	281	140
2	260	130
3	222	110
4	201	100

Regression Analysis

maintenance) had been as in Table 11.2. Here the likely relationship is difficult to tell immediately, though a check confirms that the equation $Y = 5 + 3X$ does give, pretty accurately, the relationship between costs and throughput.

TABLE 11.2

Months	Cost	Throughput
1	425	140
2	394	130
3	336	110
4	305	100

In the above examples *cost depended on throughput*. In regression analysis terminology, cost is called the *dependent* variable, and throughput the *independent* variable. The commonsense meaning of these terms corresponds very well with their technical function here in regression analysis, and from now on we shall use these more general terms.

In cases where a linear relationship (as in the examples above) is believed to exist between the dependent and independent variables, but the exact form of the equation is too difficult to determine by inspection, it is necessary to have a technique which enables one to find the equation. Simple regression analysis is the appropriate technique.

11.2 Simple regression

Suppose the sets of data are as in Table 11.3. We have moved to the more general terminology for the variables, but we have not indicated when or where the readings (and what sets of data) were taken. They may indeed be *any* readings at any time or any place (provided that the implied underlying assumption of an ongoing relationship between the variables is satisfied). Looking at the information as laid out in the data of Table 11.3, it is not at all obvious what direct relationship (if any) exists between the variables.

In both our earlier examples the relationship was expressed by way of a simple *linear algebraic equation*. In the first case the equation was $Y = 2X$, and

TABLE 11.3

Dependent Variable (Y)	Independent Variable (X)
4	2
4	3
6	6
7	8
6	10

a graph of this equation would have shown a line through the origin; in the second case the graph would have been a line meeting the vertical axis at (0, 5). In both cases the line goes almost exactly through all the points. A graph, it seems, gives a good pictorial presentation, and may help solve the problem.

Figure 11.1 shows the data as points and confirms that there is no clear and obvious line going through, or nearly through, all these points. But is there any 'line of best fit' and does such a line have any real *meaning*? These are two important questions. We need to be able to find answers, not just for these figures, but for all sets of data.

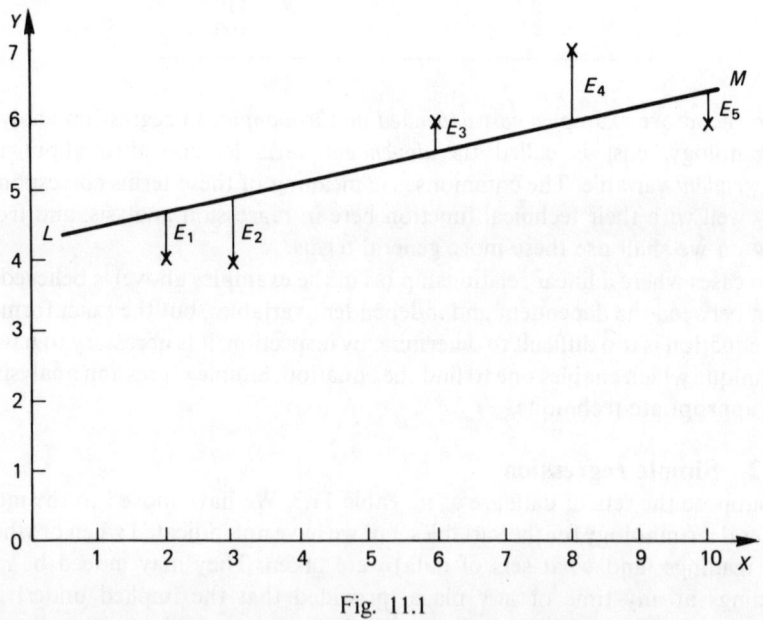

Fig. 11.1

Presumably for a line to fit the points as well as possible it must lie as closely as possible to all these points. Such a line, LM, has been drawn. If LM is a poor fit – to take a highly exaggerated example, suppose it is right off the page at the top of the diagram – then the distances of the points away from this line will be substantial. On the other hand if the line is a very good fit these distances will be small. It seems reasonable to suppose that the line of 'best fit' will be that line which makes these differences *as small as possible*.

Several problems arise in measuring these distances. They could be the perpendicular distances from the line, or distances measured horizontally (or, as in Figure 11.1, the distances measured vertically – E_1, E_2, etc.). We could try and minimise the sum of these distances, but this would cause problems with signs, since a large positive distance above the line and a large negative distance below the line would then cancel each other out. Conventionally we use the sum of the squares and arguably, there are some academic reasons for this

Regression Analysis

assumption. Anyhow this neatly gets rid of this problem of the sign, and also the mathematics is easier! Hence, we minimise:

$$S = E_1^2 + E_2^2 + E_3^2 + E_4^2 + E_5^2$$

The line that minimises this expression will give *a* line of best fit. Other formulae will give other lines. The square root of this expression – when divided by the number of points – is the standard deviation, and this itself is regarded for statistical purposes as the most useful measure of the dispersion of points about a line. Since we are trying to find how far the costs differ from the line for certain given throughputs it is usual to use the *vertical* distances in these situations.

11.3 Formulae for the line of best fit or regression line

It may, as we said, well be possible to draw a good approximation to the line of best fit by eye. However, if the points are widely scattered, and there are a large number of them this will not be easy and we shall need some formula to determine the line.

Moving to the more general case let us assume that the equation of the 'line of best fit' (or *regression line*) is:

$$Y = B_0 + B_1 X,$$

and suppose that we have n sets of past results $(X_1, Y_1) (X_2, Y_2), \ldots, (X_n, Y_n)$ giving (say) production and the corresponding cost for the last n periods. For each value (i, say) we can write:

$$Y_i = B_0 + B_1 X_i + E_i$$

(all values of i from 1 to n)
We wish to minimise S where:

$$S = \sum E_i^2 = \sum (Y_i - B_0 - B_1 X_i)^2$$

More precisely, we wish to find those values (b_0 and b_1, say) for B_0 and B_1 which will make S a minimum. S can therefore be regarded as a function of the parameters B_0 and B_1. To find the minimum we differentiate partially, first with respect to B_0 and then to B_1 and put the results equal to 0:

$$\frac{\partial S}{\partial B_0} = -2\sum (Y_i - B_0 - B_1 X_i)$$

$$\frac{\partial S}{\partial B_1} = -2\sum X_i (Y_i - B_0 - B_1 X_i)$$

$\Bigg[$ Since

$$\frac{\partial S}{\partial B} = \frac{\partial S}{\partial Z} \cdot \frac{\partial Z}{\partial B}$$

where $Z = Y_i - B_0 - B_1 X_i$ for B equal to B_0 and then B_1, and, as $S = Z^2$, then:

$$\frac{\partial S}{\partial Z} = 2Z, \quad \frac{\partial Z}{\partial B_0} = -1 \quad \text{and} \quad \frac{\partial Z}{\partial B_1} = -X_i \bigg]$$

Putting these each equal to 0, the values for b_0 and b_1 satisfy:

$$\sum(Y_i - b_0 - b_1 X_i) = 0$$
$$\sum X_i(Y_i - b_0 - b_1 X_i) = 0$$

These reduce to:

$$\sum Y_i - nb_0 - b_1 \sum X_i = 0$$
$$\sum X_i Y_i - b_0 \sum X_i - b_1 \sum X_i^2 = 0$$

That is:

$$\left. \begin{array}{c} b_0 n + b_1 \sum X_i = \sum Y_i \\ b_0 \sum X_i + b_1 \sum X_i^2 = \sum X_i Y_i \end{array} \right\} \quad (11.1)$$

These are a pair of simultaneous equations which we can solve for $b_0 b_1$. They are referred to as the *normal equations*.

From a computational point of view it is often more practical and simple to use a variation on this formula. The arithmetic means or averages are:

$$\bar{x} = \frac{\sum x}{n} \quad \bar{y} = \frac{\sum y}{n}$$

If we now express the formula in terms of deviations from these mean values, we have (changing to lower case letters, and to a new notation for the line as being $y = a + bx$, in order to show the difference):

$$a = \bar{y} - \frac{\sum(x - \bar{x})(y - \bar{y}) \times \bar{x}}{\sum(x - \bar{x})^2}$$

$$b = \frac{\sum(x - \bar{x})(y - \bar{y})}{\sum(x - \bar{x})^2}$$

11.4 Application of the formulae for the line of best fit

Now that we have a general formula which will enable us to determine the line of best fit (or 'least squares' line as it is called, since it is obtained by minimising the sum of the squares of the distances of the points above and below the line), let us see how it will work out in practice.

Table 11.4 uses the same data as Table 11.3. The first two columns have been added up and the totals divided by 5 (the number of sets of data) to give the arithmetic means – i.e., $\bar{x}$ and $\bar{y}$ as 5.8 and 5.4 respectively. The first

Regression Analysis

entry in column 3 is $(2.0 - 5.8 =) -3.8$ and in column 4 is $(4.0 - 5.4 =) -1.4$. The first entries in columns 5, 6 and 7 are $(-3.8^2 =) 14.44$, $(-1.4^2 =) 1.96$ and $(-3.8 \times -1.4 =) 5.32$. The other entries are formed in a similar way.

Adding columns 5, 6 and 7, we have:

$$\sum(x - \bar{x})^2 = 44.80$$
$$\sum(y - \bar{y})^2 = 7.20$$
$$\sum(x - \bar{x})(y - \bar{y}) = 15.40$$

Applying the formula above, and rounding off figures to two decimal places, we have,

$$a = 5.4 - \frac{15.40 \times 5.80}{44.80} = \underline{3.41}$$

$$b = \frac{15.40}{44.80} = \underline{0.34}$$

so that the equation to the line of best fit is

$$y = 3.41 + 0.34x$$

It is easy to construct this line. If we put $x = 0$ then $y = 3.41$, and if we put $x = 10$ then $y = 6.81$, consequently the line is drawn by joining the points $(0, 3.41)$ and $(10, 6.81)$. The 'correct' line of best fit is therefore more tilted than the line *LM* we drew in Figure 11.1 by eye.

In section 11.2 above we said there were two important questions – 'Is there any line of best fit, and does such a line have any real meaning?' We have answered the first of these. To consider the latter, let us take two extremes. First a line of best fit which passes through all the points. Table 11.1 will provide a good example if we alter the costs slightly so that they are all *exactly* double the throughput (i.e., the cost in month 1 is now 280 and so on for all the other months). Even if we had not known, on the other grounds, that cost should be twice throughput, we could now infer this with confidence from the figures. We could not only say that the line $(Y = 2X)$ passing through all these points was obviously the line of best fit, but also that it accurately shows the relation between costs and output, so that for other throughputs it should predict *accurately* (provided the underlying assumptions continue to apply) exactly what the cost should be. Thus for a throughput of 120 this line will give a value of 240 and we might well feel, in view of the precise way that the line fits the points, that we can rely on this prediction.

Now for the other extreme. Imagine points randomly scattered. Even these points will have some line of best fit. Any line roughly through the centre of these points will have its squares less than that of a line drawn miles away from the points. The squares of the distances will vary with the line and therefore there will be *some* line of least squares even for these points. But since they are randomly scattered there is not real relation between the points. This line of best fit is a statistical curiosity and no confidence can be attributed to it. For prediction purposes it is valueless.

TABLE 11.4

(1) Columns	(2)	(3)	(4)	(5)	(6)	(7)
x	y	$x - \bar{x}$	$y - \bar{y}$	$(x - \bar{x})^2$	$(y - \bar{y})^2$	$(x - \bar{x})(y - \bar{y})$
2	4	−3.8	−1.4	14.44	1.96	5.32
3	4	−2.8	−1.4	7.84	1.96	3.92
6	6	0.2	0.6	0.04	0.36	0.12
8	7	2.2	1.6	4.84	2.56	3.52
10	6	4.2	0.6	17.64	0.36	2.52
$\sum x = 29$	$\sum y = 27$			44.80	7.20	15.40
$\bar{x} = \dfrac{29}{5}$	$\bar{y} = \dfrac{27}{5}$					
$= 5.8$	$= 5.4$					

But what about the more usual situation which lies somewhere between these extremes? Graphically, the points will probably seem to be grouped more or less round some rather indefinite line. Already we have determined what the best expression for that line will be. All that remains to be done is to have some formula which will tell us whether the points are so close to this line that it has some significance (so that for prediction purposes we can rely on it to some extent), or not.

We can now make use of the squares of the distances from these mean values (i.e. $\bar{x}$ and $\bar{y}$) for x and y. The aim is to determine the 't' value, usually regarded as the best measure for these purposes.

A quick check on the equation shows that the line must always pass through the mean values $(\bar{x}, \bar{y})$. The 't' value for the slope (i.e., b in the formula $y = a + bx$ for the line) is—we are saying, but not proving—the most important indicator. The calculation of this t value is most easily done in two stages.

First calculate:

$$r = \frac{\sum(x - \bar{x})(y - \bar{y})}{\sqrt{\sum(x - \bar{x})^2} \cdot \sqrt{\sum(y - \bar{y})^2}}$$

then, put the value of r obtained from this, in the formula:

$$t = \frac{r\sqrt{n-2}}{\sqrt{1 - r^2}}$$

The meaning of the t value and its applications in prediction and confidence limits is susceptible to fairly detailed analysis. Here we merely use the same example to see how it works out in a simple case. From Table 11.4:

$$\sum(x - \bar{x})(y - \bar{y}) = 15.4$$
$$\sum(x - \bar{x})^2 = 44.8$$

Regression Analysis

giving

$$\sqrt{\sum(x-\bar{x})^2} = \sqrt{44.8} = 6.69$$
$$\sum(y-\bar{y})^2 = 7.2$$

giving

$$\sqrt{\sum(y-\bar{y})^2} = \sqrt{7.2} = 2.68$$

Hence

$$r = \frac{15.4}{6.69 \times 2.68} = 0.859$$

Putting in this value, we have:

$$t = \frac{0.859\sqrt{3}}{\sqrt{1-(0.859)^2}} = \frac{0.859 \times 1.732}{0.512} = 2.9 \text{ (approx.)}$$

To find how good a fit is implied by this t value we have to refer to tables. The appropriate table is Table A.4 (p. 303). This table shows 'degrees of freedom' (d.f.) on the left. The exact degree of freedom is easily found. We are trying to fit a *line* to the points. If we had two points the line would have been precisely determined since only one line can be drawn through two points. So with two points there are no degrees of freedom. With three points there is one degree: with four points, two degrees: with our five points we have $(5-2=)$ three degrees of freedom. In general, if we have n points, there are $n-2$ degrees of freedom.

Looking along the row opposite 3 d.f. we have values of 2.353 at the 0.05 probability, and 3.182 at the 0.025 probability. We have to double these probabilities since this table gives only the probability for one of the two tails. Our value therefore lies between 0.1 and 0.05. It is greater than the 0.1 probability figure, which implies that the degree of closeness of these points about the line could not have been achieved once in ten times by chance. On the other hand it is not significant at the 0.05 level. That is, if we demand a correlation between the points more reliable than that which could have been achieved once in twenty times by chance we do not have it. Possibly some reliability can be attached to the line for prediction purposes, but one cannot have any great confidence in it. It is usual to say that if the 5% level is exceeded it is 'probably' significant; if the 1% level is exceeded it is 'definitely' significant, though even these terms are conventional. A 1% probability *could* have been achieved just by chance.

11.5 Prediction and confidence limits

We have argued that the greater the reliability we can place on the line, the better it will be for prediction purposes. But the confidence we can have in the prediction will depend on one other factor as well – the relative *position* of the point.

If we choose a large value for x – again right off the page just by way of example – then any error in the slope of the line will be much magnified, and we shall not be able to have much confidence in the predicted value for y. The line passes through the point $(\bar{x}, \bar{y})$ – the mean value of the points – and the error will be at its least there. The further away from this mean value the less the confidence that can be placed on the prediction. Looked at the other way round, if we wish to know what are the limits within which we may feel sure (at some predetermined level of confidence – say, 95 %) of the predicted value, then these limits will increase the further we are away from the mean $(\bar{x}, \bar{y})$. In practice, tables are used to determine the limit for any given value of x.

11.6 Multiple regression

We can summarise what has been achieved so far by saying that if we have sets of data showing how one variable (the dependent variable) alters to correspond with changes in another variable (the independent variable) then simple regression analysis enables us to determine that linear equation which best expresses the relationship between these two. But is it usual, in the business world, for a variable to depend for its value solely on one other variable? Most business planning involves a forecast of sales: indeed we put it more strongly, and say that the starting point of almost all business planning is an estimate of the sales figures in the future. But how do we arrive at such an estimate? Do sales depend only on one factor? Probably not. There will be *internal* factors – e.g., the level of advertising expenditure – and *external* factors such as the number of people needing the product in the area in which it will be sold. In practice sales are likely to depend on a whole range of factors. This dependence on a number of factors is typical of the business situation. Clearly we need a technique to take account of this.

The graphical solution was useful when we had only one independent variable (X) to compare with the dependent variable (Y). On this basis we would need to use three-dimensional geometry if we had two independent variables X_1, X_2, and n-dimensional geometry for the general case. Clearly we need a different approach.

First we revert to the basic equation for one independent variable, that is, $Y_i = B_0 + B_1 X_i + E_i$, which holds good for all values of i from 1 to n. We can write these as:

$$Y_1 = B_0 + B_i X_1 + E_1$$
$$Y_2 = B_0 + B_i X_2 + E_2$$
$$Y_3 = B_0 + B_i X_3 + E_3$$
$$- \quad - \quad - \quad -$$
$$- \quad - \quad - \quad -$$
$$Y_n = B_0 + B_i X_n + E_n$$

Changing to matrix notation, we call Y the column vector of the values of the

Regression Analysis

dependent variable, X the matrix of the independent variables, B the column vector of the parameters, and E the column vector of the individual errors E. That is:

$$Y = \begin{pmatrix} Y_1 \\ Y_2 \\ Y_3 \\ - \\ - \\ Y_n \end{pmatrix} \quad X = \begin{pmatrix} 1 & X_1 \\ 1 & X_2 \\ 1 & X_3 \\ - & - \\ - & - \\ 1 & X_n \end{pmatrix} \quad B = \begin{pmatrix} B_0 \\ B_1 \end{pmatrix} \quad E = \begin{pmatrix} E_1 \\ E_2 \\ E_3 \\ - \\ - \\ E_n \end{pmatrix}$$

and the fact that our basic equation holds good for all the n values can be written as:

$$Y = XB + E$$

The transpose A' of a matrix A is that matrix obtained by interchanging its rows and columns. Vectors can be seen as special cases of matrices, hence:

$$E' = (E_1 \, E_2 \, E_3 \, \ldots \, E_n)$$

The multiplication:

$$E'E = (E_1^2 + E_2^2 + E_3^2 + \ldots + E_n^2)$$

Similarly:

$$Y'Y = (Y_1^2 + Y_2^2 + Y_3^2 \ldots Y_n^2)$$

$$X'X = \begin{pmatrix} 1 & 1 & 1 & \ldots & 1 \\ X_1 & X_2 & X_3 & \ldots & X_n \end{pmatrix} \begin{pmatrix} 1 & X_1 \\ 1 & X_2 \\ 1 & X_3 \\ - & - \\ - & - \\ 1 & X_n \end{pmatrix} = \begin{pmatrix} n & \sum X \\ \sum X & \sum X^2 \end{pmatrix}$$

$$X'Y = \begin{pmatrix} 1 & 1 & 1 & \ldots & 1 \\ X_1 & X_2 & X_3 & \ldots & X_n \end{pmatrix} \begin{pmatrix} Y_1 \\ Y_2 \\ Y_3 \\ - \\ Y_n \end{pmatrix} = \begin{pmatrix} \sum Y \\ \sum XY \end{pmatrix}$$

If we postmultiply $X'X$ by the column vector

$$b = \begin{pmatrix} b_0 \\ b_1 \end{pmatrix}$$

we have the left-hand side of the two normal equations (11.1). These normal equations can therefore be written in matrix form as:

$$X'Xb = X'Y$$

giving:

$$b = (X'X)^{-1}X'Y$$

provided the inverse exists.

We now have the layout in quite general form, so that we can extend the problem from the present position (where we have only one independent variable X) to the general regression situation where we have any number (p, say) of independent variables. Specifically,

$$Y = XB + E$$

where

Y is a $(n \times 1)$ column vector of n results;

X is an $(n \times p)$ matrix whose elements are the value of the p independent variables, each for n different results;

B is a $(p \times 1)$ column vector of the parameters – the values of which we wish to find;

and E is a $(n \times 1)$ column vector of errors.

The solution is the same – i.e., the values for B are given by:

$$b = (X'X)^{-1}X'Y$$

This is the matrix solution to the multiple regression problem. Since we know methods for matrix inversion, this is the best method to use.

When the number of parameters is small, say up to three, the 'alternative' approach by partial differentiation can be used. (In practice the equations derived are very similar.) We have already seen one example in the calculation earlier in the section when we had only one variable (X) and two parameters B_0 and B_1. We now consider the case where there are two independent variables $X_1 X_2$ and three parameters $B_0 B_1 B_2$. The basic algebraic equation will be:

$$Y = B_0 + B_1 X_1 + B_2 X_2 + E$$

The matrix solution for the parameters will be given by:

$$b = (X'X)^{-1}X'Y$$

Alternatively, we can write:

$$S = \sum E^2 = \sum (Y_i - B_0 - B_1 X_{1i} - B_2 X_{2i})^2$$

and differentiate partially with respect to B_0 B_1 and B_2, giving the algebraic form of the normal equations:

Regression Analysis

$$b_0 + b_1\sum X_{1i} + b_2\sum X_{2i} = \sum Y_i$$
$$b_0\sum X_{1i} + b_1\sum X_{1i}^2 + b_2\sum X_{1i}X_{2i} = \sum X_{1i}Y_i$$
$$b_0\sum X_{2i} + b_1\sum X_{2i}X_{1i} + b_2\sum X_{2i}^2 = \sum X_{2i}Y_i$$

Solving these simultaneous equations will give the required values of $b_0\ b_1\ b_2$ for the parameters $B_1\ B_2\ B_3$. As a *sales example*, suppose we have available past records which show us how a dependent variable (sales) has altered with changes in two other variables (advertising expenditure and population density). Table 11.5 shows this record.

TABLE 11.5

Sales (£m)	Advertising expenditure (£000)	Population density (100,000)
227.7	116.6	121.0
242.0	122.1	138.6
231.0	122.1	124.3
237.6	126.5	113.3
245.3	132.0	112.2
256.3	136.4	113.3
260.7	138.6	107.8
220.0	110.0	110.0
226.6	114.4	108.9

Since we have two independent variables, the equation will be of the form:

$$Y = a + b_1X_1 + b_2X_2$$

where X_1 and X_2 are the two independent variables which together determine the values of the dependent variable Y (using a in place of b_0 to tie in with the computer package).

We can now use one of the two approaches (partial differentiation or matrix inversion) to determine the values for a, b_1, b_2. There are a large number of standard computer packages which do all the hard work. These programs not only print out the values for a, b_1, b_2, etc. but they also give the various statistical checks which indicate whether reliance can be placed on these values.

The principal checks which indicate what reliance can be placed on the values for b_1 and b_2 are the 't' tests. These t values were briefly discussed earlier (Section 11.4).

In multiple regression analysis, just as with the one independent variable case, we have to refer to the table of t values (Table A.4, p. 303), and compare our t values (as printed out by the computer program) with the figures given there. With only one independent variable, the number of degrees of freedom for n sets of data was $n-2$, with one extra independent variable there is one less degree of freedom, so that if there are n sets of data the appropriate degrees of freedom are $n-3$.

Regression analysis is such a widely-used tool that the appropriate computer programs have been devised for great ease of application. After selecting the standard regression program all that is required is that the data is fed straight in. For our example the figures read in will be 227.7, 116.6, 121.0, 242.0, etc. The only other inputs required are overall guidelines such as the number of independent variables (2 in our case) and the number of sets of data (9 in our case). The print out for these figures is shown in Table 11.6.

11.7 Interpretation of the statistical data

Even with simple regression analysis, confirmation as to whether the 'line of best fit' obtained did, or did not, show some real correlation between the (one) independent variable and the dependent variable was not easy. With multiple regression analysis the tests are more complicated.

Fortunately the same approach can be used; all that is needed is to extend it to multiple regression. There are a large number of statistical tests, etc. that can be applied. We shall briefly mention the main ones and show how they can be used for our example. (For more details consult Johnstone, 1963).

1. *The regression coefficients* – These are the values for a, b_1 and b_2 which determine our regression equation as:

$$Y = 55.725 + 1.364X_1 + 0.114X_2$$

2. *Degrees of freedom* – Since we have 9 sets of data and three regression coefficients, the formula shows that there are $(9-3=)$ 6 degrees of freedom.
3. *The t values* – The printout shows t values for the regression coefficients of:

$$9.530 \quad \text{for } b_1$$
$$0.794 \quad \text{for } b_2$$

Comparing these figures with those given in the table of t values (p. 000), we see that the value of:

b_1 is significant even at the 0.01 level
b_2 is not significant even at the 0.2 level

(These interpretations are a good example of the very rough and ready rule that t values over 2 are meaningful, and the higher the value the better.) Clearly b_1 contributes a great deal, and b_2 hardly at all. Indeed it can be argued that X_2 should be left out of the equation (which reduces it to simple regression) but this point is not followed up here.
4. *Coefficient of multiple determination* – The t values show how well *each* of the coefficients contribute, but do not give an *overall* indication of how good the fit of the equation is to the sets of data. The coefficient of multiple

Regression Analysis 233

TABLE 11.6

Multiple regression analysis

Regression coefficients	t values–6 degrees of freedom
55.725	
1.364	9.530
0.114	0.794

Coefficient of multiple determination = 0.939

Observed Y values	Calculated Y values	Residuals
227.700	228.561	−0.861
242.000	238.070	3.930
231.000	236.440	−5.440
237.600	241.187	−3.587
245.300	248.564	−3.264
256.300	254.691	1.609
260.700	257.065	3.635
220.000	218.305	1.695
226.600	224.181	2.419

Correlation matrix

	Sales	Advertising	Population
Sales	1.000	0.965	−0.087
Advertising		1.000	−0.173
Population			1.000

determination, referred to as R^2, gives this indication. The R^2 figure can be anything from 0 (showing a random scatter of points with no correlation) to 1 (the perfect fit). The nearer to 1 the better. An R^2 of 0.939, as in our case, is an indication of a good fit.

5. *Observed and calculated values, and residuals* – Multiple regression analysis started off as a technique used in scientific work. The sets of data were the results of observations made in scientific experiments. The term 'observed' for the actual values of the sets of data is still used today, though the applications of regression analysis have extended far beyond the original scientific field.

The term 'observed Y values' therefore means the actual values for the Y (i.e. dependent) variable. To see how near the regression equation is to this actual value we need only to put in the values of the appropriate independent variable in this equation. Take, for example, the first set of data, i.e.:

$$227.700 \quad 116.600 \quad 121.000$$

The regression equation is:

$$Y = 55.725 + 1.364X_1 + 0.114X_2$$

Putting in the values of 116.600 for X_1 and 121.000 for X_2 the 'calculated value' for Y is:

$$55.725 + 1.364 \times 116.600 + 0.114 \times 121.000$$
$$= 228.561$$

This is the calculated figure for Y for these values of the independent variables. The corresponding observed, or actual, value is 227.700. The difference between these is the 'residual'. For each of the sets of data there will be a residual. Obviously the smaller the residuals the more closely fitting is the regression equation. With only one independent variable – that is, reverting to simple regression analysis – the residuals are the distances E_1, E_2, etc., the sum of whose squares we were trying to minimise. The regression equation was then to a line – the line of least squares.

6. *Autocorrelation* – Let us take here a simple regression example and suppose that we are comparing overhead costs with throughput. The sets of data relate to the four quarters of the last year. Table 11.7 gives these sets of data, and in Figure 11.2 the position is set out diagrammatically.

TABLE 11.7

Periods	Overhead costs	Throughput
1	350	60
2	350	100
3	650	100
4	650	60

From Figure 11.2 it is obvious that the cost has been varying over the last year in a cyclical or seasonal form, and in a manner roughly represented by the dotted line. If the same relationship continues between costs and throughput the pattern will presumably continue in much the same way. But will regression analysis indicate this? It is hardly necessary to calculate the line of best fit. Clearly it is the line *PQ*. All the points are entirely symmetrical about this line. If we rely on regression analysis for our prediction it will always give the same value (of 500). In other words, it irons out or ignores the 'true' cyclical pattern.

How can we check whether there is any serial or cyclical relationship in more complex situations? One way is to look at the residuals, and see if there is any overriding pattern in their signs and their magnitude as we go from one row to the next one above or below. This is quite effective for simple regression analysis but for multiple regression analysis we require a more sophisticated way of checking on this factor – termed *autocorrelation*.

Regression Analysis

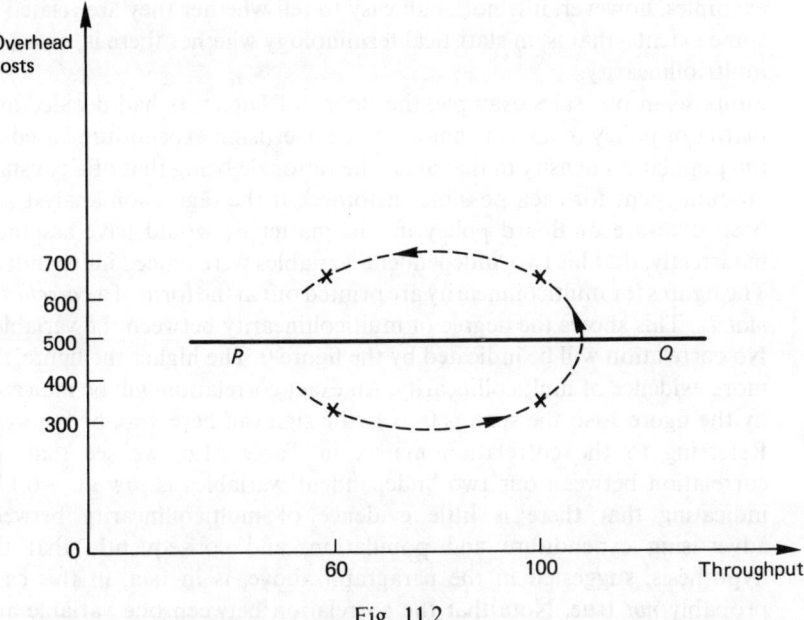

Fig. 11.2

The Durban–Watson statistic is widely accepted as the appropriate check. This figure unfortunately was not printed out by the regression program used for our regression example though many regression programs do print it out. The figure can vary from 0 to 4 and the nearer it is to the value of 2 the less evidence there is of this overriding relationship from one reading (or set of data) to another. Frequently, as in the example above, the overriding factor, if it is present, is connected with *time*.

7. *Multicollinearity and the correlation matrix* – In the preceding section we discussed the possibility of a relationship between one set of data and the one immediately above or below it, that is, between one *row* and the ones below or above it. In this section we consider artificial relationships between one independent variable and another. With the usual layout for sets of data this means between one *column* and the next. Again an example may make the position clear. Let us suppose we wish to find how to arrive at the total cost before delivery of an optical instrument. This product is made in one factory and its rather expensive special container in another. If we were very naive we might take as the two 'independent' variables the numbers of optical instruments and the number of containers. But are these 'independent' variables really independent? The answer is obvious. They are not. Each product must have its own container. There will be a one to one correspondence between the two. In this ridiculously simple example it is evident from commonsense that the two 'independent' variables are not independent of each other. In many more complex

examples, however, it is not at all easy to tell whether they are related to some extent – that is, in statistical terminology whether there is, or is not, multicollinearity.

Suppose, in our sales example, the Board of Directors had decided as a matter of policy to allocate amounts to advertising expenditure based on the population density in that area (the rationale being that of a constant amount spent for each possible customer). If the regression analyst had been unaware of Board policy in this matter he would have assumed, incorrectly, that his two 'independent' variables were indeed independent. The figures for multicollinearity are printed out in the form of a *correlation matrix*. This shows the degree of multicollinearity between the variables. No correlation will be indicated by the figure 0. The higher the figure, the more evidence of multicollinearity. An exact correlation will be indicated by the figure 1, so the span is 0 to 1 (the sign can here, too, be ignored). Referring to the correlation matrix in Table 11.6, we see that the correlation between our two 'independent' variables is low at -0.173, indicating that there is little evidence of multicollinearity between advertising expenditure and population, and consequently that the hypothesis, suggested in the paragraph above, is in fact, in this case, probably *not* true. Note that the correlation between one variable and itself is naturally exact, and is shown as 1 in the matrix.

11.8 Application of the multiple regression formulae

We have now not only arrived at the regression equation:

$$Y = 55.725 + 1.364X_1 + 0.114X_2$$

but also convinced ourselves, as a result of applying a whole battery of checks, that (with some reservations) this equation does give a good indication of how sales depend on the two independent variables (advertising expenditure and population).

We can therefore use it with some confidence for predicting sales in the future. If we wish to determine the likely values for sales next year, for example, all that is required is an estimate of the expected advertising expenditure for that year, and the estimated population density figure. The Marketing or Finance Director (or whoever has been responsible for estimates of advertising expenditure in the past) will supply the first figure; the second will be obtained from the appropriate government demographic analysis. If we suppose that these figures are £135,000 and 11 million respectively then sales will be obtained by putting these figures into the regression equation. Estimated sales for next year will therefore be:

$$55.725 + 1.364 \times 135 + 0.114 \times 110$$
$$= £252.4 \text{ million approx.}$$

Regression Analysis

A full statistical analysis would typically also include information on *Confidence limits*.

11.9 Extensions of the regression model

We have argued that in the business situation it often seems intuitively obvious that the model can be represented by an equation which is linear in the independent variables. Sometimes, however, the linear assumption is clearly not justified. For instance in a closed market situation where the total sales revenue obtainable for a product is theoretically a constant sum c (say) regardless of the quantity sold then the relationship between x (the unit price) and y (the quantity sold) is, we know, given by $xy = c$ or $y = c/x$. Another example we have met is that of the rate of production of a new item x months after the beginning of a production run. Frequently it is found to be fairly accurately described by the learning curve equation

$$y = A(1 - e^{-bx})$$

where A and b are constants.

Neither of these equations is linear in x, and consequently regression analysis would, *prima facie*, not appear to be a technique capable of dealing with them. However, if we write $Z = 1/x$ – i.e., we use the reciprocal of x as the independent variable – we have the equation $y = cZ$ for our closed market model. This is linear in Z and therefore regression analysis can be used.

In the second case, if we write $y_1 = \{(A-y)/A\}$ so that $y_1 = e^{-bx}$, take logarithms of both sides, so that $\log y_1 = -bx$, and finally write y_2 for $\log y_1$, we have $y_2 = -bx$. This, too, is now in linear form. Reverting to the original variables, the learning curve is $\log\{(A-y)/A\} = -bx$. This means that the columns for the independent variable will show the individual values of x, but the columns for the dependent variable will show the individual values of $\log\{(A-y)/A\}$. However, we should add that statistical problems can arise when using logarithms in regression analysis.

In scientific models it is not uncommon for the relationship to be a polynomial in one independent variable. That is, expressed as an equation of order n:

$$y = B_0 + B_1 x + B_2 x^2 + \ldots + B_n x^n$$

Such models are 'intrinsically linear' models – using this term in the sense that they can be expressed in linear terms by suitable transformations. All we have to do is write:

$$Z_1 = x, \; Z_2 = x^2, \; Z_3 = x^3 \ldots Z_n = x^n$$

and the equation is transformed into:

$$y = B_0 + B_1 Z_1 + B_2 Z_2 + B_3 Z_3 + \ldots + B_n Z_n$$

which is linear in the variables $Z_1, Z_2, Z_3 \ldots, Z_n$. It is therefore in a suitable form for dealing with by regression analysis.

The special case of the second order model:

$$y = B_0 + B_1 x + B_2 x^2$$

will be transformed into:

$$y = B_0 + B_1 Z_1 + B_2 Z_2$$

which is a linear model with two independent variables.

This particular model has many applications in the field of science. One typical example would be when a ball is thrown in the air. Ignoring the drag effect of the air's resistance, the vertical height (y) at any time (t) is given by the equation $y = ut - \frac{1}{2}gt^2$ where u is the initial vertical velocity and g is the gravitational constant. Management models are not so common, though one can occur when we look at the relationship between volume of production (y) from a factory, and the size of the workforce (x). We noted earlier, too, that in some industrial situations the equation $y = B_0 + B_1 x + B_2 x^2$ (where B_0 B_1 and B_2 are positive constants, and B_2 is fairly small) has been found to represent the relationship fairly exactly between total cost and throughput.

11.10 Dummy variables

Sometimes data may arise from several sources – for example, they may come from various factories, machines and operators or task forces all of whom are doing much the same thing. If we wish to use the information available from all these sources in determining the regression line we really should allow for possible innate differences between these sources.

This can be done by the introduction of *dummy variables* which reflect these differences between one source and another. Again reverting to our basic example for the relationship between the cost (y) and the output of (say) two basic products (X_3, X_4), both of which are produced at three separate factories, we now need to introduce two dummy variables X_1 and X_2 to reflect the differences that the data from any two of the plants may have when compared with the data from the third plant.

The basic equation will be:

$$Y = B_0 + B_1 X_1 + B_2 X_2 + B_3 X_3 + B_4 X_4$$

and X_1 and X_2 will be used for plant identification for individual data readings in accordance with this rule:

X_1	X_2	for data from plant no:
1	0	1
0	1	2
0	0	3

Regression Analysis 239

The other data readings for X_3 and X_4 will be entered in the usual way. We now have a series of data readings for four independent variables $(X_1\ X_2\ X_3\ X_4)$, and so the regression 'line' can be determined.

Dummy variables can also be used when there is a once and for all basic change in the structure. For instance, all the data in the example above, might have come from one plant over a period of time. At some stage, however, there may have been some substantial change or improvement in the operation of the plant, due perhaps to the introduction of some more sophisticated machinery in one section. To allow for the effect of this, one dummy variable will need to be used which will have a value of 0 for periods before the change, and 1 for periods after the change.

Where there are trends or periodic movements in significant variables that are omitted from the regression equations, the residuals, we noted, may be *autocorrelated*. One way of dealing with this is to introduce *time* as one of the independent variables. For dealing with quarterly seasonal variations the best method is to use three dummy variables (000, 100, 010, 001 for the quarters).

11.11 Business uses of regression analysis

We have already shown the application of regression analysis to two of the main areas of use in business:

1. Cost estimation.
2. Sales forecasting.

Both are important. The cost estimation method has been mentioned in the text several times and a short summary of this use may be helpful. This approach may be used for determining how one cost (e.g., manufacturing cost) varies with changes in throughput, or it may be concerned with showing how total cost varies with throughput. In the latter (and more usual) case, certain assumptions are made. These were indeed implied when we dealt briefly with the total cost function in Chapter 8. More rigorously stated, they are:

(a) That costs can be split into those costs that vary with the level of activity (i.e., 'variable' costs), and those costs that do not vary at all (i.e., 'fixed' costs). It is assumed, too, that costs that vary with volume, but less than proportionately (that is, semi-variable costs) can themselves be split up into variable and fixed elements.

(b) That throughout the relevant range 'variable' costs do remain variable with volume (or some other measure of activity) and fixed costs remain fixed.

Examples of typical variable costs are raw materials, direct labour costs, direct selling commission and some utility expenses such as power. Examples of typical fixed costs are depreciation of buildings and equipment, rent and rates, most salaries, most insurance expenses, etc. We say 'typical' in each case because the exact split between fixed and variable will change slightly from one business to another.

When past cost data go back some period in time, and inflation has been a material factor during this period, we have the problem of 'drift'–that is, costs may have risen over the years, at least partly, as a result of rises in money levels. In extreme cases the rising level of costs may merely reflect inflation, and have little or nothing to do with changes in volume. The solution in such circumstances is to eliminate the effect of inflation by putting in all costs in real values (in practice this means using some year as a base).

The limited range assumption is important (as indeed elsewhere in regression analysis work). Once the limit of normal production is exceeded, fixed costs such as depreciation of plant or buildings will jump in steps–as when another machine is brought in, or an extension to the building is made.

Adding the revenue line (that is, the line of sales revenue against output) to a graph of the total cost line gives a break even chart, profit/volume chart, or profitgraph. Such a chart shows how profits vary with the level of output. Some firms have high fixed costs, others low. The level of fixed asset cost is reflected in the operating leverage. Air line firms, for instance, have very high fixed costs. Additional passengers' fares represent an almost equal increase in operating profits.

11.12 Discriminant analysis

Discriminant analysis is a technique which is similar to regression analysis. Its principal business application is in evaluating trade credit applicants.

The assumption in such a situation is that a suitable combination of weighted financial ratios (and possibly other business data) will allow a business to discriminate between potential good and bad payers.

Suppose we start with two evaluation ratios only–let us say these are the liquidity ratio and the payout (i.e. dividend cover) ratio. To try them out we check on the values of these ratios for all new credit applicants for a trial period of perhaps six months or a year. After this period we classify all these accounts into good payers and bad payers according to some predetermined objective rule.

Using as axes the liquidity and payout ratios we now plot the position of all these credit applicants as a scatter diagram. Figure 11.3 shows a typical set. The circles represent bad accounts, the crosses those classified as good accounts according to our rule.

The objective is to find that discriminant boundary line which best separates the good and bad accounts. Let us suppose that its equation is:

$$c = a_1 X + a_2 Y$$

In Chapter 8 we noted that for any line (such as $a_1 X + a_2 Y - c = 0$) the value of the expression for points not on the line will be positive for points on one side and negative on the other. Provided we can choose suitable values for a_1, a_2, c,

Regression Analysis

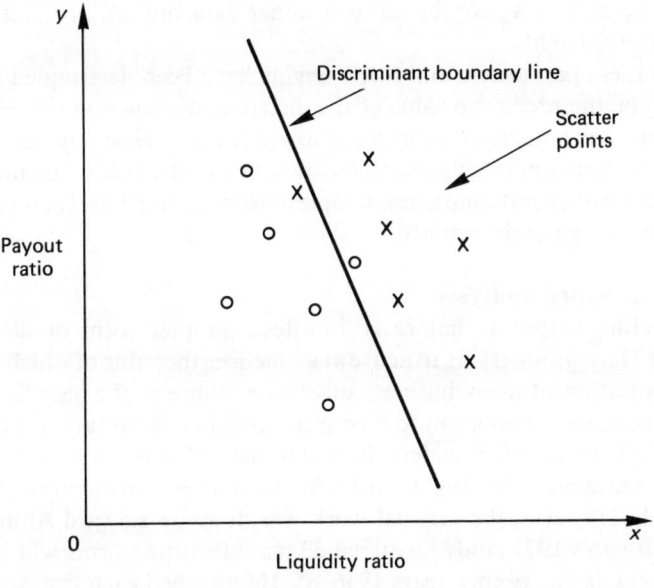

Fig. 11.3

so that all (or most) of the good and bad payers are on opposite sides of the line, we shall have a suitable discriminant line.

Though we do not prove it, a_1 and a_2 can be computed statistically. The formula is:

$$a_1 = \frac{d_x S_{yy} - d_y S_{xy}}{S_{xx} S_{yy} - (S_{xy})^2} \quad a_2 = \frac{d_y S_{xx} - d_x S_{xy}}{S_{xx} S_{yy} - (S_{xy})^2}$$

where S_{xx} and S_{yy} are the variances of x and y respectively

S_{xy} is the covariance of x and y

d_x is the mean of good payers liquidity ratios minus the mean of bad payers ratios

d_y is the same for the payout ratios.

To find the value of c we calculate its values for the good and the bad payers and then choose that value which *best discriminates between the two*. There is usually a grey area in the middle where classification is uncertain. (Fig. 11.3 shows one misclassification each side of our boundary line). In practice payers falling into this area will usually be subjected to further analysis.

The example we have used has been a simple one using only two ratios. In the more general case, where we have many ratios (or other relevant financial or economic data) the discriminant 'line' will be of the form:

$$c = a_1 X_1 + a_2 X_2 + a_3 X_3 + \ldots + a_n X_n$$

Where $X_1 X_2 \ldots X_n$ are the ratios or other data, and $a_1, a_2 \ldots a_n$ are their appropriate weights.

Since these parameters and their weights have been determined from past credit data, the predictive value of the discriminant function will be effective only if current potential customers, who are being screened by this approach, retain the same underlying characteristics. Once this fails to be the case the system will not operate and a new sample must be selected, and new parameters (and their weights) determined.

11.13 Z score analysis

Predicting corporate failure is, in effect, another form of discriminant analysis. Here we are trying to arrive at a function, the value of which will give a good indication of likely business success or failure in the near future. Such early prediction is important to creditors, investors, bond rating agencies and would-be bank or other lenders. Indeed in the UK it is the joint stock banks (and in particular the 'Big Four') who have been most interested in this approach. However, the original work was done by Edward Altman in the USA. Altman's 1977 study identified 33 manufacturing firms which filed for bankruptcy in the twenty years 1946–65. He matched each firm with a non-failed firm of similar asset size in the same industry.

The function which he found predicted failure most accurately was:

$$Z = 0.012\ (X_1) + 0.014\ (X_2) + 0.033\ (X_3) + 0.006\ (X_4) + 0.999\ (X_5)$$

where X_1 = working capital to total assets
X_2 = retained earnings to total assets
X_3 = earnings before interest and tax (i.e., trading profit) to total assets
X_4 = market value of equity to book value of total debt
X_5 = sales to total assets.

Altman found that firms with a Z score of 3.00 or more came fairly clearly in the 'non-fail' category; while those with Z scores less than 1.80 came in the 'fail' category. In the grey area between 1.80 and 3.00 Altman concluded that the Z score which best discriminated between 'fail' and 'non-fail' firms was around 2.675.

In the UK the most important work has been that done by Richard Taffler. His 1982 UK-based study found that financial leverage as well as profitability contributed substantially towards an effective prediction. Some recent work has indicated that there may have been too heavy a reliance on purely financial ratios. David Myddleton has suggested audit qualifications, delay in publishing accounts (or increase in delay compared with earlier years) and the occurrence of large extraordinary expense items as possible factors. More modern ratios such as the defensive interval, ratios incorporating cash and near cash and perhaps some employees' ratios may also eventually be found to be helpful, in combination with others, as good predictors.

Regression Analysis

11.14 Exercises
Worked answers are provided in Appendix C to questions marked with an asterisk.

1. Accountants use a variety of aids to assist in statistical calculations, including the following:

    (i) desk/hand calculators;
    (ii) mathematical and statistical tables;
    (iii) computers.

    (a) What are the advantages and disadvantages of each of these aids?
    (b) Explain briefly how you would use *either* a calculator *or* a computer to produce the equation of a regression line.

    (*Institute of Cost and Management Accountants*, May 1985)

2.* (a) Draw scatter diagrams of about 10 points to illustrate the following degrees of linear association:

    (i) weak, positive correlation
    (ii) approximately zero correlation
    (iii) $r = -1$
    (iv) fairly strong, negative correlation.

    (b) You are involved in a planning exercise, part of which requires an assessment of the financial implications of future output levels. As part of the preliminary work you have been asked to investigate the production of product A for a whole industry. Since 1975 production (million tonnes) of A for the industry has been as follows:

Year	Production
1975	19
1976	24
1977	28
1978	33
1979	35
1980	41
1981	45
1982	47
1983	52

    You are required to:

    (i) find the least squares regression equation of production on time;
    (ii) predict production for 1985, explaining any assumptions or limitations.

    (*Institute of Cost and Management Accountants*, May 1984)

3.* A company has found that the *trend* in the quarterly sales of its furniture is well described by the regression equation

$$Y = 150 + 10X$$

where Y = quarterly sales (£000)
$X = 1$ represents the first quarter of 1980
$X = 2$ represents the second quarter of 1980
$\vdots$
$X = 5$ represents the first quarter of 1981, etc.

It has also been found that, based on a multiplicative model, that is:

$$\text{Sales} = \text{Trend} \times \text{Seasonal} \times \text{Random}$$

the mean seasonal quarterly index for its furniture sales is as follows:

Quarter	1	2	3	4
Seasonal index	80	110	140	70

You are required:

(a) to explain the meaning of *this* regression equation, and *this* set of seasonal index numbers;
(b) using the regression equation, to estimate the trend values in the company's furniture sales for each quarter of 1985;
(c) using the seasonal index, to prepare sales forecasts for the company's quarterly furniture sales in 1985;
(d) to state what factors might cause your sales forecasts to be in error.
(*Institute of Cost and Management Accountants*, May 1985)

4. (i) A hospital cost model is based on the numbers of patients treated. For a given year a sample of 102 hospitals is selected and for hospital i its costs in thousands of pounds sterling (Y_i), and number of patients treated in thousands (P_i) are obtained. The model hypothesised is:

$$Y_i = a + bP_i + \varepsilon_i$$

Preliminary analysis of the data gives:

$\bar{Y} = 12{,}000$, $\bar{P} = 18$, $\Sigma y_i^2 = 125$ million, $\Sigma p_i^2 = 100$,

$\Sigma p_i y_i = 50{,}000$ ($y_i = Y_i - \bar{Y}$, $p_i = P_i - \bar{P}$).

For the hypothesised model:
(a) Give the least-squares estimate of a and b.
(b) Test whether the sample estimate of b is significantly different from zero at the 1% level.

Regression Analysis

(ii) The model is extended to include the number of beds (Q_i at hospital i) as follows:

$$Y_i = \alpha + \beta P_i + \gamma Q_i + \varepsilon_i$$

The results from running this model on a multiple regression package were:

Variable	Coefficient	Standard error	t	Probability
Constant	1000	100	10	0.000
P	100	200	0.5	0.617
Q	20	4	5	0.000

Correlation matrix:

	Y	P	Q
Y	1	0.45	0.7
P	0.45	1	0.9
Q	0.7	0.9	1

Discuss the following statements:
(a) The average cost of treating a case is £100.
(b) The number of cases treated has no significant effect on hospital costs.
(c) The number of beds is a better prediction of costs than the number of cases.

5. Eastern Finance Ltd uses discriminant analysis as one criterion for assessing applicants for loans. The equation which Eastern uses is
$Y = 0.2X_1 + 1.25X_2$
where X_1 = current assets/current liabilities
X_2 = net profit/total assets.
Wanda Tregorran has applied to Eastern for a loan to expand her Computer Software business. Her last accounts disclose:

Current assets = £30,000
Current liabilities = £50,000
Net profit = £51,000
Total assets = £105,000

Eastern has learnt from experience that 80% of its debtors are good credit risks. Variable costs are 55% and the contribution margin is 45%. Using the discriminant analysis requirement should Eastern grant a loan to Ms Tregorran?

Chapter 12
Linear Programming

Robert Ashford

12.1 Introduction

Quantitative methods are becoming an increasingly valuable part of industrial and commercial organisation. The trend for a more numerate management and cheaper computing resources will ensure that they will become more widespread in the future. Any quantitative technique requires some mathematical model of all or part of a commercial organisation, its competitors, customers or supplies, or some aspect of the environment within which it operates. These models may be very simple or highly complex, and the construction and study of them is becoming increasingly important. One that has already achieved widespread popularity is that of linear programming. It is a technique that is prescriptive rather than descriptive and its prescriptions can yield substantial dividends. Reliable computer software exists to exploit it and this has become so efficient that models with five thousand decisions can be used routinely. Thus it is an exciting and powerful tool of business analysis.

This chapter will introduce you to the ideas and concepts of linear programming, and explain the main techniques for solving or optimising linear programming models. Methods of analysing the solution are explained both in mathematical and in practical terms.

12.2 A simple problem

The following problem, which is very simple, serves to introduce some of the major concepts of linear programming in a straightforward way.

A joinery makes two different sizes of boxwood chess sets. The smaller requires three hours of machining on a lathe and the larger two hours (the smaller ones are more fiddly). The joinery has ten lathes with skilled operators and works a 40-hour week. Boxwood is scarce. The smaller set requires a hundredth of a cubic metre of boxwood and the larger three hundredths. The joinery can procure at most two cubic metres of boxwood each week. The profit contribution per chess set made is £5 for the smaller one and £20 for the

Linear Programming

larger. How many sets of each type should the joinery make each week to yield the most profit?

If we let x_1 and x_2 denote the number of small and large chess sets made each week respectively, then the weekly additional profit is:

$$£5x_1 + 20x_2$$

400 hours of machining time is available each week and the number actually used is

$$3x_1 + 2x_2$$

So we have the first restriction on the joinery's production:

$$3x_1 + 2x_2 \leqslant 400$$

The boxwood actually used each week, in cubic metres is:

$$x_1/100 + x_2 3/100$$

and this cannot be greater than the available supply of 2 cubic metres/week, so we have the second production restriction:

$$x_1/100 + 3x_2/100 \leqslant 2$$

or

$$x_1 + 3x_2 \leqslant 200.$$

Of course, it makes no sense to manufacture negative numbers of chess sets, so we must have:

$$x_1 \geqslant 0 \text{ and } x_2 \geqslant 0$$

Thus we can state the joinery's problem as one of

Maximising	$5x_1 + 20x_2$	(profit)
subject to	$3x_1 + 2x_2 \leqslant 400,$	(machining)
	$x_1 + 3x_2 \leqslant 200,$	(boxwood)
and	$x_1 \geqslant 0, x_2 \geqslant 0.$	(non-negativity)

We can represent the weekly production level of chess sets (x_1, x_2) as a point on a graph. And we can represent the set of production levels satisfying the machining limitation as those points lying to the left of the line $3x_1 + 2x_2 = 400$.

Similarly, we can represent those production levels not using too much boxwood as lying to the left of the line $x_1 + 3x_2 = 200$, and those satisfying the non-negativity restrictions as lying in the non-negative quadrant (see Figure 12.1).

In addition, we can draw in lines on which the weekly profit is constant: it is £500 on line (1), £1,000 on line (2) and £1,333 on line (3).

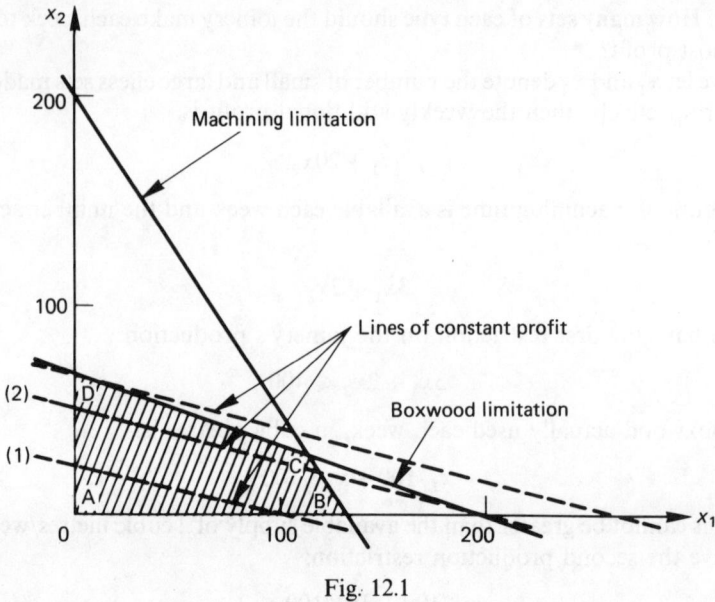

Fig. 12.1

All production levels (x_1, x_2) which lie in the shaded figure, $ABCD$ are possible in that they satisfy the given restrictions whereas all points outside that region are not. To determine the production levels which yield the most profit, you can think of those on a line of constant profit lying within $ABCD$ (say, line (1)) and then move that line in the direction of improving profitability (say, to line (2)) until no points within $ABCD$ lie on the line if it is moved any further (line (3)). Point D on the graph thus represents the best production schedule: $166\frac{2}{3}$ large chess sets per week, and no small chess sets, which yields a weekly profit of £1,333.33.

12.3 Some basic concepts

In our joinery example, we identified aspects of the production process which were under our control (the weekly production levels of the two sizes), and those that were not (profit contributions, lathe availability, boxwood supply, and the technology of the production process). Those aspects of the process which we can control are called *decision variables*. We optimised – in this case maximised – a linear function of the decision variables called the *objective function* (or objective, for short) subject to a number of linear restrictions or *constraints* upon them. In other circumstances, we may wish to minimise the objective function – if it represents running costs, for example. In either case we say that the objective function is to be optimised, and if the decision variables are subject only to linear constraints then our problem is a *linear program*. All linear programs require a linear objective function to be

Linear Programming

optimised subject to a set of linear constraints. We will discuss the general form of linear programs later; for the moment, let us return to the joinery example.

The set of possible or practical decisions – those which satisfy the constraints – is known as the *feasible region*, any decision within it is said to be *feasible*. In our example the feasible region is the quadrangle *ABCD*. In general the feasible region is always a *polyhedron*, which is a region bounded by lines – or, more generally, (hyper)planes.

Notice two things about it. Firstly it need not be closed (see Figure 12.2). So it may be possible for our decision variables to be moved in a direction which improves our objective (for example, increase the profit) indefinitely. In this case the linear program is said to be *unbounded*. This does not, of course, mean in practice that we can make an infinite amount of money. Usually it means that we have made a mistake in our interpretation of the problem as a linear program and forgotten some critical constraint!

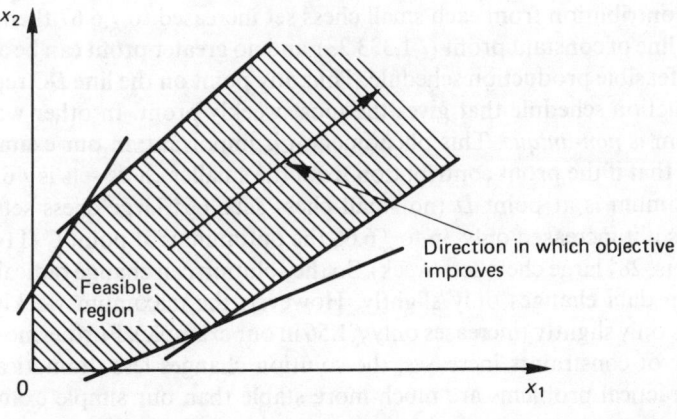

Fig. 12.2

Secondly notice that there may be no feasible decisions which satisfy the constraints. This would be the case in our joinery example if we had to fulfil an order for 100 large chess sets within a week. You can see by looking at Figure 12.1 that this is impossible. In this case the feasible region is empty and the linear program is said to be *infeasible*. Again, in practice, this usually means that we have made a mistake in our modelling rather than that it is impossible to run our plant.

There is one further point to notice about the feasible region. That is that any point on a straight line joining two points within it is also in the feasible region. In other words, the feasible region is *convex*. Examples of convex and non-convex shapes are shown in Figure 12.3.

It is intuitively clear – and it can be shown mathematically – that if an optimum solution exists then there is always one at a vertex or 'corner' of the feasible polyhedron. So in our search for the best solution we need only

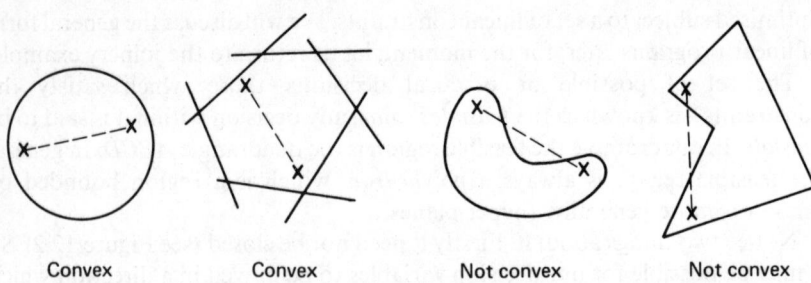

Convex   Convex   Not convex   Not convex

Fig. 12.3

consider those which lie on *vertices of the feasible region*. This fact will be exploited later in automatic solution procedures.

Return to our chess set example. Look at Figure 12.1. You can see that if the profit contribution from each small chess set increased to £6.67 then the line *DC* is a line of constant profit (£1,333.33), and no greater profit can be achieved by any feasible production schedule. Thus any point on the line *DC* represents a production schedule that gives the most weekly profit. In other words the optimum is *non-unique*. This phenomenon is important; in our example you can see that if the profit contribution from the smaller chess sets is £6.65 then the optimum is at point *D* (no small chess sets, $66\frac{2}{3}$ large chess sets/week), whereas if it increased only 3p to £6.68 the optimum is at point *C* ($114\frac{2}{7}$ small chess sets, $28\frac{4}{7}$ large chess sets/week). So the solution can change radically if the problem data changes only slightly. However, the maximum weekly profit changes only slightly (increases only £1.56 in our example). Furthermore as the number of constraints increases, the solution changes less dramatically and most practical problems are much more stable than our simple example.

We now come to another important concept of linear programming. We have seen that if the profit contributions from smaller chess sets changed from £5 to £6.67, then it just became profitable to produce them, taking our production schedules to some point between *C* and *D*. This increase of £1.67 is known as the *reduced cost* of x_1 (the production of smaller chess sets). It tells us how much more profitable (or less costly) a variable that is zero at the optimal solution needs to become before becoming positive. It would be worthwhile to make the smaller chess sets if x_1 becomes positive in the optimal solution.

There is an analogous concept for constraints that are limiting at the optimum. In the chess set example, the optimal production schedule is limited by the quantity of boxwood available. This constraint is said to be *binding*. Notice that at the optimum, *D*, we are not limited by the amount of machining time available. In this case the machining constraint is said to have *positive slack* or be *non-binding*. Suppose the joinery were able to procure an extra cubic metre of boxwood each week. The boxwood limitation constraint would rise so

Linear Programming

that point D represented 100 large chess sets/week. (The constant on the right-hand side of the boxwood constraint would change from 200 to 300.) Point D would still be the optimum and our profit would rise £666.67 to £2,000 per week. This change in profit per unit of extra resource is known as the *shadow price* of the boxwood constraint, and it tells us the most that it would be worth paying for an additional cubic metre of boxwood, assuming no other constraint becomes binding.

Both reduced costs and shadow prices are computed automatically by the most widely used solution procedure, known as the *simplex method*. You can think of them as '*forces*' holding the optimal point within the feasible region as it is pulled in the direction of most rapid improvement in the objective.

12.4 An algebraic approach

The graphical approach that we have used on the chess set example would be more difficult to use if the joinery decided to produce three different sizes of set, for then we would need a three-dimensional graph to plot values of three decision variables and for problems with four or more decision variables it would be unworkable. Since practical problems may involve a thousand or more decision variables we need an algebraic technique. We will illustrate one on the chess set example and then formalise it into the simplex method in the following sections.

To begin with, suppose we add extra non-negative decision variables, x_3 and x_4 called *slack* variables) to the machinery and boxwood constraints to turn them from inequality relations into equalities, so x_3 represents the unused hours of machining time available and x_4 represents the quantity of boxwood which is available for use but is unused (in hundredths of a cubic metre). If we let x_0 denote the weekly profit, the problem becomes:

Maximise $\quad x_0 = 5x_1 + 20x_2$
subject to $\quad 3x_1 + 2x_2 + x_3 = 400$
and $\quad\quad\quad x_1 + 3x_2 + x_4 = 200$
where $\quad\quad x_1, x_2, x_3$ and $x_4 \geq 0$

Rearranging the equalities, we can write:

$$x_0 = \quad\quad 0 + 5x_1 + 20x_2$$
$$x_3 = 400 - 2x_1 - \quad x_2$$
$$x_4 = 200 - \quad x_1 - 3x_2$$

We can think of these as expressing the *dependent* variables x_0, x_3 and x_4 on the left-hand side in terms of the *independent* variables x_1 and x_2 on the right-hand side. For any given values of the independent variables the corresponding values of the dependent variables are immediate. In linear programming terminology, the dependent variables are called *basic* and the independent variables *non-basic*. If we assign the value zero to non-basic variables then those of the basic variables are simply the constant terms on the right-hand side above. In fact, any feasible set of values of the decision variables obtained by setting the non-basic ones to zero always corresponds to a vertex of the feasible region. The one above corresponds to vertex A in Figure 12.1.

We can use the last equation above to express x_2 in terms of x_1 and x_4:

$$x_2 = \frac{200}{3} - \frac{1}{3}x_1 - \frac{1}{3}x_4$$

And we can use this to replace x_1 in the first two equations by x_1 and x_4 to yield:

$$x_0 = \frac{4000}{3} - \frac{5}{3}x_1 - \frac{20}{3}x_4$$

$$x_3 = \frac{1000}{3} - \frac{5}{3}x_1 + \frac{1}{3}x_4$$

$$x_2 = \frac{200}{3} - \frac{1}{3}x_1 - \frac{1}{3}x_4$$

Setting x_1 and x_4 to zero we have a representation of point D on Figure 12.1 which is the optimum, $x_1 = 0, x_2 = 200/3$. We have a new set of basic variables, x_0, x_3 and x_2, and a new set of non-basic variables, x_1 and x_4. Calling the set of basic variables the *basis* we say that we have taken x_4 out of the basis and put x_2 into it.

The x_0 row of the two sets of equations deserves special notice. That of the first set:

$$x_0 = 0 + 5x_1 + 20x_2$$

tells us the rate of change to x_0 as x_1 or x_2 is varied. As x_2 is increased from zero to enter the basis at 200/3, we see that x_0 increases by 20 per unit change in x_2. You can regard this as the rate of increase of our objective, x_0, as we proceed from point A to point D in Figure 12.1. Similarly we can see that the rate of change of x_0 with respect to x_1 is $+5$: this is the rate of increase of x_0 if we were to proceed from point A to point B in Figure 12.1

In other words, the coefficients of the non-basics in the x_0 (objective) row are the rates of change of x_0 as the corresponding non-basic is increased from zero. They are *intermediate reduced costs*. In going from the first to second set of equations (from points A to D) we chose to move in the direction of the fastest improvement to x_0 by bringing the non-basic with largest reduced cost (x_2) into the basis.

The x_0 row of the second set of equations:

$$x_0 = 4000/3 - 5/3x_1 - 20/3x_4$$

tells us that the solution would get worse if x_1 or x_4 were introduced into the

Linear Programming

basis, since their reduced costs are both negative. Therefore the solution is optimal. Notice that the reduced cost of x_1, $-5/3$, agrees with that found in the previous section and the reduced cost of the slack on the boxwood constraint, x_4, is $-20/3$ which is minus its scaled shadow price. The shadow price discussed in the previous section related to the original constraint before we scaled it by multiplying it by 100.

Returning to our change of basis between the two sets of equations, we can see why x_2 rather than x_1 was introduced into the basis. But which variable leaves the basis? The leaving variable, x_4, is the first basic to be reduced to zero as x_1 increased from zero. Geometrically this means hitting the boxwood constraint (on which x_4 is the slack) before any other constraint as we proceed from A along the x_2 axis.

We have thus seen an algebraic way of solving the chess set problem. We now discuss it more formally for general linear programs.

12.5 The simplex method

We can formally state the general linear programming problem as:

Maximise (or minimise) $\quad \sum_{j=1}^{n} c_j x_j$

subject to $\quad \sum_{j=1}^{n} a_j x_j \begin{Bmatrix} \leqslant \\ = \\ \geqslant \end{Bmatrix} b_i \quad$ for $i = 1, 2, \ldots, m$

and $\quad x_j \geqslant 0 \quad$ for $j = 1, 2, \ldots, n$

where the ith row may be an equality or a '$\geqslant$' or '$\leqslant$' inequality. For convenience, we suppose that each b_i is non-negative (otherwise just multiply the entire constraint by -1).

We can turn each '$\leqslant$' inequality into an equality by adding a slack variable x_{n+i} (say) to give:

$$\sum_{j=1}^{n} a_{ij} x_j + x_{n+i} = b_i, \; x_{n+i} \geqslant 0$$

and we can turn each '$\geqslant$' inequality into an equality by subtracting the slack x_{n+i}:

$$\sum_{j=1}^{n} a_{ij} x_j - x_{n+i} = b_i, \; x_{n+i} \geqslant 0$$

Both types of inequality can be turned into equalities by adding or subtracting slack variables and the problem may therefore be written in the *standard form*:

Maximise (or minimise) $\quad x_0 = \sum_{j=1}^{n} c_j x_j$

subject to $$\sum_{j=1}^{n} a_{ij}x_j = b_i \quad \text{for } i = 1, \ldots, m$$

and $$x_j \geq 0 \quad \text{for } j = 1, \ldots, n$$

Of course, the total number of variables (decision variables plus slacks), n, may well have increased, and some of the a_{ij} may be zero. We will now discuss the simplex method as applied to problems in standard form.

Let B be a subset of $\{0, 1, 2, \ldots, n\}$ containing zero and exactly m other numbers and let N be the set of those left over, so that

$$B \cup N = \{0, 1, 2, \ldots, n\}$$

Then a set of decision variables, $\{x_i : i \in B\}$ is said to be a *basis* if we can write

$$x_i = \alpha_{i0} - \sum_{j \in N} \alpha_{ij} x_j \quad \text{for each } i \in B \tag{12.1}$$

for some numbers α_{ij}. A solution in which $x_j = 0$ whenever $j \in N$ and x_i is α_{i0} whenever $i \in B$ is said to be *basic*. If each α_{i0} is non-negative it is said to be *feasible*. All basic feasible solutions represent vertices of the feasible region.

Suppose, for the moment, that we have a basic feasible solution to the linear program and we know the α_{ij}'s in the above equations expressing the *basic variables* $\{x_i : i \in B\}$ in terms of the *non-basic variables* $\{x_j : j \in N\}$. We will discuss how to achieve such a basic feasible solution later in this section.

Now, if we are maximising and each α_{0j} is positive then the solution is optimal, as it is if we are minimising and each α_{0j} is negative. Otherwise suppose we are maximising and let α_{0q} be the most negative of the α_{0j}'s. Then the objective, x_0, will increase faster with respect to an increase in x_q than in any other non-basic variable. If we increase x_q from zero to δ then the value of the ith basic variable becomes

$$\alpha_{i0} - \alpha_{iq}\delta$$

because all other non-basic variables are set to zero. As δ increases the first basic to become zero will be given by

$$\text{Minimum } \{\alpha_{i0}/\alpha_{iq} : \alpha_{iq} > 0 \text{ and } i \geq 1\}$$

Let the minimum be achieved by α_{p0}/α_{pq}. Then x_p is the variable to leave the basis. This is known as the *ratio test*.

If there is no positive α_{iq} then no basic variable will diminish in value and so x_q may be increased without limit. In this case the objective value also improves without limit, being $\alpha_{00} - \alpha_{0q}x_q$, and the problem is *unbounded*.

We can express x_q in terms of the new non-basics, by the pth equation of (12.1):

$$x_q = \alpha_{i0}/\alpha_{pq} - \sum_{j \in N\backslash q}(\alpha_{pj}/\alpha_{pq})x_j - (1/\alpha_{pq})x_p$$

Linear Programming

where $N\backslash q$ (see Chapter 4) is the set of indices of the non-basic variables excluding q.

This can be substituted into the other equations (12.1):

$$x_i = \alpha_{i0} - \sum_{j \in N\backslash q} \alpha_{ij} x_j - \alpha_{iq} x_q$$

so that the other basic variables can be expressed in terms of the new non-basics:

$$x_i = (\alpha_{i0} - \alpha_{iq}\alpha_{p0}/\alpha_{pq}) - \sum_{j \in N\backslash q} (\alpha_{ij} - \alpha_{iq}\alpha_{pj}/\alpha_{pq})x_j + (\alpha_{iq}/\alpha_{pq})x_p$$

for $i \in B\backslash p$, where $B\backslash p$ is the set of indices of basic variables excluding p.

Write the new basics in terms of the new non-basics as:

$$x_i = \alpha'_{i0} - \sum_{\substack{j \in N\backslash q \\ j = p}} \alpha'_{ij} x_j \quad \text{for } i \in B\backslash p \text{ and } i = q$$

where the sum is over all $j \in N$ but excluding q, and over $j = p$. Then the representation of the new basic variables in terms of the new non-basic variables can be obtained by exchanging x_p and x_q and setting:

$$\alpha'_{ij} \text{ to } \alpha_{ij} - \alpha_{iq}\alpha_{pj}/\alpha_{pq} \quad \text{for all } i \neq p \text{ and } j \neq q$$
$$\alpha'_{iq} \text{ to } -\alpha_{iq}/\alpha_{pq} \quad \text{for all } i \neq p$$
$$\alpha'_{pj} \text{ to } \alpha_{pj}/\alpha_{pq} \quad \text{for all } j \neq q$$
$$\text{and} \quad \alpha'_{pq} \text{ to } 1/\alpha_{pq}.$$

This operation is known as *pivoting* and enables us to go mechanically from one basis to the next, increasing the objective by an amount

$$-\alpha_{0q}\alpha_{p0}/\alpha_{pq}$$

If we were minimising we would choose the non-basic variable, x_j, whose objective row coefficient, α_{0j} is most positive, and thus the objective value decreases. Except for the special case of *degeneracy* (which is discussed below), the objective value strictly improves at each basis change or *iteration* provided that the problem is not discovered to be unbounded.

Since there are only a finite number of possible bases or sets of basic variables (actually $n!/(n-m)!m!$) and the objective value strictly improves at each iteration (excepting degeneracy), no basis will be repeated so an optimal vertex will be found in a finite number of iterations.

We may therefore solve the linear program, starting from a basic feasible solution by the simplex method as follows:

STEP 1: *For maximisation* – If every coefficient of the non-basic variables in the objective row is positive then the solution is optimal and we terminate. Otherwise select the most negative α_{0j}. Call it α_{0q}.

STEP 1: *For minimisation* – If every coefficient of the non-basic variables in the objective row is negative then the solution is optimal and we terminate. Otherwise select the most positive α_{0j}. Call it α_{0q}.

STEP 2: (*Ratio test*) – If every α_{iq}, for $i > 0$ and $i \in B$ is non-positive then the problem is unbounded and we terminate. Otherwise select the smallest:

$$\alpha_{i0}/\alpha_{pq} \text{ where } \alpha_{iq} > 0$$

Call it α_{p0}/α_{pq}.

STEP 3: (*Pivoting*) – Form a new set of α coefficients, by setting:

α_{ij}	to	$\alpha_{ij} - \alpha_{iq}\alpha_{pj}/\alpha_{pq}$	if $i \neq p$ and $j \neq q$,
α_{iq}	to	$-\alpha_{iq}/\alpha_{pq}$	if $i \neq p$,
α_{pj}	to	α_{pj}/α_{pq}	if $j \neq q$
α_{pq}	to	$1/\alpha_{pq}$	

Form a new basis by interchanging x_p and x_q – i.e., remove x_p from the basis and add x_q. The α coefficients in the new x_p column are now α_{iq} (so call them α_{ip}) and the α coefficients in the new x_q row are now α_{pj} (so call them α_{qi}). Go to step 1.

If any of the basic variables have a value of zero – in other words, $\alpha_{i0} = 0$ for some i, in a row which has a positive coefficient, α_{iq}, in the column corresponding to the incoming variable x_q – then such a variable is bound to be chosen to leave the basis by the ratio test. In this event α_{p0} will be zero and the solution will not improve. This condition is known as *degeneracy* and the basis is said to be *degenerate*. It can happen that the basis remains degenerate through a sequence of simplex iterations which repeat themselves or *cycle*. If it does happen, perturbation procedures can be employed to break the cycle. However, these are beyond the scope of this chapter and although cycling is of theoretical interest, it happens rarely in practice.

We now turn to the problem of obtaining a basic feasible solution in the first place. Of course, if none exists then the problem is *infeasible*: no set of values for the decision variables will satisfy all the constraints. We need to be able to detect this as well as finding a basic feasible solution when it exists.

We may assume that we have formulated the problem in the general form in which all the original right-hand side coefficients, b_i, are positive. Consider, first of all a ' $\leq$ ' inequality row:

$$\sum_{j=1}^{n} a_{ij}x_j \leq b_i$$

If we add the non-negative *slack variable*, s_i, so that:

$$\sum_{j=1}^{n} a_{ij}x_j + s_i = b_i$$

Linear Programming

we have immediately that:

$$s_i = b_i - \sum_{j=1}^{n} a_{ij}x_j$$

so we can choose v_i to be a basic variable and set (with reference to equation (12.1)):

$$\alpha_{i0} = b_i \quad \text{and} \quad \alpha_{ij} = \alpha_{ij} \text{ for each } j$$

This works because $b_i \geqslant 0$.

Now consider an equality row:

$$\sum_{j=1}^{n} a_{ij}x_j = b_i$$

Then we can add a special non-negative variable, called an *artificial variable*, a_i, (not to be confused with the coefficients a_{ij}), so that:

$$\sum_{j=1}^{n} a_{ij}x_j + a_i = b_i$$

and

$$a_i = b_i - \sum_{j=1}^{n} a_{ij}x_j$$

We start with such a a_i being basic in the same way that we used the slack s_i above. This construction yields a basic solution which will later become feasible if we can remove a_i from the basis.

Lastly, consider an '$\geqslant$' inequality row:

$$\sum_{j=1}^{n} a_{ij}x_j \geqslant b_i$$

We add both a non-negative slack variable, s_i, and a non-negative artificial variable, a_i to this row so that:

$$\sum_{j=1}^{n} a_{ij}x_j - s_i + a_i = b_i$$

The constraint will be satisfied if $w_i = 0$ and we can write:

$$a_i = b_i - \left\{ \sum_{j=1}^{n} a_{ij}x_j - s_i \right\}$$

Again, we can start with the artificial w_i being basic in the same way as for equality constraints.

Noting that the objective row is simply:

$$x_0 = 0 - \sum_{j=1}^{n} (-c_j)x_j$$

we see that a basic solution has been obtained. And we even have all the α's if we assign an objective row coefficient of zero to the slack and artificial variables. However, this solution will not be feasible if we have used any artificial variables. It would, of course, become feasible if we could remove them from the basis by the steps of the simplex method that we have already encountered. Notice that the initial basic solution is not feasible because the artificials may take non-zero values, not because any of the α_{i0}'s are negative (they are all non-negative, being the b_i's). There are two ways of achieving feasibility by the simplex method.

Firstly we can give each artificial a very large penalty cost, so that its initial objective row coefficient is $-M$ for maximisation or M for minimisation, where M is a very large positive number. We must choose M to be so large that the optimal solution cannot possibly have any artificial variables at a positive level unless the problem is infeasible. This is known as the *big-M technique*.

The other method is to give each artificial variable a unit cost or penalty and ignore the costs or profit contributions of all other variables. The costs of the artificial variables can then be minimised. If this minimum turns out to be zero then a feasible solution has been achieved. If it is positive then the problem is infeasible. The linear program is thus solved in two phases: firstly that of obtaining feasibility by minimising the sum of the artificials' costs, and secondly that of optimising the real objective function having dropped the artificial variables, once this has been achieved. This is known as the *two-phase* or *Phase I/phase II technique*.

We still need a method of obtaining the x_0 row coefficients of the non-basic variables, α_{0j}'s, if the artificials have a cost of M, or if they have a cost of unity, or at the start of phase II of the two-phase method. Suppose that the jth decision or artificial variable has an original objective function coefficient of c_j. In the big-M method c_j for each decision variable is as given in the problem, but for each artificial variable it is $-M$ if we are maximising or $+M$ if we are minimising. At the start of Phase I of the two-phase method the original objective function coefficient of each decision variable will be zero and that of each artificial variable will be unity. At the start of Phase II of the two-phase method that of each decision variable will be as given and the artificial variables will have been removed from the problem. We know all the α's apart from those in the x_0 row so we can compute:

$$x_i = \alpha_{i0} - \sum_{j \in N} \alpha_{ij} x_j$$

for each $i \in B \backslash 0$, the set of the indices of all basic variables excluding x_0. Now x_0 can be expressed in terms of all the other variables by:

$$x_0 = \sum_{j=1}^{n} c_j x_j = \sum_{i \in B \backslash 0} c_i x_i + \sum_{j \in N} c_j x_j$$

Linear Programming

and we can substitute for x_i in the first sum above to express x_0 in terms of the non-basic variables only:

$$x_0 = \sum_{i \in B \setminus 0} c_i \left(\alpha_{i0} - \sum_{j \in N} \alpha_{ij} x_j \right) + \sum_{j \in N} c_j x_j$$

$$= \left(\sum_{i \in B \setminus 0} c_i \alpha_{i0} \right) - \sum_{j \in N} \left(\sum_{i \in B \setminus 0} c_i \alpha_{ij} - c_j \right) x_j$$

$$= \alpha_{00} - \sum_{j \in N} \alpha_{0j} x_j$$

Thus the coefficients of the non-basic variables in the x_0 row, α_{0j}, can be computed by:

$$\alpha_{0j} = \sum_{i \in B \setminus 0} c_i \alpha_{ij} - c_j$$

We will now look at a convenient way of representing the αs and see how the method works on a more realistic example than that of the chess sets.

12.6 The simplex tableau

A cattle feed manufacturer blends soya extract, maize and fish meal into a form suitable for feeding cattle in winter. The final product must contain at least 15% protein, 32.5% carbohydrate and no more than 20% fat. The cost per ton of each raw material and their composition is tabulated below.

	Raw material		
	Soya	Maize	Fishmeal
Cost/ton	£200	£100	£150
Protein	20%	10%	30%
Carbohydrate	30%	40%	10%
Fat	10%	10%	40%

Assuming that no protein, carbohydrate or fat is lost or added in the blending process find the minimum cost blend which satisfies the requirements.

Let x_1, x_2 and x_3 denote the quantities (in tons) of soya, maize and fishmeal used per ton of cattle feed.

Then the unit blending cost is:

$$x_0 = 200 x_1 + 100 x_2 + 150 x_3$$

and if this is in £/ton we must have:

$$x_1 + x_2 + x_3 = 1$$

Now the percentage of protein in the blend is:

$$20 x_1 + 10 x_2 + 30 x_3$$

and this must not be less than 15%, so:
$$20x_1 + 10x_2 + 30x_3 \geq 15$$
Similarly to satisfy the carbohydrate requirement, we must have:
$$30x_1 + 40x_2 + 10x_3 \geq 65/2$$
And the percentage of fat in the blend is:
$$10x_1 + 10x_2 + 40x_3$$
which must not exceed 20%, so:
$$10x_1 + 10x_2 + 40x_3 \leq 20$$

Clearly we cannot use negative quantities of raw material, so x_1, x_2 and x_3 must be non-negative.

The problem of finding the minimum cost blend is therefore the linear program:

Minimise $x_0 = 200x_1 + 100x_2 + 150x_3$

subject to
$$\begin{aligned}
x_1 + x_2 + x_3 &= 1 \quad \text{(material balance)} \\
20x_1 + 10x_2 + 30x_3 &\geq 15 \quad \text{(protein)} \\
30x_1 + 40x_2 + 10x_3 &\geq \frac{65}{2} \quad \text{(carbohydrate)} \\
10x_1 + 10x_2 + 40x_3 &\leq 20 \quad \text{(fat)}
\end{aligned}$$

and $x_1, x_2, x_3 \geq 0$

If we add the slack variables s_2, s_3 and s_4 to the protein, carbohydrate and fat constraints and the artificial variables a_1, a_2 and a_3 to the material balance, protein and carbohydrate constraints we obtain the problem of

Minimising $x_0 = 200x_1 + 100x_2 + 150x_3 + Ma_1 + Ma_2 + Ma_3$

subject to
$$\begin{aligned}
x_1 + x_2 + x_3 \quad\quad\quad\quad + a_1 \quad\quad\quad\quad &= 1 \\
20x_1 + 10x_2 + 30x_3 - s_2 \quad\quad + a_2 \quad\quad &= 15 \\
30x_1 + 40x_2 + 10x_3 \quad\quad - s_3 \quad\quad + a_3 &= 65/2 \\
10x_1 + 10x_2 - 40x_3 \quad\quad\quad + s_4 \quad\quad\quad\quad &= 20
\end{aligned}$$

and $x_1, x_2, x_3, s_2, s_3, s_4, a_1, a_2, a_3 \geq 0$,

where M is a large positive number.

We start by taking x_0, a_1, a_2, a_3 and s_4 to be the basic variables and x_1, x_2, x_3, s_2 and s_3 to be the non-basic ones. Thus we write:

$$\begin{aligned}
a_1 &= 1 \quad - \{\; x_1 + \;\; x_2 + \;\; x_3\} \\
a_2 &= 15 \;\; - \{20x_1 + 10x_2 + 30x_3 - s_2\} \\
a_3 &= 65/2 - \{30x_1 + 40x_2 + 10x_3 \quad\quad\quad - s_3\} \\
s_4 &= 20 \;\; - \{10x_1 + 10x_2 + 40x_3\}
\end{aligned}$$

Linear Programming

The x_0 row can now be computed. The constant term is:

$$M + 15M + \tfrac{65}{2}M = \tfrac{97}{2}M,$$

the coefficient of x_1 is:

$$-\{M + 20M + 30M - 200\} = -\{51M - 200\}$$

that of x_2 is:

$$-\{M + 10M + 40M - 100\} = -\{51M - 100\}$$

and that of x_3 is:

$$-\{M + 30M + 10M - 150\} = -\{41M - 150\}$$

The coefficients of both the slack variables, s_2 and s_3 are both $-M$, so we can write the x_0 row as:

$$x_0 = \tfrac{97}{2}M - \{(51M - 200)x_1 + (51M - 100)x_2 \\ + (41M - 150x_3) - Ms_2 - Ms_3\}$$

thus expressing the basic variables in terms of the non-basic ones we have:

$$x_0 = \tfrac{97}{2}M - \{(51M - 200)x_1 + (51M - 100)x_2 + (41M - 150)x_3 - Ms_2 - Ms_3\}$$
$$a_1 = 1 \quad -\{ \quad 1x_1 + \quad 1x_2 + \quad 1x_3 \qquad \qquad \}$$
$$a_2 = 15 \quad -\{ \quad 20x_1 + \quad 10x_2 + \quad 30x_3 - 1s_2 \quad \}$$
$$a_3 = \tfrac{65}{2} \quad -\{ \quad 30x_1 + \quad 40x_2 + \quad 10x_3 \qquad - 1s_3\}$$
$$s_4 = 20 \quad -\{ \quad 10x_1 + \quad 10x_2 + \quad 40x_3 \qquad \qquad \}$$

These equations are more conveniently represented in a table known as the *simplex tableau* as follows:

	x_1	x_2	x_3	s_2	s_3	s_4	a_1	a_2	a_3	RHS
x_0	$51M-200$	$51M-100$	$41M-150$	$-M$	$-M$	0	0	0	0	$\tfrac{97}{2}M$
a_1	1	1	1	0	0	0	1	0	0	1
a_2	20	10	30	-1	0	0	0	1	0	15
a_3	30	㊵	10	0	-1	0	0	0	1	$\tfrac{65}{2}$
s_4	10	10	40	0	0	1	0	0	0	20

The constant terms have been put in a column labelled 'Right-Hand Side' (or 'RHS', for short) because they appear on the right-hand side of the constraints in the original formulation. The columns between the dotted lines (labelled s_4, a_1, a_2 and a_3) correspond to the basic variables and are not needed if a record is kept of which rows correspond to which basic variables (i.e., the terms appearing on the left-hand side in the above equations). We have, in fact, done this in the leftmost column. Notice that the entries in these columns are zero except for that in the row corresponding to the same basic variable when it is unity.

We could have an extra column, labelled x_0, with a 1 in the x row and zeros elsewhere.

It is worth remarking that you could interpret the tableau rows as *equations*. For each row the sum of the product of each coefficient and the variable in whose column it lies is equal to the RHS coefficients. Taking the third row as an example we have that:

$$20x_1 + 10x_2 + 30x_3 - s_2 + a_2 = 15$$

We now turn to solving the problem represented in the above simplex tableau. Since we are minimising, the solution will be optimal only if all the coefficients of the non-basics in the x_0 row are *negative*. Clearly they are not and that of x_2 is the most positive ($51M - 100$) so the objective improves faster as x_2 is raised from zero to enter the basis than it would if any other non-basic variable were used. Performing the ratio test we see that:

$$\text{Minimum } \{1/1, 15/10, 65/80, 20/10\} = 65/80$$

so a_3 is the variable to leave the basis and we pivot on the element in column x_2 and row a_3. All the elements in the columns corresponding to the non-basic variables are updated by the simplex pivot rules (for example, the coefficient of x_1 in the a_1 row becomes $1 - 30 \times 1/40 = 1/4$, that of x_3 in the a_1 row becomes $1 - 1 \times 10/40 = 3/4$ and that of x_1 in the s_4 row becomes $10 - 30 \times 10/40 = 5/2$) and x_2 and a_3 are interchanged to yield the second tableau:

	x_1	a_3	x_3	s_2	s_3	s_4	a_1	a_2	x_2	RHS
x_0	$\frac{51M}{4} - 125$	$\frac{5}{2} - \frac{51M}{40}$	$\frac{113M}{4} - 125$	$-M$	$\frac{11M}{40} - \frac{5}{2}$	0	0	0	0	$\frac{113}{16}M + \frac{325}{4}$
a_1	$\frac{1}{4}$	$-\frac{1}{40}$	$\frac{3}{4}$	0	$\frac{1}{40}$	0	1	0	0	$\frac{3}{16}$
a_2	$\frac{25}{4}$	$-\frac{1}{4}$	$\frac{55}{2}$	-1	$\frac{1}{4}$	0	0	1	0	$\frac{55}{8}$
x	$\frac{3}{4}$	$\frac{1}{40}$	$\frac{1}{4}$	0	$-\frac{1}{40}$	0	0	0	1	$\frac{13}{16}$
s_4	$\frac{5}{2}$	$\frac{1}{4}$	$\frac{75}{2}$	0	$\frac{1}{4}$	1	0	0	0	$\frac{95}{8}$

(Pivot element $\frac{3}{4}$ circled in the a_1 row, x_3 column.)

The non-basic artificial variable, a_3, will never be re-introduced into the basis and so the column labelled a_3 can be dropped. Nothing is lost by this since shadow price information can be gained from s_3. This would not have been the case if a_1 were removed from the basis since there is no slack on the first constraint of the problem.

Linear Programming

The second tableau is still not optimal as we introduce x_3 into the basis, since it has the largest x_0 row coefficient.

However, the ratio test yields a tie between a_1 and a_2 ($\frac{3}{16} \div \frac{3}{4} = \frac{55}{8} \div \frac{55}{2}$). We can choose either a_1 or a_2 to leave the basis. It does not really matter which we choose; however, whichever remains will stay in the basis with a value of zero. In other words, the solution will become *degenerate*. The degeneracy persists in this case throughout the remaining iterations of the simplex method and will have important consequnces for the interpretation of the final solution as we shall see. Choosing a_1 to leave the basis, the third tableau is obtained:

	x_1	a_1	s_2	s_3	s_4	x_3	a_2	x_2	RHS
x_0	$\frac{10M}{3} - \frac{250}{3}$	$\frac{500}{3} - \frac{113M}{3}$	$-M$	$\frac{5}{3} - \frac{2}{3}M$	0	0	0	0	$\frac{225}{2}$
x_3	$\frac{1}{3}$	$\frac{4}{3}$	0	$\frac{1}{30}$	0	1	0	0	$\frac{1}{4}$
a_2	$\boxed{\frac{10}{3}}$	$-\frac{110}{3}$	-1	$-\frac{2}{3}$	0	0	1	0	0
x_2	$\frac{2}{3}$	$-\frac{1}{3}$	0	$-\frac{1}{30}$	0	0	0	1	$\frac{3}{4}$
s_4	-10	-50	0	-1	1	0	0	0	$\frac{5}{2}$

This solution is feasible since the objective value which is the term in the RHS column of the x_0 row does not contain any term involving M. Artificials a_1 and a_3 are non-basic and although artificial a_2 is basic it has value zero. If all the artificial variables were non-basic we could drop the terms involving M and simply never consider any as candidates to enter the basis. But this is not the case yet, and so we choose x_1 to enter the basis since it has the largest positive x_0 row coefficient for M sufficiently large. Of course, a_2 leaves, to yield the fourth simplex tableau:

	a_2	a_1	s_2	s_3	s_4	x_3	x_1	x_2	RHS
x_0	$25 - M$	$750 - M$	-25	-15	0	0	0	0	$\frac{225}{2}$
x_3	$-\frac{1}{10}$	5	$\frac{1}{10}$	$\frac{1}{10}$	0	1	0	0	$\frac{1}{4}$
x_1	$\frac{3}{10}$	-11	$-\frac{3}{10}$	$-\frac{1}{5}$	0	0	1	0	0
x_2	$-\frac{1}{5}$	7	$\frac{1}{5}$	$\frac{1}{10}$	0	0	0	1	$\frac{3}{4}$
s_4	-3	-160	3	-3	1	0	0	0	$\frac{5}{2}$

The a_2 column can be dropped from the tableau without loss of information. Notice that the objective function value remains unchanged at 225/2 and the solution is still degenerate. All the x_0 row coefficients of non-basic variables eligible to enter the basis are negative, so the solution is optimal.

The interpretation is that the minimum cost cattle feed mix is £112.50 per ton, composed of no soya, 75% maize and 25% fishmeal. Furthermore the fat constraint is not binding and so does not influence the optimal mix. This is evident from the RHS column of the fourth tableau. The shadow prices of the constraints are contained in the x_0 row. These reveal that an extra 1% of protein would cost £25.00 per ton in the final mix (from the coefficient of s_2) and an extra 1% of carbohydrate would cost £15.00 (from the coefficient of s_3). Or does it?

The above interpretation of the shadow prices is the usual one, but it needs to be considered carefully. Compare the solution of the fourth simplex tableau with that of the third. The two solutions are actually the same ($x_2 = \frac{3}{4}$, $x_3 = \frac{1}{4}$, $s_4 = \frac{5}{2}$, all other variables zero); they are merely different ways of representing it. Notice that the x_0 row coefficients (apart from that in the RHS column) are radically different. This condition arises only if one of the basic variables has a value of zero and the solution is degenerate. But when it does, the shadow prices and reduced costs are not unique. So which should be used for marginal costing information? The answer is that each reduced cost or shadow price is valid over some range, which may well be of zero width. The only way to tell is to use a form of sensitivity analysis called *ranging*. This is described in chapter 13; such a facility is almost always available in any computer code which solves linear programs. It should be used whenever the final basis is degenerate and marginal costing information is required. The range over which the reduced costs and shadow prices is valid, amongst other information, is always given. In the case of our example, the range over which the reduced costs and shadow prices is valid in the third tableau is, in fact, of zero or null length and those of the fourth tableau are 'correct' from the point of view of their usual interpretation.

The final point to notice about this example is that although we have used the 'big M' method of achieving a feasible solution we have not actually used any particular value for M. If the problem were being solved by a computer program then a particular value, say 1500, would have to be used. But in our case we have always been able to choose the x_0 row coefficient in the tableau which is larger than all others *for sufficiently large M* to determine the variable to enter the basis. Thus the only real difference between this and the phase I/phase II method is that we always keep track of the 'phase II' x_0 row coefficients, those not involving M. Thus we do not have to calculate them at the start of phase II and we can tie-break between terms whose coefficient of M is the same (we did so in choosing x_2 rather than x_1 to enter the basis in the first tableau).

This concludes our discussion of the use of the simplex tableau as a formal

Linear Programming

means of solving linear programs. We build on this same example in our work in the next chapter.

12.7 Exercises

Worked answers are provided in Appendix C to questions marked with an asterisk.

1.* A company needs to purchase a number of of small printing presses, of which there are two types, X and Y.

    Type X costs £4,000, requires two operators and occupies 20 square metres of floor space. Type Y costs £12,000, also requires two operators but occupies 30 square metres.

    The company has budgeted for a maximum expenditure on these presses of £120,000. The print shop has 480 square metres of available floor space, and work must be provided for at least 24 operators.

    It is proposed to buy a combination of presses X and Y that will maximise production, given that type X can print 150 sheets per minute and type Y, 300 per minute.

    You are required to:

    (a) write down all the equations/inequalities which represent the cost and space conditions. The labour conditions are given by $2X + 2Y \geqslant 24$;

    (b) draw a graph to represent this problem, shading any unwanted regions;

    (c) use the graph to find the numbers of presses X and Y the company should buy to achieve its objective of maximum production;

    (d) state the figure of maximum production and the total cost of the presses in this case.

    (*Institute of Cost and Management Accountants*, May 1984)

2. The groundsman of the local tennis club proposes to treat the grass courts with two types of fertilizer. Both types use the same three basic ingredients, in the following proportions:

    Ingredients

Fertilizer	I_1	I_2	I_3
F_1	6	4	2
F_2	1	3	5

    The groundsman buys:

tons	£ per ton
6 I_1	3
12 I_2	4
10 I_3	5

Given these purchased quantities, how should he mix the ingredients to make the best use of his available stock?

If he finds that he can apply 10% less of F_2 than F_1 in order to achieve comparable results, which fertilizer would you advise him to concentrate on for economy?

(*Institute of Chartered Accountants in England and Wales*, November 1985)

3.* (a) Hitech, a division of Sunrise plc, produces and sells three products:

*HTO*1, *HTO*2 and *HTO*3

The following details of prices and product costs have been extracted from Hitech's cost accounting records:

	Product		
	*HTO*1	*HTO*2	*HTO*3
Price per unit	£150	£200	£220
Costs per unit:			
Direct labour at £4/hr	£100	£120	£132
Direct material at £20/kg	£20	£40	£40

Direct labour is regarded as a variable product cost.

A regression analysis has been carried out in order to estimate the relationship between overhead costs and production of the three products. Expressed in weekly terms the results of the analysis show:

$$Y = 4{,}000 + 0.5x_1 + 0.7x_2 + 0.8x_3$$

where Y = total overhead cost per week
x_1 = *HTO*1, weekly direct labour hours
x_2 = *HTO*2, weekly direct labour hours
x_3 = *HTO*3, weekly direct labour hours

The company operates a 46-week year.

You are required to compute the total variable product costs for each of *HTO*1, *HTO*2 and *HTO*3.

(b) The material used by Hitech is also used in a wide variety of other applications and is in relatively limited supply. As business conditions improve in general, there will be pressure for the price of this material to rise, but strong competition in Hitech's sector of the market would make it unlikely that increased material costs can be passed on to customers in higher product prices. The position on material supplies is that Hitech can obtain 20,000 kg at current prices.

Linear Programming

Further, reductions in the skilled labour force made during the recession mean that the number of available direct labour hours is estimated at no more than 257,600 hours for the next year.

Demand for each product over the year is forecast to be:

	Units
HTO1	16,000
HTO2	10,000
HTO3	6,000

You are required to formulate a linear program from the above data in order to obtain the annual production/sales plan which will maximise Hitech's contribution earnings and profit. (You are *not* required to solve the problem.)

(c) The following are the final tableau and its immediately preceding tableau, obtained as a result of running the linear program:

Final tableau

HTO1	HTO2	HTO3	S_1	S_2	S_3	S_4	S_5	RHS
0.0	0.0	1.0	0.0	0.0	0.0	0.0	1.0	6,000.0
1.0	1.2	1.3	0.0	0.0	0.0	0.0	0.0	10,304.0
0.0	−1.2	−1.3	−0.0	0.0	1.0	0.0	0.0	5,696.0
0.0	1.0	0.0	0.0	0.0	0.0	1.0	0.0	10,000.0
0.0	0.8	1.7	−0.0	1.0	0.0	0.0	0.0	9,696.0
0.0	2.0	1.5	0.7	0.0	0.0	0.0	0.0	180,320.0

Preceding tableau

HTO1	HTO2	HTO3	S_1	S_2	S_3	S_4	S_5	RHS
0.0	−0.5	0.0	0.0	−0.6	0.0	0.0	1.0	228.6
1.0	0.6	0.0	0.1	−0.8	0.0	0.0	0.0	2,685.7
0.0	−0.6	0.0	−1.0	0.8	1.0	0.0	0.0	13,314.3
0.0	1.0	0.0	0.0	0.0	0.0	1.0	0.0	10,300.0
0.0	0.5	1.0	−0.0	0.6	0.0	0.0	0.0	5,771.4
0.0	1.3	0.0	0.7	−0.9	0.0	0.0	0.0	171,663.0

where $S_1 \ldots S_5$ are the slack variables for labour, materials, HTO1, HTO2 and HTO3 respectively.

You are required to:

(i) provide as complete an interpretation of the final tableau as you can; compare the outcome of the final with that of the preceding tableau and give an estimate of the final net profit figure;

(ii) state, if material prices were to rise, the order in which production of each of Hitech's products would cease to be worthwhile.

(*Institute of Cost and Management Accountants*, May 1984)

4.* A company manufactures four products *A, B, C* and *D* which are marketed in tubs. Of its total of 25 machines, ten are suitable for manufacturing all four products whilst the remaining 15 are unsuitable for products *A* and *D*.

Each machine is in production for 48 weeks per year and each is used on a given product in terms of full weeks and not in fractions of weeks. The company has no problem in obtaining adequate supplies of labour and raw materials.

Marketing policy is that all four products should be sold, and the minimum annual production is 500 tubs of each. Fixed costs are budgeted at £820,000 for the year.

Information on production, market price and direct costs is given below:

	Product			
	A	B	C	D
Production (tubs per machine per week)	14	4	3	6
	£	£	£	£
Market price (per tub)	390	390	450	570
Direct costs:				
Process 1: Direct materials (per week)	238	108	96	156
Process 1: Direct labour (per week)	448	304	186	264
Process 2: Direct materials (per tub)	10	10	10	10
Process 2: Direct labour (per tub)	80	72	100	120
Transport (per tub)	130	130	100	240

Because the demand for products *A* and *D* is increasing, the company is considering converting to all-product machines those which, at present, are unsuitable for products *A* and *D*. The cost quoted for this work is

Linear Programming

£70,000 per machine. The company expects a 12% return over a three-year period for this type of expenditure.

Market research indicates that the company's expected sales over the next three years for products A and D would be 7,000 tubs and 2,100 tubs per annum respectively.

You are required, as management accountant, to:

(a) calculate the profit that the company would earn if it worked on the most profitable basis of allocation of productive time to the various products (within the constraints outlined above);
(b) recommend the maximum number of machines to be converted into all-purpose machines, giving supporting calculations.

Ignore tax and inflation.

(*Institute of Cost and Management Accountants*, May 1985)

Note that the best approach here is to calculate contribution per product.

5. A small furniture manufacturer makes two speciality products, tables and chairs. Three stages in the manufacturing process are required – machining, assembling and finishing. The number of minutes required for each unit is shown below:

	Machining	Assembling	Finishing
Table	4	10	12
Chair	12	5	12

Each day the machining equipment is available for six hours (360 minutes), the assembly shop for six hours and the finishing equipment for eight hours. The contribution is £3 per table and £5 per chair. All equipment can be used for the production of either tables or chairs at all times it is available.

You are required to:

(a) state all the equations/inequalities (constraints) which describe the production conditions;
(b) graph these constraints;
(c) find how many tables and chairs the manufacturer should produce to maximise contribution;
(d) calculate this maximum contribution and include any comments or reservations you have about this analysis generally.

(*Institute of Cost and Management Accountants*, May 1986)

Chapter 13
Sensitivity Analysis, Duality and the Transportation Algorithm

Robert Ashford

In this chapter we examine the solution of linear programs in more detail, consider concepts of duality and look at a special form of the algorithm for solving transportation problems.

13.1 Sensitivity analysis or ranging

We have already seen how small changes to the cost coefficients and right-hand-side values alter the optimal solution in a fairly informal way. In this section we will study how these changes perturb the solution much more systematically; it is important to ascertain how large these changes can be before the basis ceases to be optimal and the calculated rate of change of the objective value ceases to be valid. For consistency, suppose that the linear program is in standard form:

$$\text{Maximise} \quad x_0 = \sum_{j=1}^{n} c_j x_j$$

$$\text{subject to} \quad \sum_{j=1}^{n} a_{ij} x_j = b_i \quad \text{for } i = 1, \ldots, m$$

$$\text{and} \quad x_j \geq 0 \quad \text{for } j = 1, \ldots, n,$$

and we have the optimal solution:

$$x_i = \alpha_{i0} - \sum_{j \in N} \alpha_{ij} x_j \quad \text{for } i \in B$$

where B is the index set of the basic variables including x_0, and N is that of the non-basic variables. Notice that for feasibility we must have:

$$\alpha_{i0} \geq 0 \quad \text{for each } i \in B \backslash 0$$

Sensitivity Analysis, Duality, Transport Algorithm

where $B\backslash 0$ is the index set of the basic variables excluding x_0, and for optimality we must have:
$$\alpha_{0j} \geq 0 \quad \text{for each } j \in N$$

If the problem were one of minimisation, the above inequality would, of course, be reversed.

Firstly, consider a small change in one of the objective coefficients, c_k to $c_k + \delta$. There are two cases that must be considered separately, namely when x_k is basic and when x_k is non-basic, but in either case it is important to understand the nature of the reduced costs, the α_{0j}'s. We saw in the last chapter that the objective row coefficients of the non-basic variables can be computed from:
$$\alpha_{0j} = \sum_{i \in B\backslash 0} c_i \alpha_{ij} - c_j \quad \text{for each } j \in N$$

and this is valid at any basic solution, including the optimal one. Now consider increasing the original coefficient of the non-basic variable x_k, c_k, by δ. The only item to change is the reduced cost of x_k, α_{0k} which becomes:
$$\sum_{i \in B\backslash 0} c_i \alpha_{ik} - c_k - \delta = \alpha_{0k} - \delta$$

in other words it simply decreases by δ. So if the solution is to remain optimal this reduced cost must stay non-negative and δ must be such that:
$$\delta \leq \alpha_{0k}$$

If we were minimising then δ would have to be no greater than α_{0k}. The maximum permissible *increase or decrease* to the cost coefficients of the non-basics is given by their reduced costs or coefficients in the x_0 row of the final simplex tableau according to whether we are maximising or minimising, respectively. Notice that if we were minimising, the reduced costs would be negative and therefore we would obtain a lower bound on the value of the non-basic cost coefficient before the basis changes. If the non-basic's cost changes by more than these limits then it would enter the basis in the optimal solution. If the variable represented production of some item, this tells us how much more profitable or less costly it would have to become before being worthwhile to produce.

Changes to the objective coefficients of basic variables are slightly more complicated to analyse. Suppose x_k is basic and its objective coefficient is increased from c_k to $c_k + \delta$.

Then the reduced costs of all the non-basics will change; that of the jth becomes:
$$\sum_{i \in B\backslash 0} c_i \alpha_{ij} - c_j + \delta a_{kj} = \alpha_{0j} + \delta \alpha_{kj}$$

And these must all remain non-negative, so:
$$\delta \geq -\alpha_{0j}/\alpha_{kj}, \quad \text{for all } j: \alpha_{kj} > 0$$
$$\delta \leq -\alpha_{0j}/\alpha_{kj}, \quad \text{for all } j: \alpha_{kj} < 0$$

If we were minimising instead of maximising, the above inequalities would be reversed. In either case however, the final basis and the optimal value of the decision variable will not change provided δ remains within these specified limits, and the objective function will change by $x_k \delta$.

For example, consider the earlier fertilizer problem. Suppose that the cost of maize were increased by δ £ per ton from £100/ton to £$(100+\delta)$ per ton. Inspection of the final simplex tableau reveals that the solution would remain optimal provided

$$\delta \leqslant -(-25)/\tfrac{1}{5} = 125$$
$$\text{and} \quad \delta \leqslant -(-15)/\tfrac{1}{10} = 150$$

The solution would therefore remain optimal provided the cost of maize is less than £225 per ton. It can, of course, be anything less than £100 per ton as well.

Now suppose that the cost of soya were increased by δ £/ton from £200/ton to £$(200+\delta)$/ton. Again, inspection of the final simplex tableau reveals that the solution would remain optimal provided

$$\delta \geqslant -(-25)/(-\tfrac{3}{10}) = -250/3$$
$$\text{and} \quad \delta \geqslant -(-15)/(-\tfrac{1}{5}) = -75$$

In other words, the optimal solution would not change provided the cost of soya were greater than £125 per ton.

We now consider changes to the right-hand side coefficients. To do this we must consider the constraints of the standard form linear program in more detail. We may suppose that the first m variables, $x_1, \ldots, x_m$ are slacks and artificials used to convert the linear program in general form to one in standard form. In the case of the ith constraint being a '$\geqslant$' inequality originally, suppose that x_i is the slack, the artificial appearing in the remaining $n-m$ variables. Thus we may write the ith constraint as:

$$(\pm 1)x_i + \sum_{j=m+1}^{n} a_{ij} x_j = b_i$$

where the coefficient of x_i is -1 if the original constraint was a '$\geqslant$' inequality and is otherwise unity.

Firstly, consider inequality constraints in which the slack variable is basic in the optimal solution. If the right-hand-side coefficient was b_k and is changed to $b_k + \delta$, then the solution will not alter as long as the slack x_k, originally equal to α_{k0}, remains non-negative. It will change to $\alpha_{k0} + \delta$, so the solution remains unchanged provided:

$$\delta \geqslant -\alpha_{k0} \text{ if the constraint was '}\leqslant\text{'}$$
$$\delta \leqslant \alpha_{k0} \text{ if the constraint was '}\geqslant\text{'}$$

and the objective value remains unchanged. If the artificial variable of an equality constraint is basic (necessarily at zero) in the optimal solution then we cannot easily deduce the effect of a change to its right-hand-side coefficient from the optimal simplex tableau.

In our cattle feed example, the solution will remain unchanged if the maximum fat content is diminished by $2\frac{1}{2}\%$ to $17\frac{1}{2}\%$ from 20%.

Now consider '$\leqslant$' constraints in which the slack variable x_k is non-basic in the optimal solution. If the original right-hand-side coefficient is changed from b_k to $b_k + \delta$, the values of all the basic variables may change. We can explore the effect of this by allowing the slack to change in value from zero to $-\delta$, *even if this makes it negative*, for by doing so the left-hand-side of the kth constraint can take the value $b_k + \delta$. That is,

$$\sum_{j=1}^{n} a_{kj} x_j = b_k + \delta,$$

and the equation

$$x_k + \sum_{j=1}^{n} a_{kj} x_j = b_k$$

is preserved if $x_k = -\delta$.

The ith row of the optimal solution is

$$x_i = \alpha_{i0} - \sum_{j \in N \setminus k} \alpha_{ij} x_j - \alpha_{ik} x_k$$

where $\{x_j : j \in N \setminus k\}$ is the set of non-basic variables excluding x_k. Since these variables are set to zero, the value of the basic variable x_i changes from α_{i0} to

$$\alpha_{i0} - \alpha_{ik} x_k = \alpha_{i0} + \alpha_{ik} \delta.$$

In particular, from the x_0 (or objective) row of the optimal solution, we can see that the value of the objective function, x_0, changes from α_{00} to $\alpha_{00} + \alpha_{0k}\delta$. So the reduced cost of the slack variable x_k, or *shadow price* of the kth constraint, α_{0k}, gives the marginal change to the objective function value with respect to a change in b_k.

In our cattle feed example we can see that our profit diminishes by £25 per percent of protein required in the final blend, and by £15 per percent of carbohydrate, since the shadow prices on the protein and carbohydrate constraints are -25 and -15 respectively in the optimal solution.

Returning to the change in the ith basic variable from α_{i0} to $\alpha_{i0} + \alpha_{ik}\delta$, we can see that the solution will remain feasible provided:

$$\delta \geqslant -\alpha_{i0}/\alpha_{ik} \quad \text{if } \alpha_{ik} > 0$$

and

$$\delta \leqslant -\alpha_{i0}/\alpha_{ik} \quad \text{if } \alpha_{ik} < 0$$

for each basic variable x_i (except x_0). *This is the range over which the marginal costing information provided by the shadow prices is valid.* So if, in our cattle feed example, the protein requirement changes from 15% to $(15+\delta)\%$, this will cost an extra £25 provided:

$$\delta \geq -1/4 \div 1/10 = -5/2$$
$$\delta \leq -0 \div -3/10 = 0$$
$$\delta \geq -3/4 \div 1/5 = -15/4$$
$$\delta \geq -5/2 \div 3 = -5/6$$

i.e. provided $-5/2 \leq \delta \leq 0$. So we cannot deduce what the cost of an extra 1% in the protein requirement should be, but know that a drop of $2\frac{1}{2}$% would save £25 × $2\frac{1}{2}$ = £$62\frac{1}{2}$. If the carbohydrate requirement changed from $32\frac{1}{2}$% to $(32\frac{1}{2}+\delta)$%, then this would cost £15δ, provided $-5/2 \leq \delta \leq 0$ (this is only coincidentally the same as that for the protein requirement).

Equality ('=') constraints in which the artificial variable is non-basic may be treated in exactly the same way as '$\leq$' constraints above, except that x_k is an *artificial* rather than a slack variable.

For '$\geq$' constraints in which the slack variable is non-basic, the analysis of changes to the right-hand-side coefficients proceeds in a similar fashion to that for '$\leq$' constraints, except that a change from b_k to $b_k + \delta$ necessitates a change of the slack variable x_k from zero to $+\delta$. So the shadow price gives minus the rate of change to the objective value with respect to b_k, and this marginal costing is valid provided the solution remains feasible, which will be the case if:

$$\delta \leq \alpha_{0k}/\alpha_{ik} \quad \text{if } \alpha_{ik} > 0$$

and

$$\delta \geq \alpha_{0k}/\alpha_{ik} \quad \text{if } \alpha_{ik} < 0$$

for each basic variable x_i except x_0.

13.2 Duality

We have seen how to formulate a practical problem as a linear program and solve it by the simplex method. The problem has been posed in terms of finding the best possible values for a number of structural or decision variables, subject to a number of linear constraints upon them. We have seen how the simplex method tells us the marginal price we have to pay for each active or binding constraint. The problem can in fact be posed the other way around – to determine the best possible set of marginal prices for each constraint, subject to a set of linear constraints on those prices, one for each decision variable. This is called the *dual* problem and if we apply the simplex method to it the marginal price of each constraint tells us the optimal value of each decision variable in the original or *primal* problem. This is very useful for two reasons:

1. It helps us to 'explore' the problem, analysing the sensitivity of the solution to changes in input data, and enhance our model of the practical situation.
2. It can be used within different solution methods, especially the transportation algorithm which will be discussed later.

Sensitivity Analysis, Duality, Transport Algorithm

Earlier we stated the linear program in standard form as:

Maximise $x_0 = \sum_{j=1}^{n} c_j x_j$ over $x_1, \ldots, x_n$

subject to $\sum_{j=1}^{n} a_{ij} x_j = b_i$ for $i = 1, \ldots, m$

and $x_j \geq 0$ for $j = 1, \ldots, n$

The dual to this primal problem is defined to be:

Minimise $y_0 = \sum_{i=1}^{m} b_i y_i$ over $i = 1, \ldots, m$

subject to $\sum_{i=1}^{m} a_{ij} y_i \geq c_j$ for $j = 1, \ldots, n$

Notice that to each primal constraint there corresponds a dual variable and to each primal variable there corresponds a dual constraint. There is a dual problem associated with every linear program, not just those in standard form, and this constraint–variable correspondence always applies. With this in mind, we can now state the rules by which a dual problem may be constructed for any linear program.

1. If we maximise in the primal, we minimise in the dual and vice-versa.
2. If a primal constraint is an equality then the corresponding dual variable is unrestricted in sign; if it is a '$\leq$' inequality the corresponding dual variable is non-positive if we are minimising in the primal and non-negative if we are maximising; if it is a '$\geq$' inequality the corresponding dual variable is non-negative if we are minimising in the primal and non-positive if we are maximising.
3. If a primal variable is non-negative and the primal problem one of maximisation, the corresponding dual constraint is a '$\geq$' inequality, whereas if the primal problem is one of minimisation the corresponding dual constraint is a '$\leq$' inequality. If a primal variable is non-positive then the corresponding dual constraint is the other way around – '$\leq$' for maximisation in the primal and '$\geq$' for minimisation. If a primal variable is unrestricted in sign, then the corresponding dual constraint is an equality.

You can verify that these relationships follow from the definition of the dual to the linear program in standard form by expressing the general linear program (with constraints of all three types and variables which are non-negative, non-positive and unrestricted in sign) in standard form and then looking at the resultant dual. Notice that to express a variable unrestricted in sign (say x_j) in terms of non-negative variables, we can write:

$$x_j = x_j^+ - x_j^-$$

where both x_j^+ and x_j^- are non-negative. To change a maximisation problem into a minimisation one (or vice-versa) replace each objective coefficient c_j by $-c_j$.

Furthermore, you can see that rules (1), (2) and (3) make intuitive sense in terms of the interpretation given to the dual at the beginning of this section. For example, if we are maximising in the primal then the 'marginal price' of a '$\leqslant$' constraint – in other words, the rate of increase in the optimal objective value with respect to a change in the right-hand side coefficient – cannot be negative. This is because we enlarge the feasible region as the right-hand side coefficient is increased and so the objective function must improve – in other words, increase. Hence the corresponding dual variable must be non-negative. On the other hand, if we were minimising then the marginal price cannot be positive because an improvement in objective value is a decrease. Hence the corrresponding dual variable must be non-positive. For an equality constraint in the primal, an increase in right-hand side value may make the optimal objective function value increase or decrease. Hence the corresponding dual variable is unrestricted in sign.

It is reassuring to observe that the dual of the dual program is the original primal one, so primal and dual programs exhibit a form of symmetry. However, they conform to the duality concept discussed in Chapter 4.

Let us construct the dual to the blending problem that we solved earlier. The original or primal problem was:

Minimise $\quad x_0 = 200 x_1 + 100 x_2 + 150 x_3 \quad$ (cost)
subject to $\quad\quad\quad x_1 + \quad x_2 + \quad x_3 = 1 \quad$ (material balance)
$\quad\quad\quad\quad\quad 20 x_1 + 10 x_2 + 30 x_3 \geqslant 15 \quad$ (protein)
$\quad\quad\quad\quad\quad 30 x_1 + 40 x_2 + 10 x_3 \geqslant 32\tfrac{1}{2} \quad$ (carbohydrate)
$\quad\quad\quad\quad\quad 10 x_1 + 10 x_2 + 40 x_3 \leqslant 20 \quad$ (fat)
and $\quad\quad\quad\quad x_1, x_2, x_3 \quad\quad\quad\quad\quad \geqslant 0$

The dual problem is one in four variables, y_1, y_2, y_3 and y_4, corresponding to the material balance, protein carbohydrate and fat constraints respectively. The dual variable y_1 is unrestricted in sign, y_2 and y_3 are non-negative, and y_4 is non-positive. The dual has three constraints, one corresponding to each primal variable and these are all '$\leqslant$' inequalities. So the dual problem is to:

maximise $\quad Y_0 = y_1 + 15 y_2 + 32\tfrac{1}{2} y_3 + 20 y_4$
subject to $\quad y_1 + 20 y_2 + 30 y_3 + 20 y_4 \leqslant 200$
$\quad\quad\quad\quad y_1 + 10 y_2 + 40 y_3 + 10 y_4 \leqslant 100$
$\quad\quad\quad\quad y_1 + 30 y_2 + 10 y_3 + 40 y_4 \leqslant 150$

and y_1, unrestricted in sign, $y_2 \geqslant 0$, $y_3 \geqslant 0$, and $y_4 \leqslant 0$

You can think of this as the problem of maximising the total marginal value of

Sensitivity Analysis, Duality, Transport Algorithm   277

the quality restrictions subject to constraints arising from the raw material costs.

We can now state the crucial relationships between primal and dual linear programs. The first set is known as the *weak duality* results. Assume that the primal problem is in standard form. Then:

1. If $x_1, \ldots, x_n$ and $y_1, \ldots, y_m$ are feasible solutions to the primal and dual problem respectively, then:

$$\sum_{j=1}^{n} c_j x_j \leq \sum_{i=1}^{m} b_i y_i$$

   In other words, any feasible primal objective value is not more than any feasible dual objective value. If the two objective values are equal then the solutions are optimal.

2. If the primal problem is unbounded, then the dual is infeasible and conversely if the primal is infeasible then the dual is unbounded.

The first weak duality result is easy to show. Since $x_1, \ldots, x_n$ is a feasible solution to the primal, we must have:

$$b_i = \sum_{j=1}^{n} a_{ij} x_j$$

So
$$\sum_{i=1}^{m} b_i y_i = \sum_{i=1}^{m} \left(\sum_{j=1}^{n} a_{ij} x_j \right) y_i$$

$$= \sum_{j=1}^{n} \left(\sum_{i=1}^{m} a_{ij} y_i \right) x_j \geq \sum_{j=1}^{n} c_j x_j$$

because $y_1, \ldots, y_m$ is feasible for the dual and the x's are all non-negative. Furthermore if the two objective values are equal, then:

$$x_0 = \sum_{i=1}^{m} b_i y_i = \sum_{j=1}^{n} c_j x_j$$

$$\geq \sum_{j=1}^{n} c_j \tilde{x}_j$$

for any other primal feasible solution $\tilde{x}_1, \ldots, \tilde{x}_n$. So $x_1, \ldots, x_n$ is an optimal primal solution, there being none with a larger objective function value. Similarly, it follows that $y_1, \ldots, y_m$ is optimal for the dual program. To see the second weak duality results notice that if a feasible dual solution exists then the dual objective value provides an immediate upper bound on that of the primal by 1, and so the primal cannot be unbounded. Hence if the primal is unbounded then the dual cannot be feasible. The converse result follows from a similar argument.

Thus a connection between the primal and dual programs and their solutions has been established. However this is not all. The major result is that of *strong duality* which we now state:

> If the primal problem has an optimal solution, then so has the dual and their objective values are equal.

The proof is beyond the scope of this chapter: if you are concerned to see it then consult any standard textbook on the advanced theory of linear programming, for example Vajda (1974).

It can also be shown that if the primal problem is in standard form and has a unique non-degenerate optimal solution, then so has the dual. Moreover the shadow prices of the primal constraints are the optimal dual variable values and vice-versa.

There is a correspondence between shadow prices for general linear programs and optimal dual variable values: the shadow prices of equality or '$\leq$' inequality constraints are the same as the corresponding optimal dual variable values, whereas for '$\geq$' inequality constraints they are *minus* the optimal dual values. For this reason, shadow prices are not usually displayed on output from most linear programming computer codes, but dual variable values are displayed instead.

An important corollary to the above duality results is a property called *complementary slackness*. This says that at an optimal solution to the primal and dual problems (for each primal variable and corresponding dual constraint) either the primal variable is zero or the slack on the dual constraint is zero, or both. In terms of the linear program in standard form and its dual, which we defined earlier, we can express this mathematically as

$$x_j \left(c_j - \sum_{i=1}^{m} a_{ij} y_i \right) = 0$$

whenever $x_1, \ldots, x_n$ and $y_1, \ldots, y_m$ are optimal solutions to the primal and dual programs respectively.

This result is very easy to show, for by strong duality we must have:

$$\sum_{j=1}^{n} c_j x_j = \sum_{i=1}^{n} b_i y_i$$

$$= \sum_{i=1}^{m} \left(\sum_{j=1}^{n} a_{ij} x_j \right) y_i$$

$$= \sum_{j=1}^{n} \left(\sum_{i=1}^{m} a_{ij} y_i \right) x_j$$

So:

$$\sum_{j=1}^{n} \left(c_j - \sum_{i=1}^{m} a_{ij} y_i \right) x_j = 0$$

Since $x_j \geq 0$ for all j and $c_j - \sum_{i=1}^{m} a_{ij} y_i \leq 0$ for all j, each term in the above sum must be non-positive, and so:

$$\left(c_j - \sum_{i=1}^{n} a_{ij} y_i \right) x_j = 0 \quad \text{for each } j = 1, \ldots, n$$

Since the dual of the dual program is the original primal one, it follows that at the optimum of the primal and dual, either the dual variable is zero or the slack on the corresponding primal constraint is zero, or both.

Complementary slackness makes obvious sense from an intuitive point of view. If you think of each primal variable value being the marginal worth of its corresponding dual constraint, then that worth (and hence the primal variable value) will be zero if the constraint is not binding – in other words, has positive slack.

13.3 The transportation algorithm

Many practical applications of linear programming are to problems which involve moving goods or commodities from places of procurement, manufacture or storage, to places of disposal, consumption or further storage. There is often a cost associated with moving a unit of the commodity from one place to another depending (for example) on distance, and considerable freedom in choice over which routes to use.

In most practical applications such commodity movement forms only part of a more comprehensive linear programming business planning model. However, in this section we study such commodity movement or transportation models in their pure form without additional variables or constraints for the modelling of other business functions. There are two reasons for this. Firstly, we will obtain an intuitive grasp of the way in which these transportation models behave. This is important for the construction and development of more realistic practical models. Many other assignment, allocation and network problems have the same structure as the simple transportation model, so the domain of applicability is larger than it may first appear.

Secondly, transportation models tend to involve a very large number of variables and without special solution techniques they may not be soluble at all.

Suppose we wish to plan the transport of fertilizer for a company which has factories at Reading, Birmingham and Manchester and warehouses at Norwich, Plymouth and Hastings. The costs in pounds per ton of moving fertilizer between each factory and each warehouse is given by the table:

	Warehouse	1	2	3
Factory		Norwich	Plymouth	Brighton
1	Reading	20	25	9
2	Birmingham	13	28	22
3	Manchester	24	36	32

There are orders for 100, 70 and 50 tons of fertilizers from Norwich, Plymouth and Brighton respectively and there is available 80, 150 and 60 tons of fertilizer at Reading, Birmingham and Manchester. The problem is to satisfy all the orders from the warehouses from the factories at minimum cost.

You can think of this problem as one of *optimal routing* through the simple network:

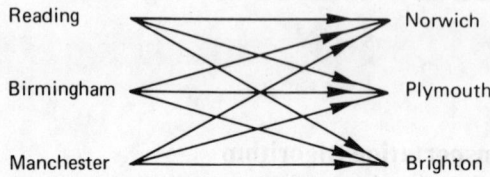

If we number each factory, 1, 2 and 3 and each warehouse 1, 2 and 3 so that factory 2 is at Birmingham and warehouse 3 is at Brighton, then we may let x_{ij} represent the tonnage of fertilizer to be moved from factory i to warehouse j, we may write our problem as the linear program:

Minimize $x_0 =$
$$20x_{11} + 25x_{12} + 9x_{13} + 13x_{21} + 28x_{22} + 22x_{23} + 24x_{31} + 36x_{32} + 32x_{33}$$

subject to

$$
\begin{aligned}
x_{11} + x_{12} + x_{13} &\leq 80 \\
x_{21} + x_{22} + x_{23} &\leq 150 \\
x_{31} + x_{32} + x_{33} &\leq 60 \\
x_{11} + x_{21} + x_{31} &\geq 100 \\
x_{12} + x_{22} + x_{32} &\geq 70 \\
x_{13} + x_{23} + x_{33} &\geq 50
\end{aligned}
$$

and $x_{11}, x_{12}, x_{13}, x_{21}, x_{22}, x_{23}, x_{31}, x_{32}, x_{33} \geq 0$

This problem has a special structure, in common with all transportation problems, which we shall exploit. The structure is perhaps more apparent if we express the fertilizer problem algebraically.

Denote the cost of £/ton of transporting fertilizer between factory i and warehouse j by c_{ij}. We may denote the available supply in tons at factory i by a_i

Sensitivity Analysis, Duality, Transport Algorithm

and the demand in tons from warehouse j by d_j. Then we have the problem:

Minimise
$$x_0 = \sum_{i=1}^{3} \sum_{j=1}^{3} c_{ij} x_{ij}$$

subject to
$$\sum_{j=1}^{3} x_{ij} \leq a_i \quad \text{for} \quad i = 1, 2, 3$$
(we cannot move more than the available supply)

$$\sum_{i=1}^{3} x_{ij} \geq d_j \quad \text{for} \quad j = 1, 2, 3$$
(we must satisfy the demand)

and
$$x_{ij} \geq 0 \quad \text{for} \quad i = 1, 2, 3 \text{ and } j = 1, 2, 3$$

Clearly a possible routing or feasible solution will exist provided demand does not exceed supply, i.e.:

$$\sum_{i=1}^{3} a_i \geq \sum_{j=1}^{3} d_j$$

This is the case in our example above.

It is convenient to ensure that the demand is exactly equal to the available supply. If this is not already the case, it can be achieved by introducing a 'fictitious' or dummy destination (say warehouse 4 in the fertilizer example) with a demand requirement equal to the difference between the total available supply and the total actual demand, to which goods can be sent at zero cost. When there is no excess of supply, all the supply and all the demand constraints must be satisfied with equality. In general, suppose that there are m factories or *sources* of goods and n warehouses or *destinations* for them. Then if c_{ij} is the unit transport cost between source i and destination j, a_i is the available supply at source i and d_j is the demand requirement at destination j, we may write the general transportation problem as the linear program:

Minimise
$$x_0 = \sum_{i=1}^{m} \sum_{j=1}^{n} c_{ij} x_{ij} \qquad \text{(cost)}$$

subject to
$$\sum_{j=1}^{n} x_{ij} = a_i \quad \text{for} \quad i = 1, \ldots, m \qquad \text{(supply)}$$

$$\sum_{i=1}^{m} x_{ij} = d_j \quad \text{for} \quad j = 1, \ldots, n \qquad \text{(demand)}$$

and
$$x_{ij} \geq 0 \quad \text{for} \quad i = 1, \ldots, m \text{ and } j = 1, \ldots, n$$

where x_{ij} is the amount of goods to be transported between source i and destination j.

You can see that this program has mn variables and $m + n$ constraints. In fact, one of the constraints is redundant, for if we miss one out it can be inferred from the remaining $m+n-1$ constraints. To see this, suppose we omit the last demand constraint (the one for destination n). Then from the remaining $m+n-1$ constraints we have that:

$$\sum_{j=1}^{n} d_j = \sum_{i=1}^{n} a_i \qquad \text{(since demand equals supply)}$$

$$= \sum_{i=1}^{m} \sum_{j=1}^{n} x_{ij} \qquad \text{(from the supply constraints)}$$

$$= \sum_{j=1}^{n-1} \left(\sum_{i=1}^{m} x_{ij} \right) + \sum_{i=1}^{m} x_{in}$$

So $\sum_{i=1}^{m} x_{in} = \sum_{j=1}^{n-1} \left(d_j - \sum_{i=1}^{m} x_{ij} \right) + d_n = d_n$

(using the $n-1$ demand constraints)

and this is the last demand constraint which we omitted.

We will use the properties of the dual to the transportation problem to derive an efficient algorithm for its solution. If we let v_i denote the dual variable corresponding to the ith supply constraint and let w_j be that corresponding to the jth demand constraint, then the dual program is:

Maximise $\sum_{i=1}^{m} a_i v_i + \sum_{j=1}^{n} d_j w_j$

subject to

$$v_i + w_j \leq c_{ij} \quad \text{for} \quad i=1,\ldots,m \text{ and } j=1,\ldots,n$$

and v_i, w_j unrestricted in sign

If $\{x_{ij}\}$ is an optimal solution to the primal problem and $\{v_i, w_j\}$ is an optimal solution to the dual, then a set of reduced costs for the primal solution is given by:

$$v_i + w_j - c_{ij}$$

for each non-basic variable x_{ij}.

To see this, recall that the reduced cost of a *particular* non-basic x_{ij} is minus the rate of change of the objective function as it is raised from zero, in other words, $\partial x_0/\partial x_{ij}$. Suppose x_{ij} is increased by a small amount ∂_{ij}. Then the objective, which is the total cost will rise an amount $c_{ij}\delta_{ij}$. But the solution will have become infeasible unless the right-hand side coefficients of the ith availability and jth demand are raised to $a_i + \delta_{ij}$ and $d_j + \delta_{ij}$ respectively. So if x_{ij} is increased from zero to δ_{ij} and if the supply a_i is increased to $a_i + \delta_{ij}$ and if the demand d_j is increased to $d_j + \delta_{ij}$, the total cost increases an amount $c_{ij}\delta_{ij}$. But we are not interested in changing the right-hand side coefficients. So we reduce the ith supply from $a_i + \delta_{ij}$ back to a_i and the jth demand from $d_j + \delta$

back to d_j. This will *reduce* the total cost by

δ_{ij} (shadow price of ith supply constraint
 + shadow price of jth demand constraint)
$= \delta_{ij}(v_i + w_j)$

since the dual variable values are equal to the shadow prices on equality constraints. Thus the objective value will increase an amount:

$$\delta_{ij}(c_{ij} - v_i - w_j)$$

as non-basic variable x_{ij} is increased from zero to δ_{ij}. Since the reduced cost of x_{ij} is minus the rate of change of the objective function with respect to x_{ij}, this is:

$$v_i + w_j - c_{ij}$$

Notice that these reduced costs are not unique; the primal program is degenerate (having one redundant constraint) and thus the optimal basic solution will always have a basic variable at zero level, and the dual solution will have multiple optima. We can calculate the reduced costs knowing only which of the x_{ij}'s are non-zero at the optimum by using the complementary slackness condition. This shows us that:

$$v_i + w_j = c_{ij} \quad \text{whenever } x_{ij} > 0$$

Since there will be $m+n-1$ non-zero x_{ij}'s in any basic solution (one of the basics is zero) this yields $m+n-1$ equations in $m+n$ dual variables and this fact will be exploited later. We have already remarked that the dual solution is not unique, in fact the optimal dual variables are determined up to a constant in that any given constant can be added to each of the v_is and subtracted from each of the w_js without affecting the dual feasibility or the value of the dual objective function – try it and see. So we can set one of the dual variables to zero and use the above $m+n-1$ equations to solve for the remaining $m+n-1$ dual variables and hence calculate reduced costs for the non-basic x_{ij}'s.

For any primal solution (not just the optimal one) there is corresponding dual solution. This will not satisfy the dual constraints unless the primal solution is optimal, but it will satisfy the complementary slackness condition and give the x_0 or objective row coefficients of the x_{ij}'s in the primal simplex tableau. And we can calculate these in the same way as we can the reduced costs at the optimum.

This immediately suggests a solution method, known as the *transportation algorithm*, once an initial basic feasible solution has been found.

STEP 1 Set $v_1 = 0$ and use the $m+n-1$ equations $v_i + w_j = c_{ij}$ whenever $x_{ij} > 0$ to solve for the remaining dual variables. (Recall that at most $m+n-1$ of the x_{ij}'s can be non-zero.

STEP 2 Calculate the reduced costs of the non basic variables, $v_i + w_j - c_{ij}$. If these are all negative stop; the solution is optimal. Otherwise choose the non-basic x_{ij} with largest negative reduced cost to raise from zero and enter the basis.

STEP 3   Find the basic variable to leave the basis and re-calculate the values of the remaining basic variables and calculate the value at which the new variable enters the basis. Go to step 1.

Step 3 above can be done by the 'stepping stone' method which is best explained with reference to an actual example. We shall do this using the fertilizer example below.

As with the standard simplex method, we need some way of obtaining an initial basic feasible solution. Fortunately, for the transportation model this can be done *by inspection*. There are several methods available for doing this – we shall describe one called the *Northwest corner rule* with reference to the fertilizer example.

To begin with, it is convenient to represent the problem in the form of a *transportation table*. Like the simplex tableau this enables the initial problem and the intermediate calculations to be displayed in a compact methodical way. The table has one row for each supply point and one column for each destination (including a dummy one, if necessary). Cells of the table are used to contain the current values of the x_{ij}'s and their cost coefficients (the c_{ij}'s). In addition, we use two extra columns; one contains the availabilities at each supply point (the a_i's) and the other the calculated dual variable values for the supply constraints (the v_i's). Also we have two extra rows, one for the requirements at each destination (the d_j's) and the other for the calculated dual variable values for the demand constraints (the w_j's). For the fertilizer example, the first such table is:

i \ j	1	2	3	4	a_i	v_i
1	20 −θ 80	25 0 10	9 +θ 0 20	0 0 7	80 8̶0̶ 0	0
2	13 +θ 20	28 70	22 −θ 50	0 10	150 1̶5̶0̶ 1̶3̶0̶ 60 0	−7
3	24 0 −11	36 0 −8	32 0 −10	0 60	60 6̶0̶ 0	−7
d_j	100 1̶0̶0̶ 20 0	70 7̶0̶ 0	50 5̶0̶ 0	70 7̶0̶ 0	290	
w_j	20	35	29	9		

Sensitivity Analysis, Duality, Transport Algorithm

The top left-hand corner of each cell contains an item of problem data: c_{ij}, a_i or d_j. Notice that the cell in the supply column on the demand row contains the *total supply*. This is convenient for calculating the dummy demand requirement, which ensures that total supply and total demand are equal. Initially we do not enter any v_i, w_j or x_{ij} values.

To obtain an initial basic feasible solution by the Northwest corner rule, we allocate the largest possible value to x_{11} in the top right-hand cell. This is the smaller of a_1 and d_1, in this case 80. Either the first demand requirement will be satisfied or the first available supply exhausted and we enter the remaining supply and unsatisfied demand in the supply column and demand row. In this case it is the supply that is 'used up'; we enter 20 in the first cell of the demand row, and zero in first cell of the supply column. In the former case, no goods are needed from any other supply point to satisfy the first demand, so zero x_{i1}s can be entered in the remaining cells of the first column. In the latter case, no goods are available from the first supply point to satisfy any other demand so zero x_{1j}s can be entered in the remaining cells of the first row. This is the case in our example. We can now repeat the process using the *remaining* unsatisfied demands and available supplies in the cells where the x_{ij}'s have not been allocated, updating these in the supply row and demand column. We finish with an initial allocation of values to the x_{ij}'s. Notice that there should always be exactly $m+n-1$ of these non-zero, unless the problem is still further degenerate, in which case an initial basic feasible solution cannot be obtained by inspection. In this case one or more of the a_i's and d_j's should be perturbed slightly until the extra degeneracies are removed and exactly $m+n-1$ of the x_{ij}'s are positive. In our example there are 6 non-zero x_{ij}'s: x_{11}, x_{21}, x_{22}, x_{23}, x_{24}, and x_{34}, and these are in our initial feasible basis.

We can now set the first dual variable, v_1, to zero and use the relations

$$v_i + w_j = c_{ij} \text{ whenever } x_{ij} > 0$$

to solve for the remaining $m+n-1$ dual variables. In our example, we use:

$$v_1 + w_1 = 20$$
$$v_2 + w_1 = 13$$
$$v_2 + w_2 = 28$$
$$v_2 + w_3 = 22$$
$$v_2 + w_4 = 0$$

and
$$v_3 + w_4 = 0$$

Simple elimination reveals that these relations are satisfied by:

$$v_1 = 0 \qquad w_1 = 20$$
$$v_2 = -7 \qquad w_2 = 35$$
$$v_3 = -7 \qquad w_3 = 29$$
$$w_4 = 7$$

These can now be entered in the transportation table.

We can now calculate the reduced costs of the non-basic variables, those that are currently zero, by:

$$v_i + w_j - c_{ij}$$

and enter it in the bottom right-hand corner of the cells for the non-basics.

Not all non-basics have negative reduced costs, so that solution is not optimal. We choose that with the largest positive reduced cost to enter the basis. Of course, we can make at most only one non-basic positive at each iteration. In our example, it is x_{13}. Increase it from zero to θ (say). Correspondingly reduce two basic x_{ij}'s: one in the column containing the entering non-basic variable and the other in the row containing the entering non-basic variable. In our example we decrease x_{23} and x_{11} by θ. This preserves the supply and demand balance in the row and column containing the entering non-basic. Now we need to find other basic variables whose values when increased or decreased by θ preserve the balance on the remaining supply and demand constraints. In our example, we can achieve this by increasing x_{21} by θ. We can think of the basics whose value we are adjusting as 'stepping-stones', hence this particular technique is called the *stepping stone method*. The value of θ is now increased until one of the existing basic variables descends to zero and so leaves the basis. In our example this is x_{23}. The maximum value of θ is that at which the non-basic enters the basis and the values of the basic variables that we have used as stepping stones can now be adjusted. It is convenient to write these values afresh into the next transportation table, although if you use a pencil and a sufficiently efficacious eraser all the values can be updated on the same table. For the purpose of exposition we write out the transportation table afresh. Notice that the supply column and demand row are no longer required, and can therefore be omitted.

j \ i	1	2	3	4	v_i

1	20 $-\theta$
	    30
		25 $+\theta$
		        10
			9
			  50
				0
				          7
					0

2	13 $+\theta$
	    70
		28 $-\theta$
		    70
			22
			       -20
				0
				      10
					-7

3	24
	    -11
		36
		    -8
			32
			    -30
				0
				      60
					-7

w_j	20	35	9	7	

Sensitivity Analysis, Duality, Transport Algorithm

We can now repeat the process, re-calculating the values of the dual variables and the reduced costs of the non-basics. There are still those with positive reduced costs; the most positive is now x_{12} which we choose to enter the basis. The stepping stone method reveals that x_{11} leaves and x_{12} enters at a value of 30, yielding the next transportation table:

i \ j	1	2	3	4	v_i
1	20 −10	25 30	9 50	0 −3	0
2	13 100	28 40	22 −10	0 10	3
3	24 −11	36 −8	32 −20	0 60	3
w_j	10	25	9	−3	

Calculation of the dual variable values and hence the reduced costs of the non-basics reveals that this solution is optimal since all the reduced costs are negative. Thus the optimal solution to the fertilizer problem is:

> Transport 30 tons from Reading to Plymouth
> Transport 50 tons from Reading to Brighton
> Transport 100 tons from Birmingham to Norwich
> Transport 40 tons from Birmingham to Plymouth

The total cost is

$$30 \times £25 + 50 \times £9 + 100 \times £13 - 40 \times £28 = £3,620$$

We can use the final transportation table to explore the sensitivity of the solution to changes in the input data, both the unit cost of each route, the c_{ij}s, and to changes in the qualities available for supply, the a_is, and to the demand requirements, the d_js.

Firstly, consider changes to the *unit cost of each route*. We have to treat the case in which the route is basic and non-basic separately. Consider an increase of δ to the cost of the non-basic route x_{ij}. The current solution will remain optimal and the minimum total cost will remain unchanged, provided the reduced cost of that route:

$$v_i + w_j - (c_{ij} + \delta)$$

remains non-positive. In other words, we can reduce the cost by as much as minus the reduced cost of that route before the optimal solution changes. In our example this means that we can reduce the cost of the routes

and
Reading to Norwich by £10/ton
Birmingham to Brighton by £10/ton
Manchester to Norwich by £11/ton
Manchester to Plymouth by £8/ton
Manchester to Brighton by £20/ton

before the optimal solution changes.

Changes to the *costs of basic routes* are a little more complicated, because if we change any one of them, the reduced costs of all the non-basic routes may change. Suppose the cost of the route (1, 2) from Reading to Plymouth were increased by δ £/ton (from 25 to $25 + \delta$). This changes the values of the dual variables. Since our setting of v_1 to zero was simply a device to obtain values of the $m+n$ dual variables from $m+n-1$ equations, it can remain at zero. So the value of w_2 must change from 25 to $25 + \delta$, and so the value of v_2 must change from 3 to $3 - \delta$. This forces w_1 to change from 10 to $10 + \delta$ and w_3 to change from -3 to $-3 + \delta$, and the latter change forces v_3 to change from 3 to $3 - \delta$. These implications are displayed in the table below, as are the reduced costs of the non-basic routes, this time including terms in δ:

i \ j	1	2	3	4	v_i
1	20 $-10+\delta$	25 30	9 50	0 $-3+\delta$	0
2	13 100	28 40	22 $-10-\delta$	0 10	$3-\delta$
3	24 -11	36 -8	32 $-20-\delta$	0 60	$3-\delta$
w_j	$10+\delta$	$25+\delta$	9	$-3+\delta$	

Sensitivity Analysis, Duality, Transport Algorithm

The current solution will remain optimal provided the reduced costs of the non-basic routes remain non-positive. In other words provided:

$$-10 + \delta \leq 0 \quad \text{or} \quad \delta \leq 10$$
$$-3 + \delta \leq 0 \quad \text{or} \quad \delta \leq 3$$
$$-10 - \delta \leq 0 \quad \text{or} \quad \delta \geq -10$$
$$\text{and} \quad -20 - \delta \leq 0 \quad \text{or} \quad \delta \geq -20$$

All these conditions are satisfied provided

$$-10 \leq \delta \leq 3$$

So the current solution remains optimal if c_{12} lies within the range 15 to 28. Of course, this time as c_{12} changes so does the total minimum cost, by an amount £$x_{12}\delta$ or £30δ.

The rate of change of the objective function with respect to changes in supply availabilities or demand requirements is given by the shadow prices on the corresponding rows. We have already seen how these are related to the dual variable values, but we need to take special care here. Firstly notice that such changes will not make any sense unless the dummy destination is dropped. Otherwise they will just make the problem infeasible. Secondly (as we have already observed), the dual variables are not unique: we can add any constant to the v_is provided we subtract it from the w_js. So we need to find the constant which gives the correct shadow prices in terms of the original problem without the dummy destination. Applying complementary slackness to the original formulation, we see that the dual variables corresponding to a '$\leq$' supply constraint must be zero whenever there is an excess of supply from that source, in other words, whenever there is positive slack in the final solution. In our fertilizer example we see that there is an excess of supply at Birmingham ($i = 2$) and Manchester ($i = 3$). By subtracting 3 from the v_is and adding it to the w_js we achieve $v_2 = v_3 = 0$. The other dual variables values become $v_1 = -3, w_1 = 13, w_2 = 28, w_3 = 12, w_4 = 0$. Recalling that the shadow prices equal the dual variable values on the supply constraints and are minus the dual variable values on the demand constraints (they are '$\geq$' inequalities, and we are minimising) we have that the shadow prices of the supply constraint is:

$$\begin{array}{ll} & -3 \text{ for supply from Reading} \\ & -13 \text{ for demand from Norwich} \\ & -78 \text{ for demand from Plymouth} \\ \text{and} & -12 \text{ for demand from Brighton} \end{array}$$

So if an extra ton of fertilizer were required at Norwich, Plymouth or Brighton it would cost £13, £28, or £12, to transport respectively and if an extra ton were available at Reading, it would save £3 on the transport bill. Of course, marginal changes to the supply at Birmingham or Manchester will make no difference to the total cost, since there is already more than is required available at these sources.

13.4 Integer programming

Throughout this chapter and Chapter 12 we have assumed that the decision variables are *continuous* in the sense of being able to take any values between some lower limit (usually zero) and some upper limit (which may be infinite). In practice, of course, this assumption may seem unrealistic. It makes no sense to employ half a labourer or buy half a lorry. In such circumstances it is natural to use *integer* variables (which can take only the values 0, 1, 2, 3, . . .) to model these decision variables. Linear programs in which some of the variables are thus constrained are known as *mixed integer programs* (or MIPs, for short) and a technique known as *integer programming* may be used to solve them. Integer programming is still rather a 'black art'; the optimisation process often requires the modeller's guidance and is not automatic in the manner of linear programming. Moreover, even small models can be very difficult to solve: the solution effort required is more a function of their structure than size, and it may take more than 50 times the computational effort to solve a MIP program than the same model without integer restrictions. A full discussion of integer programming is beyond the scope of this book and the interested reader should refer to Beale (1968) or Williams (1978).

Fortunately, it is rarely necessary to use integer programming to model decision variables such as labourers to hire or lorries to use. It turns out that with a little common sense rounding to the nearest integer variable is usually adequate when the number of units concerned is more than about 20. The most important use of integer programming is for modelling YES/NO decisions, start-up costs and non-linear functions.

13.5 Exercises

Worked answers are provided in Appendix C to question(s) marked with an asterisk.

1.* Mega Manufacturers has three product lines: A, B and C. There are four constraints on its operations, indicated below. Management has set up the following linear programme to find the optimal production plan:

Maximise: $6.0A + 12.0B + 18.0C$
Subject to: $0.75A + 1.5B + 0.37C \leqslant 1{,}500$ (skilled labour hours)
$3.0A + 0.3B + 1.5C \leqslant 8{,}500$ (machine hours)
$4.5A + 3.0B + 1.1C \leqslant 7{,}500$ (despatch handling units)
$\quad\quad\quad + 0.25B + 2.0C \leqslant 4{,}000$ (cubic foot storage capacity)

(Row 1 is the objective row.)
The solution to this problem is given as:
Objective function value: 42,080.0

Sensitivity Analysis, Duality, Transport Algorithm

Variable	Value	Reduced cost
A	1,013.33	0.00
B	0.00	1.88
C	2,000.00	0.00

Row	Slack or surplus	Dual prices
2	0.00	8.00
3	2,460.00	0.00
4	740.00	0.00
5	0.00	7.52

Ranges in which the basis is unchanged:

Variable	Current coefficient	Allowable increase	Allowable decrease
A	6.00	30.49	0.97
B	12.00	1.88	Infinity
C	18.00	Infinity	15.04

Right-hand-side ranges (RHS):

Row	Current RHS	Allowable increase	Allowable decrease
2	1,500.00	123.33	760.00
3	8,500.00	Infinity	2,460.00
4	7,500.00	Infinity	740.00
5	4,000.00	4,108.11	1,321.43

You are required, with suitable explanation, to state:

(a) the optimal production plan, the contribution margin it is expected to earn and the amounts of any unused resources;

(b) the effect on the contribution margin if Mega produces one unit of B;

(c) the effect on the contribution margin if Mega can obtain one more

hour of skilled labour time at a premium of £1.00 per hour over existing rates;
(d) whether the company should acquire further storage space that can be obtained at a cost of £5.00 per unit;
(e) the effect on the contribution margin if Mega, due to site renovation, loses 1,000 units of its own storage capacity;
(f) what increase in despatch handling facilities there can be before the solution changes;
(g) how far despatch handling facilities may contract before the solution changes;
(h) the allowable range for product A's contribution margin such that the solution will not change and give conclusions about product A and its costing that you would draw from this;
(j) whether more than one question at a time can be investigated with respect to the initial solution; for example, the combined effect of changes in both skilled labour and machine hours.

(*Institute of Cost and Management Accountants*, November 1985)

2. A large conurbation in the North of Britain stockpiles salt and sand in five locations for spreading on roads during winter frosts and snowstorms. When frost or snow is forecast by the meteorological office all salt and sand is distributed from shelters in these five locations to five different zones in the city.

The weather forecasts are not always reliable and often give only short notice of an impending frost. It has therefore been decided that the salt and sand should be transported from the shelters to the centres of the five zones in the shortest time possible. It is thought that the most effective way of doing this is to minimise the overall distance travelled by lorry loads from their sources to the five destinations. The table below summarises the distance (in miles) between the sand and salt shelters and the centre of each zone. It also shows the number of lorry loads typically required to salt and sand all the streets in each zone and the number of lorry loads which can be provided from each shelter without restocking. Some of the shelters are unable to supply some of the zones because of geographical barriers (e.g., one-way streets) or because it would be impractical.

Shelter	1	2	Zone 3	4	5	Lorry loads available
1	1.2	—	2.2	—	—	40
2	3.8	4.5	5.2	2.3	3.2	50
3	—	6.4	5.0	4.2	1.8	30
4	2.9	3.1	5.2	4.1	3.8	60
5	1.5	—	4.0	6.1	2.2	20
Lorry loads required	30	40	60	25	25	

Sensitivity Analysis, Duality, Transport Algorithm

Required:

(a) Solve the problem by the transportation method to find the optimal allocation of lorry loads from the shelters to the zones.

(b) If the operating cost per mile for each lorry is £0.50 find the minimum transport cost for distributing salt and sand to the zones.

(c) In your optimal allocation which of the shelters will have excess salt and sand and how many lorry loads in excess will they have?

(*Chartered Institute of Public Finance and Accounting*, November 1984)

3. The Kaolene Co. Ltd has six different products all made from fabricated steel. Each product passes through a combination of five production operations: cutting, forming, drilling, welding and coating.

   Steel is cut to the length required, formed into the appropriate shapes, drilled if necessary, welded together if the product is made up of more than one part, and then passed through the coating machine. Each operation is separate and independent, except for the cutting and forming operations, when, if needed, forming follows continuously after cutting. Some products do not require every production operation.

   The output rates from each production operation, based on a standard measure for each product, are set out in the tableau below, along with the total hours of work available for each operation. The contribution per unit of each product is also given. It is estimated that three of the products have sales ceilings and these are also given below:

Products	X_1	X_2	X_3	X_4	X_5	X_6
Contribution/unit (£)	5.7	10.1	12.3	9.8	17.2	14.0
Output rates/hour						
Cutting	650	700	370	450	300	420
Forming	450	450	—	520	180	380
Drilling	—	200	380	—	300	—
Welding	—	—	380	670	440	720
Coating	500	—	540	480	600	450
Maximum sales units (000)	—	—	150	—	20	70

	Cutting	Forming	Drilling	Welding	Coating
Production hours available	12,000	16,000	4,000	4,000	16,000

The production and sales for the year were found using a linear programming algorithm. The final tableau is given below:

	X_1	X_2	X_3	X_4	X_5	X_6	X_7	X_8	X_9	X_{10}	X_{11}	X_{12}	X_{13}	X_{14}	Variable in basic solution	Value of variable in basic solution
	1	0	−1.6	−.22	−.99	0	10.8	0	3.0	(18.5)	0	0	0	0	X_1	(43,287.0) units
	0	0	−.15	−.02	.12	0	−1.4	1	−3.0	.58	0	0	0	0	X_8	(870.3) hours
	0	1	.53	0	.67	0	0	0	3.33	0	0	0	0	0	X_2	13,333.3 units
	0	0	1.9	1.08	1.64	1	0	0	0	12	0	0	0	0	X_6	48,019.2 units
	0	0	.06	.01	0	0	−1.3	0	.37	.63	0	0	0	0	X_{11}	4,404.6 hours
	0	0	1	0	0	0	0	0	0	0	1	0	0	0	X_{11}	150,000.0 units
	0	0	0	0	0	0	0	0	0	0	0	1	0	0	X_{12}	20,000.0 units
	0	0	−1.9	−1.0	−1.6	0	0	0	0	−12	0	0	1	0	X_{13}	
	0	0	(10.0)	4	6.83	0	0	0	0	−12	0	0	0	1	X_{14}	21,980.8 units
	0	0	10.0	4	6.83	0	61.7	0	(16.0)	62.1	0	0	0	0	$(Z_i - C_i)$	(£1,053,671.4)

Sensitivity Analysis, Duality, Transport Algorithm 295

Variables X_7 to X_{11} are the slack variables relating to the production constraints, expressed in the order of production. Variables X_{12} to X_{14} are the slack variables relating to the sales ceilings of X_3, X_5 and X_6 respectively.

After analysis of the above results, the production manager believes that further mechanical work on the cutting and forming machines costing £200 can improve their hourly output rates as follows:

	X_1	X_2	X_3	X_4	X_5	X_6
Cutting	700	770	410	500	330	470
Forming	540	540	—	620	220	460

The optimal solution to the new situation indicates the shadow prices of the cutting, drilling and welding sections to be £59.3, £14.2 and £75.1 per hour respectively.

Requirements

(a) Explain the meaning of the six items ringed in the final tableau.
(b) Show the range of values within which the following variables or resources can change without changing the optimal mix indicated in the final tableau:
   (i) c_4 – contribution of X_4.
   (ii) b_5 – available coating time.
(c) Formulate the revised linear programming problem taking note of the revised output rates for cutting and forming.
(d) Determine whether the changes in the cutting and forming rates will increase profitability.
(e) Using the above information discuss the usefulness of linear programming to managers in solving this type of problem.

(*Institute of Chartered Accountants, England and Wales*, December 1985)

4. Decibels plc manufactures classical and electric guitars. It operates a 48-week year, divided into 4-week periods.
   The models, demand and selling prices are:

Model	Expected demand per period (units)	Selling price (£)
Folk	200	110
Junior Electric	150	90
Mastersound	130	130
Mightysound	50	240

Each of two department stores has contracted to purchase 50 units per period of the Folk and Junior Electric guitars.

Decibels plc has three production departments, each with a largely skilled workforce, working 40 hours a week, with productive time amounting to 75% of total attendance time. The guitars are tested in the company laboratories, where staff work a 35-hour week, also with effective time being 75% of attendance time.

Details relating to the production of guitars are:

Department	No. of operators	Folk (hr)	Junior Electric (hr)	Master-sound (hr)	Mighty-sound (hr)
Soundbox	10	3.0	1.5	2.0	3.5
Neck and Fingerboard	10	2.0	2.0	3.5	4.0
Assemble	15	3.5	3.0	4.5	6.0
Laboratory test	3	0.5	0.5	1.0	1.5

Fixed costs are running at £14,000 per month and the contribution margins are £40, £30, £25 and £80 per unit respectively for Folk, Junior Electric, Mastersound and Mightysound.

Decibels plc. produced a linear programme to estimate monthly income from the above data. At *Appendix A* to this question is the final tableau resulting from running the programme. The slack variables are arranged in the following order: 2 to 5 relate to production and laboratory test department operator constraints; 6 and 7 represent the stores' contracts and 8 to 11 represent the estimated demands per month. Row 1 is the objective row.

You are required, from the information given, to:

(a) formulate the initial equations of the linear programme;
(b) interpret the results if the linear programme, including the dual (shadow) prices;
(c) state and explain whether it would be worthwhile to employ another member of staff in the test laboratory or permit overtime working of up to 8 hours per man per week. (Overtime is paid at $1\frac{1}{2}$ times normal rates, the latter being £3.50 per hour.)

QUESTION 4: APPENDIX A LINEAR PROGRAMME FINAL TABLEAU

Row	Folk	Junior Electric	Master-sound	Mighty-sound	2	3	4	5	Slack variables 6	7	8	9	10	11	by
1	0.0	0.0	0.0	0.0	0.0	0.0	0.0	25.0	0.0	0.0	27.5	17.5	0.0	42.5	18,125
2	0.0	0.0	0.0	0.0	1.0	0.0	0.0	−2.0	0.0	0.0	−2.0	−0.5	0.0	−0.5	70.0
3	0.0	0.0	0.0	0.0	0.0	1.0	0.0	−3.5	0.0	0.0	−0.25	−0.25	0.0	1.25	72.5
4	0.0	0.0	0.0	0.0	0.0	0.0	1.0	−4.5	0.0	0.0	−1.25	−0.75	0.0	0.75	57.5
5	0.0	0.0	1.0	0.0	0.0	0.0	0.0	1.0	0.0	0.0	−0.5	−0.5	0.0	−1.5	65.0
6	0.0	0.0	0.0	0.0	0.0	0.0	0.0	0.0	1.0	0.0	1.0	0.0	0.0	0.0	150.00
7	0.0	0.0	0.0	0.0	0.0	0.0	0.0	0.0	0.0	1.0	0.0	1.0	0.0	0.0	100.00
8	1.0	0.0	0.0	0.0	0.0	0.0	0.0	0.0	0.0	0.0	1.0	0.0	0.0	0.0	200.00
9	0.0	1.0	0.0	0.0	0.0	0.0	0.0	0.0	0.0	0.0	0.0	1.0	0.0	0.0	150.00
10	0.0	0.0	0.0	0.0	0.0	0.0	0.0	−1.0	0.0	0.0	0.5	0.5	1.0	1.5	65.00
11	0.0	0.0	0.0	1.0	0.0	0.0	0.0	0.0	0.0	0.0	0.0	0.0	0.0	1.0	50.00

(*Institute of Cost and Management Accountants, May 1985*)

Appendix A
Tables

TABLE A.1 LOGARITHMS

	0	1	2	3	4	5	6	7	8	9	1 2 3 4 5 6 7 8 9
10	0000	0043	0086	0128	0170						4 9 13 17 21 26 30 34 38
						0212	0253	0294	0334	0374	4 8 12 16 20 24 28 32 36
11	0414	0453	0492	0531	0569						4 8 12 15 19 23 27 31 35
						0607	0645	0682	0719	0755	4 7 11 15 19 22 26 30 33
12	0792	0828	0864	0899	0934						3 7 11 14 18 21 25 28 32
						0969	1004	1038	1072	1106	3 7 10 14 17 20 24 27 31
13	1139	1173	1206	1239	1271						3 7 10 13 16 20 23 26 30
						1303	1335	1367	1399	1430	3 7 10 13 16 19 22 25 29
14	1461	1492	1523	1553	1584						3 6 9 12 15 19 22 25 28
						1614	1644	1673	1703	1732	3 6 9 12 15 17 20 23 26
15	1761	1790	1818	1847	1875						3 6 9 11 14 17 20 23 26
						1903	1931	1959	1987	2014	3 6 8 11 14 17 19 22 25
16	2041	2068	2095	2122	2148						3 5 8 11 14 16 19 22 24
						2175	2201	2227	2253	2279	3 5 8 10 13 16 18 21 23
17	2304	2330	2355	2380	2405						3 5 8 10 13 15 18 20 23
						2430	2455	2480	2504	2529	2 5 7 10 12 15 17 20 22
18	2553	2577	2601	2625	2648						2 5 7 9 12 14 16 19 21
						2672	2695	2718	2742	2765	2 5 7 9 11 14 16 18 21
19	2788	2810	2833	2856	2878						2 4 7 9 11 13 16 18 20
						2900	2923	2945	2967	2989	2 4 6 8 11 13 15 17 19
20	3010	3032	3054	3075	3096	3118	3139	3160	3181	3210	2 4 6 8 11 13 15 17 19
21	3222	3243	3263	3284	3304	3324	3345	3365	3385	3404	2 4 6 8 10 12 14 16 18
22	3424	3444	3464	3483	3502	3522	3541	3560	3579	3598	2 4 6 8 10 12 14 15 17
23	3617	3636	3655	3674	3692	3711	3729	3747	3766	3784	2 4 6 7 9 11 13 15 17
24	3802	3820	3838	3856	3874	3892	3909	3927	3945	3962	2 4 5 7 9 11 12 14 16
25	3979	3997	4014	4031	4048	4065	4082	4099	4116	4133	2 3 5 7 9 10 12 14 15
26	4150	4166	4183	4200	4216	4232	4249	4265	4281	4298	2 3 5 7 8 10 11 13 15
27	4314	4330	4346	4362	4378	4393	4409	4425	4440	4456	2 3 5 6 8 9 11 13 14
28	4472	4487	4502	4518	4533	4548	4564	4579	4594	4609	2 3 5 6 8 9 11 12 14
29	4624	4639	4654	4669	4683	4698	4713	4728	4742	4757	1 3 4 6 7 9 10 12 13
30	4771	4786	4800	4814	4829	4843	4857	4871	4886	4900	1 3 4 6 7 9 10 11 13
31	4914	4928	4942	4955	4969	4983	4997	5011	5024	5038	1 3 4 6 7 8 10 11 12
32	5051	5065	5079	5092	5105	5119	5132	5145	5159	5172	1 3 4 5 7 8 9 11 12
33	5185	5198	5211	5224	5237	5250	5263	5276	5289	5302	1 3 4 5 6 8 9 10 12
34	5315	5328	5340	5353	5366	5378	5391	5403	5416	5428	1 3 4 5 6 8 9 10 11
35	5441	5453	5465	5478	5490	5502	5514	5527	5539	5551	1 2 4 5 6 7 9 10 11
36	5563	5575	5587	5599	5611	5623	5635	5647	5658	5670	1 2 4 5 6 7 8 10 11
37	5682	5694	5705	5717	5729	5740	5752	5763	5775	5786	1 2 3 5 6 7 8 9 10
38	5798	5809	5821	5832	5843	5855	5866	5877	5888	5899	1 2 3 5 6 7 8 9 10
39	5911	5922	5933	5944	5955	5966	5977	5988	5999	6010	1 2 3 4 5 7 8 9 10
40	6021	6031	6042	6053	6064	6075	6085	6096	6107	6117	1 2 3 4 5 6 8 9 10
41	6128	6138	6149	6160	6170	6180	6191	6201	6212	6222	1 2 3 4 5 6 7 8 9
42	6232	6243	6253	6263	6274	6284	6294	6304	6314	6325	1 2 3 4 5 6 7 8 9
43	6335	6345	6355	6365	6375	6385	6395	6405	6415	6425	1 2 3 4 5 6 7 8 9
44	6435	6444	6454	6464	6474	6484	6493	6503	6513	6522	1 2 3 4 5 6 7 8 9
45	6532	6542	6551	6561	6571	6580	6590	6599	6609	6618	1 2 3 4 5 6 7 8 9
46	6628	6637	6646	6656	6665	6675	6684	6693	6702	6712	1 2 3 4 5 6 7 7 8
47	6721	6730	6739	6749	6758	6767	6776	6785	6794	6803	1 2 3 4 5 5 6 7 8
48	6812	6821	6830	6839	6848	6857	6866	6875	6884	6893	1 2 3 4 4 5 6 7 8
49	6902	6911	6920	6928	6937	6946	6955	6964	6972	6981	1 2 3 4 4 5 6 7 8

TABLE A.1 LOGARITHMS

	0	1	2	3	4	5	6	7	8	9	1 2 3	4 5 6	7 8 9
50	6990	6998	7007	7016	7024	7033	7042	7050	7059	7067	1 2 3	2 4 5	6 7 8
51	7076	7084	7093	7101	7110	7118	7126	7135	7143	7152	1 2 3	3 4 5	6 7 8
52	7160	7168	7177	7185	7193	7202	7210	7218	7226	7235	1 2 2	3 4 5	6 7 7
53	7243	7251	7259	7267	7275	7284	7292	7300	7308	7316	1 2 2	3 4 5	6 6 7
54	7324	7332	7340	7348	7356	7364	7372	7380	7388	7396	1 2 2	3 4 5	6 6 7
55	7404	7412	7419	7427	7435	7443	7451	7459	7466	7474	1 2 2	3 4 5	5 6 7
56	7482	7490	7497	7505	7513	7520	7528	7536	7543	7551	1 2 2	3 4 5	5 6 7
57	7559	7566	7574	7582	7589	7597	7604	7612	7619	7627	1 2 2	3 4 5	5 6 7
58	7634	7642	7649	7657	7664	7672	7679	7686	7694	7701	1 1 2	3 4 4	5 6 7
59	7709	7716	7723	7731	7738	7745	7752	7760	7767	7774	1 1 2	3 4 4	5 6 7
60	7782	7789	7796	7803	7810	7818	7825	7832	7839	7846	1 1 2	3 4 4	5 6 6
61	7853	7860	7868	7875	7882	7889	7896	7903	7910	7917	1 1 2	3 4 4	5 6 6
62	7924	7931	7938	7945	7952	7959	7966	7973	7980	7987	1 1 2	3 3 4	5 6 6
63	7993	8000	8007	8014	8021	8028	8035	8041	8048	8055	1 1 2	3 3 4	5 5 6
64	8062	8069	8075	8082	8089	8096	8102	8109	8116	8122	1 1 2	3 3 4	5 5 6
65	8129	8136	8142	8149	8156	8162	8169	8176	8182	8189	1 1 2	3 3 4	5 5 6
66	8195	8202	8209	8215	8222	8228	8235	8241	8248	8254	1 1 2	3 3 4	5 5 6
67	8261	8267	8274	8280	8287	8293	8299	8306	8312	8319	1 1 2	3 3 4	5 5 6
68	8325	8331	8338	8344	8351	8357	8363	8370	8376	8382	1 1 2	3 3 4	4 5 6
69	8388	8395	8401	8407	8414	8420	8426	8432	8439	8445	1 1 2	2 3 4	4 5 6
70	8451	8457	8463	8470	8476	8482	8488	8494	8500	8506	1 1 2	2 3 4	4 5 6
71	8513	8519	8525	8531	8537	8543	8549	8555	8561	8567	1 1 2	2 3 4	4 5 5
72	8573	8579	8585	8591	8597	8603	8609	8615	8621	8627	1 1 2	2 3 4	4 5 5
73	8633	8639	8645	8651	8657	8663	8669	8675	8681	8686	1 1 2	2 3 4	4 5 5
74	8692	8698	8704	8710	8716	8722	8727	8733	8739	8745	1 1 2	2 3 4	4 5 5
75	8751	8756	8762	8768	8774	8779	8785	8791	8797	8802	1 1 2	2 3 3	4 5 5
76	8808	8814	8820	8825	8831	8837	8842	8848	8854	8859	1 1 2	2 3 3	4 5 5
77	8865	8871	8876	8882	8887	8893	8899	8904	8910	8915	1 1 2	2 3 3	4 4 5
78	8921	8927	8932	8938	8943	8949	8954	8960	8965	8971	1 1 2	2 3 3	4 4 5
79	8976	8982	8987	8993	8998	9004	9009	9015	9020	9025	1 1 2	2 3 3	4 4 5
80	9031	9036	9042	9047	9053	9058	9063	9069	9074	9079	1 1 2	2 3 3	4 4 5
81	9085	9090	9096	9101	9106	9112	9117	9122	9128	9133	1 1 2	2 3 3	4 4 5
82	9138	9143	9149	9154	9159	9165	9170	9175	9180	9186	1 1 2	2 3 3	4 4 5
83	9191	9196	9201	9206	9212	9217	9222	9227	9232	9238	1 1 2	2 3 3	4 4 5
84	9243	9248	9253	9258	9263	9269	9274	9279	9284	9289	1 1 2	2 3 3	4 4 5
85	9294	9299	9304	9309	9315	9320	9325	9330	9335	9340	1 1 2	2 3 3	4 4 5
86	9345	9350	9355	9360	9365	9370	9375	9380	9385	9390	1 1 2	2 3 3	4 4 5
87	9395	9400	9405	9410	9415	9420	9425	9430	9435	9440	0 1 1	2 2 3	3 4 4
88	9445	9450	9455	9460	9465	9469	9474	9479	9484	9489	0 1 1	2 2 3	3 4 4
89	9494	9499	9504	9509	9513	9518	9523	9528	9533	9538	0 1 1	2 2 3	3 4 4
90	9542	9547	9552	9557	9562	9566	9571	9576	9581	9586	0 1 1	2 2 3	3 4 4
91	9590	9595	9600	9605	9609	9614	9619	9624	9628	9633	0 1 1	2 2 3	3 4 4
92	9638	9643	9647	9652	9657	9661	9666	9671	9675	9680	0 1 1	2 2 3	3 4 4
93	9685	9689	9694	9699	9703	9708	9713	9717	9722	9727	0 1 1	2 2 3	3 4 4
94	9731	9736	9741	9745	9750	9754	9759	9763	9768	9773	0 1 1	2 2 3	3 4 4
95	9777	9782	9786	9791	9795	9800	9805	9809	9814	9818	0 1 1	2 2 3	3 4 4
96	9823	9827	9832	9836	9841	9845	9850	9854	9859	9863	0 1 1	2 2 3	3 4 4
97	9868	9872	9877	9881	9886	9890	9894	9899	9903	9908	0 1 1	2 2 3	3 4 4
98	9912	9917	9921	9926	9930	9934	9939	9943	9948	9952	0 1 1	2 2 3	3 4 4
99	9956	9961	9965	9969	9974	9978	9983	9987	9991	9996	0 1 1	2 2 3	3 3 4

TABLE A.2 PRESENT VALUE TABLES

Years hence	1%	2%	4%	6%	8%	10%	12%	14%	15%	16%	18%	20%	22%	24%	25%	26%	28%	30%	35%	40%	45%	50%
1	0.990	0.980	0.962	0.943	0.926	0.909	0.893	0.877	0.870	0.862	0.847	0.833	0.820	0.806	0.800	0.794	0.781	0.769	0.741	0.714	0.690	0.667
2	0.980	0.961	0.925	0.890	0.857	0.826	0.797	0.769	0.756	0.743	0.718	0.694	0.672	0.650	0.640	0.630	0.610	0.592	0.549	0.510	0.476	0.444
3	0.971	0.942	0.889	0.840	0.794	0.751	0.712	0.675	0.658	0.641	0.609	0.579	0.551	0.524	0.512	0.500	0.477	0.455	0.406	0.364	0.328	0.296
4	0.961	0.924	0.855	0.792	0.735	0.683	0.636	0.592	0.572	0.552	0.516	0.482	0.451	0.423	0.410	0.397	0.373	0.350	0.301	0.260	0.226	0.198
5	0.951	0.906	0.822	0.747	0.681	0.621	0.567	0.519	0.497	0.476	0.437	0.402	0.370	0.341	0.328	0.315	0.291	0.269	0.223	0.186	0.156	0.132
6	0.942	0.888	0.790	0.705	0.630	0.564	0.507	0.456	0.432	0.410	0.370	0.335	0.303	0.275	0.262	0.250	0.227	0.207	0.165	0.133	0.108	0.088
7	0.933	0.871	0.760	0.665	0.583	0.513	0.452	0.400	0.376	0.354	0.314	0.279	0.249	0.222	0.210	0.198	0.178	0.159	0.122	0.095	0.074	0.059
8	0.923	0.853	0.731	0.627	0.540	0.467	0.404	0.351	0.327	0.305	0.266	0.233	0.204	0.179	0.168	0.157	0.139	0.123	0.091	0.068	0.051	0.039
9	0.914	0.837	0.703	0.592	0.500	0.424	0.361	0.308	0.284	0.263	0.225	0.194	0.167	0.144	0.134	0.125	0.108	0.094	0.067	0.048	0.035	0.026
10	0.905	0.820	0.676	0.558	0.463	0.386	0.322	0.270	0.247	0.227	0.191	0.162	0.137	0.116	0.107	0.099	0.085	0.073	0.050	0.035	0.024	0.017
11	0.896	0.804	0.650	0.527	0.429	0.350	0.287	0.237	0.215	0.195	0.162	0.135	0.112	0.094	0.086	0.079	0.066	0.056	0.037	0.025	0.017	0.012
12	0.887	0.788	0.625	0.497	0.397	0.319	0.257	0.208	0.187	0.168	0.137	0.112	0.092	0.076	0.069	0.062	0.052	0.043	0.027	0.018	0.012	0.008
13	0.879	0.773	0.601	0.469	0.368	0.290	0.229	0.182	0.163	0.145	0.116	0.093	0.075	0.061	0.055	0.050	0.040	0.033	0.020	0.013	0.008	0.005
14	0.870	0.758	0.577	0.442	0.340	0.263	0.205	0.160	0.141	0.125	0.099	0.078	0.062	0.049	0.044	0.039	0.032	0.025	0.015	0.009	0.006	0.003
15	0.861	0.743	0.555	0.417	0.315	0.239	0.183	0.140	0.123	0.108	0.084	0.065	0.051	0.040	0.035	0.031	0.025	0.020	0.011	0.006	0.004	0.002
16	0.853	0.728	0.534	0.394	0.292	0.218	0.163	0.123	0.107	0.093	0.071	0.054	0.042	0.032	0.028	0.025	0.019	0.015	0.008	0.005	0.003	0.002
17	0.844	0.714	0.513	0.371	0.270	0.198	0.146	0.108	0.093	0.080	0.060	0.045	0.034	0.026	0.023	0.020	0.015	0.012	0.006	0.003	0.002	0.001
18	0.836	0.700	0.494	0.350	0.250	0.180	0.130	0.095	0.081	0.069	0.051	0.038	0.028	0.021	0.018	0.016	0.012	0.009	0.005	0.002	0.001	0.001
19	0.828	0.686	0.475	0.331	0.232	0.164	0.116	0.083	0.070	0.060	0.043	0.031	0.023	0.017	0.014	0.012	0.009	0.007	0.003	0.002	0.001	
20	0.820	0.673	0.456	0.312	0.215	0.149	0.104	0.073	0.061	0.051	0.037	0.026	0.019	0.014	0.012	0.010	0.007	0.005	0.002	0.001	0.001	
21	0.811	0.660	0.439	0.294	0.199	0.135	0.093	0.064	0.053	0.044	0.031	0.022	0.015	0.011	0.009	0.008	0.006	0.004	0.002	0.001		
22	0.803	0.647	0.422	0.278	0.184	0.123	0.083	0.056	0.046	0.038	0.026	0.018	0.013	0.009	0.007	0.006	0.004	0.003	0.001	0.001		
23	0.795	0.634	0.406	0.262	0.170	0.112	0.074	0.049	0.040	0.033	0.022	0.015	0.010	0.007	0.006	0.005	0.003	0.002	0.001			
24	0.788	0.622	0.390	0.247	0.158	0.102	0.066	0.043	0.035	0.028	0.019	0.013	0.008	0.006	0.005	0.004	0.003	0.002	0.001			
25	0.780	0.610	0.375	0.233	0.146	0.092	0.059	0.038	0.030	0.024	0.016	0.010	0.007	0.005	0.004	0.003	0.002	0.001	0.001			
26	0.772	0.598	0.361	0.220	0.135	0.084	0.053	0.033	0.026	0.021	0.014	0.009	0.006	0.004	0.003	0.002	0.002	0.001				
27	0.764	0.586	0.347	0.207	0.125	0.076	0.047	0.029	0.023	0.018	0.011	0.007	0.005	0.003	0.002	0.002	0.001	0.001				
28	0.757	0.574	0.333	0.196	0.116	0.069	0.042	0.026	0.020	0.016	0.010	0.006	0.004	0.002	0.002	0.002	0.001	0.001				
29	0.749	0.563	0.321	0.185	0.107	0.063	0.037	0.022	0.017	0.014	0.008	0.005	0.003	0.002	0.002	0.001	0.001	0.001				
30	0.742	0.552	0.308	0.174	0.099	0.057	0.033	0.020	0.015	0.012	0.007	0.004	0.003	0.002	0.001	0.001	0.001	0.001				
40	0.672	0.453	0.208	0.097	0.046	0.022	0.011	0.005	0.004	0.003	0.001	0.001										
50	0.608	0.372	0.141	0.054	0.021	0.009	0.003	0.001	0.001	0.001												

TABLE A.3 ANNUITY TABLES
Cumulative present value of £1 received annually for n years

Years (N)	1%	2%	4%	6%	8%	10%	12%	14%	15%	16%	18%	20%	22%	24%	25%	26%	28%	30%	35%	40%	45%	50%
1	0.990	0.980	0.962	0.943	0.926	0.909	0.893	0.877	0.870	0.862	0.847	0.833	0.820	0.806	0.800	0.794	0.781	0.769	0.741	0.714	0.690	0.667
2	1.970	1.942	1.886	1.833	1.783	1.736	1.690	1.647	1.626	1.605	1.566	1.528	1.492	1.457	1.440	1.424	1.392	1.361	1.289	1.224	1.165	1.111
3	2.941	2.884	2.775	2.673	2.577	2.487	2.402	2.322	2.283	2.246	2.174	2.106	2.042	1.981	1.952	1.923	1.868	1.816	1.696	1.589	1.493	1.407
4	3.902	3.808	3.630	3.465	3.312	3.170	3.037	2.914	2.855	2.798	2.690	2.589	2.494	2.404	2.362	2.320	2.241	2.166	1.997	1.849	1.720	1.605
5	4.853	4.713	4.452	4.212	3.993	3.791	3.605	3.433	3.352	3.274	3.127	2.991	2.864	2.745	2.689	2.635	2.532	2.436	2.220	2.035	1.876	1.737
6	5.795	5.601	5.242	4.917	4.623	4.355	4.111	3.889	3.784	3.685	3.498	3.326	3.167	3.020	2.951	2.885	2.759	2.643	2.385	2.168	1.983	1.824
7	6.728	6.472	6.002	5.582	5.206	4.868	4.564	4.288	4.160	4.039	3.812	3.605	3.416	3.242	3.161	3.083	2.937	2.802	2.508	2.263	2.057	1.883
8	7.652	7.325	6.733	6.210	5.747	5.335	4.968	4.639	4.487	4.344	4.078	3.837	3.619	3.421	3.329	3.241	3.076	2.925	2.598	2.331	2.108	1.922
9	8.566	8.162	7.435	6.802	6.247	5.759	5.328	4.946	4.772	4.607	4.303	4.031	3.786	3.566	3.463	3.366	3.184	3.019	2.665	2.379	2.144	1.948
10	9.471	8.983	8.111	7.360	6.710	6.145	5.650	5.216	5.019	4.833	4.494	4.192	3.923	3.682	3.571	3.465	3.269	3.092	2.715	2.414	2.168	1.965
11	10.368	9.787	8.760	7.887	7.139	6.495	5.937	5.453	5.234	5.029	4.656	4.327	4.035	3.776	3.656	3.544	3.335	3.147	2.752	2.438	2.185	1.977
12	11.255	10.575	9.385	8.384	7.536	6.814	6.194	5.660	5.421	5.197	4.793	4.439	4.127	3.851	3.725	3.606	3.387	3.190	2.779	2.456	2.196	1.985
13	12.134	11.343	9.986	8.853	7.904	7.103	6.424	5.842	5.583	5.342	4.910	4.533	4.203	3.912	3.780	3.656	3.427	3.223	2.799	2.468	2.204	1.990
14	13.004	12.106	10.563	9.295	8.244	7.367	6.628	6.002	5.724	5.468	5.008	4.611	4.265	3.962	3.824	3.695	3.459	3.249	2.814	2.477	2.210	1.993
15	13.865	12.849	11.118	9.712	8.559	7.606	6.811	6.142	5.847	5.575	5.092	4.675	4.315	4.001	3.859	3.726	3.483	3.268	2.825	2.484	2.214	1.995
16	14.718	13.578	11.652	10.106	8.851	7.824	6.974	6.265	5.954	5.669	5.162	4.730	4.357	4.033	3.887	3.751	3.503	3.283	2.834	2.489	2.216	1.997
17	15.562	14.292	12.166	10.477	9.122	8.022	7.120	6.373	6.047	5.749	5.222	4.775	4.391	4.059	3.910	3.771	3.518	3.295	2.840	2.492	2.218	1.998
18	16.398	14.992	12.659	10.828	9.372	8.201	7.250	6.467	6.128	5.818	5.273	4.812	4.419	4.080	3.928	3.786	3.529	3.304	2.844	2.494	2.219	1.999
19	17.226	15.678	13.134	11.158	9.604	8.365	7.366	6.550	6.198	5.877	5.316	4.844	4.442	4.097	3.942	3.799	3.539	3.311	2.848	2.496	2.220	1.999
20	18.046	16.351	13.590	11.470	9.818	8.514	7.469	6.623	6.259	5.929	5.353	4.870	4.460	4.110	3.954	3.808	3.546	3.316	2.850	2.497	2.221	1.999
21	18.857	17.011	14.029	11.764	10.017	8.649	7.562	6.687	6.312	5.973	5.384	4.891	4.476	4.121	3.963	3.816	3.551	3.320	2.852	2.498	2.221	2.000
22	19.660	17.658	14.451	12.042	10.201	8.772	7.645	6.743	6.359	6.011	5.410	4.909	4.488	4.130	3.970	3.822	3.556	3.323	2.853	2.498	2.222	2.000
23	20.456	18.292	14.857	12.303	10.371	8.883	7.718	6.792	6.399	6.044	5.432	4.925	4.499	4.137	3.976	3.827	3.559	3.325	2.854	2.499	2.222	2.000
24	21.243	18.914	15.247	12.550	10.529	8.985	7.784	6.835	6.434	6.073	5.451	4.937	4.507	4.143	3.981	3.831	3.562	3.327	2.855	2.499	2.222	2.000
25	22.023	19.523	15.622	12.783	10.675	9.077	7.843	6.873	6.464	6.097	5.467	4.948	4.514	4.147	3.985	3.834	3.564	3.329	2.856	2.499	2.222	2.000
26	22.795	20.121	15.983	13.003	10.810	9.161	7.896	6.906	6.491	6.118	5.480	4.956	4.520	4.151	3.988	3.837	3.566	3.330	2.856	2.500	2.222	2.000
27	23.560	20.707	16.330	13.211	10.935	9.237	7.943	6.935	6.514	6.136	5.492	4.964	4.524	4.154	3.990	3.839	3.567	3.331	2.856	2.500	2.222	2.000
28	24.316	21.281	16.663	13.406	11.051	9.307	7.984	6.961	6.534	6.152	5.502	4.970	4.528	4.157	3.992	3.840	3.568	3.331	2.857	2.500	2.222	2.000
29	25.066	21.844	16.984	13.591	11.158	9.370	8.022	6.983	6.551	6.166	5.510	4.975	4.531	4.159	3.994	3.841	3.569	3.332	2.857	2.500	2.222	2.000
30	25.808	22.396	17.292	13.765	11.258	9.427	8.055	7.003	6.566	6.177	5.517	4.979	4.534	4.160	3.995	3.842	3.569	3.332	2.857	2.500	2.222	2.000
40	32.835	27.355	19.793	15.046	11.925	9.779	8.244	7.105	6.642	6.234	5.548	4.997	4.544	4.166	3.999	3.846	3.571	3.333	2.857	2.500	2.222	2.000
50	39.196	31.424	21.482	15.762	12.234	9.915	8.304	7.133	6.661	6.246	5.554	4.999	4.545	4.167	4.000	3.846	3.571	3.333	2.857	2.500	2.222	2.000

TABLE A.4   PROBABILITY TABLES

d.f. (degrees of freedom)	0.10	0.05	0.025	0.01	0.005
1	3.078	6.314	12.706	31.821	63.657
2	1.886	2.920	4.303	6.965	9.925
3	1.638	2.353	3.182	4.541	5.841
4	1.533	2.132	2.776	3.747	4.604
5	1.476	2.015	2.571	3.365	4.032
6	1.440	1.943	2.447	3.143	3.707
7	1.415	1.895	2.365	2.998	3.499
8	1.397	1.860	2.306	2.896	3.355
9	1.383	1.833	2.262	2.821	3.250
10	1.372	1.812	2.228	2.764	3.169
11	1.363	1.796	2.201	2.718	3.106
12	1.356	1.782	2.179	2.681	3.055
13	1.350	1.771	2.160	2.650	3.012
14	1.345	1.761	2.145	2.624	2.977
15	1.341	1.753	2.131	2.602	2.947
16	1.337	1.746	2.120	2.583	2.921
17	1.333	1.740	2.110	2.567	2.898
18	1.330	1.734	2.101	2.552	2.878
19	1.328	1.729	2.093	2.539	2.861
20	1.325	1.725	2.086	2.528	2.845
21	1.323	1.721	2.080	2.518	2.831
22	1.321	1.717	2.074	2.508	2.819
23	1.319	1.714	2.069	2.500	2.807
24	1.318	1.711	2.064	2.492	2.797
25	1.316	1.708	2.060	2.485	2.787
26	1.315	1.706	2.056	2.479	2.779
27	1.314	1.703	2.052	2.473	2.771
28	1.313	1.701	2.048	2.467	2.763
29	1.311	1.699	2.045	2.462	2.756
30	1.310	1.697	2.042	2.457	2.750

Appendix B
Solutions to Exercises*

Exercises 1
1. 13   36   136   0
2. 19.9   42.1
3. 8.12   5.78
4. 0.296   0.025
5. 6,000   35   42
6. £210
7. 20% of cost price
8. $62\frac{1}{2}\%$ $38\frac{6}{13}\%$
9. £1,818.75
10. The *face value* is always 6 itself.
11. (a) $2468 = 2 \times 10^3 + 4 \times 10^2 + 6 \times 10 + 8 \times 1$
    (b) $54.322 = 5 \times 10 + 4 \times 1 + 3 \times 10^{-1} + 2 \times 10^{-2} + 2 \times 10^{-3}$
12. 75.298
13. (a) £2,009.80
    (b) £3,057.70 (nearly)
14. 19.5% approx: around 20% has been customary
15. £191,250   £190,000
16. 101101100
17. 365
18. $\frac{23}{24}$ $1\frac{1}{15}$
19. (a) 1.8 kilometres per hour
    (b) 1.2 kilograms
    (c) 40.5 metres; No.

Exercises 2
1. 3   4   13   9   38
2. 101   10001   110010   1111001
3. 1542
4. (a) 11 (3)   (b) 101 (5)   (c) 111 (7)
   (d) 1001 (9)   (e) 1101 (13)

* Please note that not all questions have either quantitative or worked answers.

Appendix B: Solutions to Exercises   305

5. (a) 11 (3)   (b) 1001 (9)   (c) 110 (6)
    (d) 1001 (9)
6. (a) 0 (0)   (b) 110 (6)   (c) 110 (6)
    (d) 1000001 (65)
7. (a) 11 (3)   (b) 10 (2)   (c) 101 (5)
    (d) 1010 (10)
8. 100010000111
9. 2537
10. 33.24
11. (a) 11   (b) 13   (c) 14
12. 14633
13. 57.71875
14. 0.C8
15. (1) 13   (b) E   (c) 12
16. 13B41
17. 2D2E
18. IC.B6C
19. (a) 1204B   (b) 4E.9B8

Exercises 3

1. a^{12} $\dfrac{1}{a^3}$ $\dfrac{1}{a^{10}}$ $(1.12)^{17}$
2. $b^{\frac{5}{6}}$ $\dfrac{1}{b^4}$
3. 2.31   36.35   79.66
4. 4.65   0.04728   7.906   0.1779
5. 2,000
6. −2   390
7. 4,092
8. −5,460
9. 35.06%
10. £5,830 (approx.)
11. £1,071.20
12.

NPV	A	B
10%	13,400	14,260
20%	1,780	−2,560
30%	−6,680	−13,980

   Where the curves cut the *x*-axis

13. (a) 17.903%
    (b) £11,747

14. (a) 11 cm
    (b) $643\frac{32}{81}$ cm
    (c) 675 cm
16. (a) $A = £205,675.62$
        Effective annual rate is 10.25%
    (b) $B = £55,240.93$

21. (a) 12" approx 50%
        18" approx 42%
        24" approx 33%
        30" approx 17%
        35" approx 10%
    (b) 18" is the largest size to be purchased.
23. (a) 19.56%

Exercises 4

1. (a) 80
   (b) 170
   (c) 330
2. (a) 270
   (b) 420
   (c) 210
4. No
6. (a) 4, 5, 6, 7, 8, 9, 10, 11
   (b) 2, 4, 6, 8, 10, 12, 14
   (c) No positive integers satisfy $4 + x = 3$
7. (a) $\subset$ (b) $\subset$ (c) $\not\subset$
   (d) $\subset$ (e) $\not\subset$ (f) $\subset$
   (g) $\not\subset$ (h) $\subset$
8. (a) $A \cup B = \{1, 2, 3, 4, 5, 6, 7\}$
       $A \cap B = \{4, 5\}$
   (b) $B \cup D = \{1, 3, 4, 5, 6, 7, 9\}$
       $B \cap D = \{5, 7\}$
   (c) $A \cup C = \{1, 2, 3, 4, 5, 6, 7, 8, 9\}$
       $A \cap C = \{5\}$
   (d) $A^c = \{6, 7, 8, 9\}$
   (e) $B^c = \{1, 2, 3, 8, 9\}$
   (f) $A \cap (B \cup E) = \{2, 4, 5\}$
   (g) $(B \cap F) \cup (C \cap E) = \{5, 6, 8\}$
9. (a) $(A \cap B) \cup (A \cap B^c) = A \cap U$
   (b) $(A \cup \emptyset) \cap (B \cup A) = A$

Appendix B: Solutions to Exercises 307

10. The second identity is the dual of the first.
11. (a) 2 (b) 4
12. (a) 6 (c) 60

Exercises 5

1. (a) $\begin{pmatrix} 21 & 30 \\ 11 & 18 \end{pmatrix}$ (b) $\begin{pmatrix} 11 & 26 \\ 10 & 28 \end{pmatrix}$

2. (a) 32 (b) $\begin{pmatrix} 4 & 8 & 12 \\ 5 & 10 & 15 \\ 6 & 12 & 18 \end{pmatrix}$

3. (a) $\begin{pmatrix} 1 & 4 \\ 0 & 1 \end{pmatrix}$ (b) $\begin{pmatrix} 1 & 6 \\ 0 & 1 \end{pmatrix}$

   (c) $\begin{pmatrix} 1 & 10 \\ 0 & 1 \end{pmatrix}$

4. (a) $\begin{pmatrix} 1 & 0 \\ 0 & 1 \end{pmatrix}$ (b) $\begin{pmatrix} 0 & 1 & 0 \\ 0 & 1 & 0 \\ 1 & -1 & 1 \end{pmatrix}$

5. (a) No (b) Yes
6. Yes
7. Yes

8. (a) $\begin{pmatrix} 16 & -6 \\ -2 & -5 \end{pmatrix}$ (b) $\begin{pmatrix} -40 \\ -41 \end{pmatrix}$

   (c) $\begin{pmatrix} -1 & -8 & -10 \\ 1 & -2 & -5 \\ 9 & 22 & 15 \end{pmatrix}$

9. (a) $\begin{pmatrix} 1 & 3 \\ 2 & -1 \\ 0 & 4 \end{pmatrix}$ (b) $\begin{pmatrix} 5 & 1 \\ 1 & 26 \end{pmatrix}$ (c) $\begin{pmatrix} 10 & -1 & 12 \\ -1 & 5 & -4 \\ 12 & -4 & 16 \end{pmatrix}$

10. (a) Yes (b) No
    (c) $XY = \begin{pmatrix} 11 & -6 & 14 \\ 1 & 2 & -14 \end{pmatrix}$
12. (a) 23 (b) -4 (c) $a^2 - 2b^2$
13. (a) 55 (b) 67
14. $\begin{pmatrix} -3 & 5 \\ 2 & -3 \end{pmatrix}$

Exercise 6
5. (a) Defined. A one-dimensional (or linear) array with 82 elements
   (b) Defined. A three-dimensional $(7 \times 9 \times 5)$ array with $7 \times 9 \times 5 = 315$ elements
   (c) Not defined
   (d) Defined. A two-dimensional $(N(N+1))$ array with $N(N+1) = N^2 + N$ elements

Exercises 7
1. $x = 2, y = -\frac{1}{2}$
2. $x = -1, y = 2$
3. There is no solution
4. (a) $x = \frac{3}{2}$ (b) $x = \frac{32}{9}$
   (c) There is no solution
5. (a) $x = 9$ (b) $x = 10\frac{1}{2}$
   (c) $x = 12$
6. $x = 1, y = 2, z = 3$
7. $x = 1, y = 1, z = -1$
8. $x = -1, y = -2, z = -3$

Exercises 8
3. Approx 63 units
4. Maximum occurs when $x = 10, y = 5$
9. (a) $y = 600 + 40x + x^2$

Exercises 9
1. (a) 6 (b) $4x - x$ (c) $20x^3 - 3x^2 + 2x$
   (d) $18x^2 - x^3 + x^2 + 2/x^3$ (e) $2(3x + 2/x^2) + (2x + 4)(3 - 4/x^3)$
   (f) $10(2x + 3)^4$ (g) $2/2x + 3$
   (h) $6e^{2x} - 12e^{-3x} + 2xe^x + x^2 e^x$
2. Approx 31.6

Appendix B: Solutions to Exercises

3. (a) $\dfrac{3x^2}{2} - 5x + c$ (b) $\tfrac{2}{3}x^3 - 2x^2 + 4x + c$

    (c) $\log_e(3x-5) + c$ (d) $\tfrac{2}{3}e^{3x} - e^{-x} + c$

    (e) $2e^x - \tfrac{4}{3}e^{3x} + c$ (f) $\tfrac{1}{2}x^2(\log_e x - \tfrac{1}{2}) + c$

4. $115/6$

5. (a) $\dfrac{\partial z}{\partial x} = 2 - 2x$ $\dfrac{\partial z}{\partial y} = 4 - 2y$

    (b) $\dfrac{\partial z}{\partial x} = \dfrac{\partial z}{\partial y} = (x+y)e^{x+y} + e^{x+y}$

    (c) $\dfrac{\partial z}{\partial x} = \dfrac{2x}{x^2+y^2}$ $\dfrac{\partial z}{\partial y} = \dfrac{2y}{x^2+y^2}$

    (d) $\dfrac{\partial z}{\partial x} = -\dfrac{2y}{(x-y)^2}$ $\dfrac{\partial z}{\partial y} = \dfrac{2x}{(x-y)^2}$

6. (a) $7\tfrac{1}{2}$ (b) 2   (c) -12

13. $a = 2$, $b = 288$. Speed is 12 knots. Cost of fuel is £5,800

20. $x = -163$, $y = 142$

21. $C_0 = £13{,}000$
    $C_1 = £15{,}600$
    This implies borrowing the sum of £7,000

Exercises 10

1. (a) 2/3   (b) 1/3   (c) 1   (d) 0
2. 1/36
3. (a) 1/9   (b) 5/36
4. (a) 2   (b) 6   (c) 24
5. (a) 9/80   (b) 17/80
6. (a) 2/63   (b) 2/3   (c) 1/3   (d) 19/63
7. (a) 0.7   (b) 0.275

12. (a)   EV(A) = £3,000
         EV(B) = £3,000
         VAR(A) = £2,666,667
         VAR(B) = £6,000,000
    (c)   Coeff of VAR(A) = 0.544
         Coeff of VAR(B) = 0.816
13. (a)   Highest expected profit is £9,277.50
    (b)   Value of Perfect Information is £722.50 p.a.

Exercises 11

2. (b) (i) $y = 36 + (4.05)x$ (ii) 60.3 (Assuming $x = 0$ for 1979)
5. Probably yes.

Appendix C
Worked Answers

Exercise 3

21. (a)

		12″ £000	18″ £000		24″ £000		30″ £000		36″ £000
				Cutter size					
Year 0		−20	−30		−45		−70		−100
Annual gross savings		8	10		11		8		6
Add back: Depreciation		+2	+3		+4.5		+7		+10
Net annual savings		10	13		15.5		15		16
Cost/Net Annual Savings		2	2.31		2.9		4.67		6.25
Cum. factor) 40% for 10 years) 50%		2.4 1.98	(As 12″)	30% 40%	3.09 2.4	16% 17%	4.83 4.64	9% 10%	6.42 6.14
		0.42	0.42		0.69		0.19		0.28

Rough calculation:

Difference %		$\frac{0.02}{0.42}$	$\frac{0.09}{0.42}$	$\frac{0.19}{0.69}$	$\frac{0.03}{0.19}$	$\frac{0.11}{0.28}$
		= 1%	= 21%	= 28%	= 15%	= 39%
DCF rate/yield (very approx)		50%	42%	33%	17%	10%

(b) *Present value*

15% × 10 years
= 4.97

PV =   49.7   64.61   77.04   74.55   79.52

if buy:

	PV	Capital expenditure
18" compared	64.61	30
with 12"	−49.70	−20
Extra	14.91	10 (18" preferred)
24" compared	77.04	45
with 18"	−64.61	−30
Extra	12.43	15 (18" preferred)

By inspection 30" and 36" are worse than 24"
Therefore 18" is the largest size to be purchased.

(c) *In favour of proceeding*

1. It meets the investment criteria
2. It is the best of the opportunities
3. There is a possible increase in manual labour costs over time: thus the savings may well be even greater (in opportunity cost terms)
4. DCF return of some 42% seems attractive
5. Inflation – equipment costs are increasing and may continue to do so
6. Safety – operating conditions may be safer with the new equipment as workers will be less exposed in dangerous traffic conditions.

In favour of deferment

1. If we wait there may be an improvement in the economics of a larger machine – e.g., a lower capital cost for the same savings
2. There might be some technological breakthrough or improvement for the whole range of machines
3. Alternative investment – funds might be used in alternative projects yielding greater savings or other benefits
4. Purchase of the machine might lead to redundancies with loss of job opportunities in an area of high unemployment.

Exercise 7

10. (a) (i) Let M = maintenance
L = laundry
C = catering

Then:

$$M = 56{,}000 + 0.05M + 0.05L + 0.10C$$
$$L = 250{,}000 + 0.25M + 0.15C$$
$$C = 35{,}000 + 0.10M + 0.10L + 0.05C$$

Thus:

$$0.95M - 0.05L - 0.05C = 56{,}000$$
$$-0.25M + L - 0.15C = 250{,}000$$
$$-0.10M - 0.10L + 0.95C = 35{,}000$$

This gives matrix coefficients:

$$\begin{pmatrix} 0.95 & -0.05 & -0.10 \\ -0.25 & 1 & -0.15 \\ -0.10 & -0.10 & 0.95 \end{pmatrix}$$

(ii) Let A = the coefficient matrix
s = the service department vector
C = costs
We have $\quad As = C$
Hence $\quad s = A^{-1}C$

(b)

$$\begin{pmatrix} 1.083 & 0.067 & 0.125 \\ 0.293 & 1.034 & 0.194 \\ 0.145 & 0.116 & 1.086 \end{pmatrix} \times \begin{pmatrix} 56{,}000 \\ 250{,}000 \\ 35{,}000 \end{pmatrix} = \begin{pmatrix} M \\ L \\ C \end{pmatrix}$$

Giving:

$$M = 1.083 \times 56{,}000 + 0.067 \times 250{,}000 + 0.125 \times 35{,}000$$
$$= 60{,}648 + 16{,}750 + 4{,}375$$
$$= \underline{81{,}773}$$

$$L = 0.293 \times 56{,}000 + 1.034 \times 250{,}000 + 0.194 \times 35{,}000$$
$$= 16{,}408 + 258{,}500 + 6{,}790$$
$$= \underline{281{,}698}$$

$$C = 0.145 \times 56{,}000 + 0.116 \times 250{,}000 + 1.086 \times 35{,}000$$
$$= 8{,}120 + 29{,}000 + 39{,}010$$
$$= \underline{75{,}130}$$

Exercise 8

12. (a) *Definition*

    The learning curve, also known as a cost experience curve, has been defined as 'the relationship plotted between cost per unit expressed in constant money terms, and cumulative units produced per unit—usually plotted on a double logarithmic scale'.

    (b) *Explanation of the theory*

    As an organisation repeats the manufacture of a particular product or the provision of a particular service, it becomes more adept at that manufacture or provision as a result of:

    (i) familiarity with the detailed activities in its manufacture or provision;
    (ii) exposure to, and awareness of, the problems that can arise, and consequently more experience in how to deal with those problems;
    (iii) confidence in its ability to deal with the activity in the most advantageous ways.

    This increased adeptness should lead to greater speed in carrying out the task with consequent reduction in the time taken and the costs incurred, particularly in so far as they relate to man-controlled activities.

    Improvements in productivity can be expected with each successive batch of output and the effect of the cumulative average cost at each doubled level of output can be expected to be a constant when expressed in percentage terms. On a log log scale, this would appear as a straight line. In absolute terms, of course, the drop in costs from one doubled level to the next will be progressively smaller. The more complex the activity, the greater the general possibility of improvement over time and thus the steeper the slope is likely to be. The rates at which costs fall tend to be between 75% and 90%.

    (c) *Areas of assistance in costing*

    The learning curve operates mostly where activities are person-controlled. It is of great relevance in assessing or predicting future levels of direct labour time and cost. It is useful in:

    1. setting labour standard times
    2. pricing for successive batches/contracts
    3. calculation of incentive rates in wage bargaining
    4. evaluating likely wage movements in project evaluation.

    The learning curve also influences the absorption of overhead when direct labour hour rates are used.

It is not so relevant for processes that are largely governed by machine speeds – such as where the rhythm or speed of a production line (or other machinery) governs the speed of operations.

It should be noted that the concept has been extended to relate to the whole activity of a factory or an industry. Here it is known as the 'cost experience curve' and includes such features as long-term improvements in inventiveness, product development, manufacturing equipment development (e.g., of specialised equipment for the industry). The developments in the speed/cost of producing pocket calculators over the past (say) 5–10 years, is an evident example of the operation of this curve.

(d)

Cumulative order size	Cumulative average hours	Cumulative average cost	Materials	Overhead	Total
		£	£	£	£
1	1,000	3,000	—	—	—
2	800	2,400	—	—	—
(i) = 4	640	1,920	1,800	2,000	5,720
(ii) = 8	512	1,536	1,800	1,000	4,336

Exercises 9

13. (a) Let $Y = $ cost of fuel $= 100\left(aX + \dfrac{b}{X} + 10\right)$,

    $X = $ knots

    If $X = 4$: $9{,}000 = 100\left(4a + \dfrac{1}{4b} + 10\right)$

    $9{,}000 = 400a + 25b + 1{,}000$

    $320 = 16a + b$ ... (1)

    If $X = 6$: $7{,}000 = 100\left(6a + \dfrac{1}{6b} + 10\right)$

    $7{,}000 = 600a + \dfrac{100}{6b} + 1{,}000$

    $1{,}800 = 180a + 5b$ ... (2)

    From (1) $1{,}600 = 80a + 5b$

    Subtracting: $200 = 100a$

    $a = 2$

    Similarly $b = 288$

Appendix C: Worked Answers

(b) The solution can be found by differentiation:

$$Y = 100\left(2X + \frac{288}{X} + 10\right)$$

$$Y = 200X + \frac{28,800}{X} + 1,000$$

$$\frac{dY}{dX} = 200 - \frac{28,800}{X^2}$$

To find the minimum point set $\frac{dY}{dX} = 0$:

Then $\frac{28,800}{X^2} = 200$

$$28,800 = 200X^2$$

$$X = \pm 12$$

The second derivative $\frac{d^2Y}{dX^2}$ is positive, therefore there is a minimum value of Y when $X = +12$

When $X = 12$:

$$Y = 100\left(24 + \frac{288}{12} + 10\right)$$

$$= £5,800 = \text{cost of fuel at 12 knots}$$

14. (a) The two points are (440, 13) and (300, 20)

Therefore $P - 13 = \frac{20 - 13}{300 - 440}(X - 440)$

$$P = -\frac{1}{20}(X - 440) + 13$$

$$P = 35 - \frac{1}{20}X$$

(b) Revenue = price × quantity, therefore the revenue equation is:

$$R = 35X - \frac{1}{20}X^2$$

(c) The cost equation is:

$$C = 2,000 + 10X$$

(d) The break-even points are where $R - C = 0$.

$$R - C = -\frac{1}{20}X^2 + 35X - (10X + 2{,}000) = 0$$

Giving
$$X^2 - 500X + 40{,}000 = 0$$
$$(X - 400)(X - 100) = 0$$

Therefore the two points are $X = 100$ or $X = 400$.

(e) (i) Profit: Differentiate $(R - C)$ with respect to X

$$\frac{d\,\text{Profit}}{dX} = 25 - \frac{1}{10}X$$

Maximum where $= 0$

$= 250$ units.

(second derivative $d^2\,\text{Profit}/dX^2$ is negative)

Price should be £22.50 and production 250 units to maximise profit

(ii) Revenue: Differentiate R with respect to X

$$\frac{d\,\text{Revenue}}{dX} = 35 - \frac{1}{10}X$$

Maximum where $= 0$

$= 350$ units

(second derivative $d^2\,\text{Revenue}/dX^2$ is negative)

Price should be £17.50 and production 350 units to maximise revenue

15. (a) Optimum quantity (q) is computed from the EOQ formula:

We have annual demand $= 10{,}000$ kg
cost of order $= £15.00$
cost of capital $= 0.25$
unit price $= £30.00$

$$q = \sqrt{\frac{2 \times 15 \times 10{,}000}{0.25 \times 30}}$$

$= \sqrt{40{,}000}$

$= 200$ units

Appendix C: Worked Answers 317

(b) (i) Increasing stockholding cost to 35% will cause the EOQ to be multiplied by a factor of:

$$\sqrt{\frac{0.25}{0.35}} = 0.845$$

to $200 \times 0.845 = 169$ units

(ii) Reducing price of Z to £20 per kg will cause the EOQ to be multiplied by a factor of:

$$\sqrt{\frac{30}{20}} = 1.225$$

to $200 \times 1.225 = 245$ units

(iii) Increasing consumption to 250 kg will cause the EOQ to be multiplied by a factor of:

$$\sqrt{\frac{250}{200}} = 1.118$$

to $200 \times 1.118 = 224$ units

(c) In each case, the change in EOQ is proportional to the square root of the change in the variable. Increase in demand or order cost will cause the EOQ to be increased but an increase in cost of capital or of unit cost will result in a reduction in EOQ.

Exercises 10

12. (a) Expected value of project $A = \frac{1}{3} \times (£1,000 + £3,000 + £5,000)$
$= £3,000$

Expected value of project $B = \frac{1}{3} \times (£0 + £3,000 + £6,000)$
$= £3,000$

Variance $(A) = \frac{1}{3} \times (2,000^2 + 0 + 2,000^2)$
$= \frac{1}{3} \times (4,000,000 + 4,000,000)$
$= \frac{8,000,000}{3}$
$= £2,666,667$

Variance $(B) = \frac{1}{3} \times (3,000^2 + 0 + 3,000^2)$
$= \frac{1}{3} \times (9,000,000 + 9,000,000)$
$= \frac{18,000,000}{3}$
$= £6,000,000$

(b) Project *A* is recommended: both have the same expected return but project *A* has the lesser variance.
The expected value gives a measure of the expected return and variance measures the risk.

(c) The coefficient of variation is:

$$\frac{\text{standard deviation}}{\text{mean}}$$

Standard deviation is the square root of the variance and so:

Standard deviation $A = \sqrt{2,666,667} = 1,633$
Standard deviation $B = \sqrt{6,000,000} = 2,449$

Thus the corresponding coefficients of variation are:

for A $\quad \dfrac{1,633}{3,000} = 0.544$

for B $\quad \dfrac{2,449}{3,000} = 0.816$

Coefficient of variation is a measure of dispersion without dimension which may be applied to two data sets even when measured in different units. In this example it shows that *B* is 50% more variable than *A*.

13. (a) Profit on each sale is $\quad$ (£1 − 20p =) 80p
Loss on each unsold is $\quad$ (20p − 1p =) 19p

The conditional profit table is:

Demand	40	50	60	70
Made				
40	32.00	32.00	32.00	32.00
50	30.10	40.00	40.00	40.00
60	28.20	38.10	48.00	48.00
70	26.30	36.20	46.10	56.00
Probability	.4	.3	.2	.1

(The demand of 40 occurs 100 times per year in a total of 250 working days so that the probability is $\dfrac{100}{250} = 0.40$, etc.).

Appendix C: Worked Answers

The expected profit table is:

Demand	40	50	60	70	Total
Made					
40	10.52	10.86	9.22	5.60	32.00
50	12.80	9.60	6.40	3.20	36.04
60	12.04	12.00	8.00	4.00	37.11
70	11.28	11.43	9.60	4.80	36.20

Thus the optimum policy is to make 30 units per day with an expected daily profit of £37.11 and an expected annual profit of:

$$250 \times 37.11 = £9,277.50$$

(b) Assuming perfect information the daily profit would be:

$$30 \times 0.4 + 40 \times 0.30 + 48 \times 0.20 + 56 \times 0.10$$
$$= 12.8 + 12.0 + 9.6 + 5.6$$
$$= £40 \text{ per day}$$

Annual profit would be £10,000
Value of information is £10,000 − £9,277.50 = £722.50 p.a.

Exercises 11

2. (a)

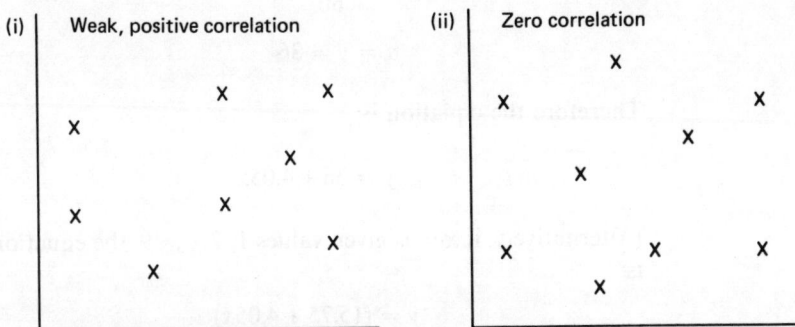

(i) Weak, positive correlation

(ii) Zero correlation

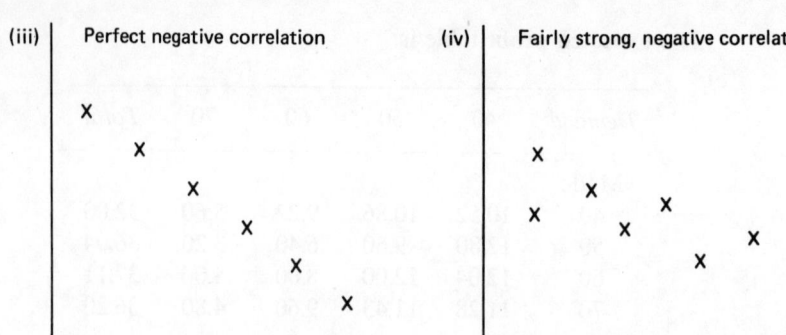

(b) (i) Let x = years; y = production.

Years	d	y	d^2	dy
1975	-4	19	16	-76
1976	-3	24	9	-72
1977	-2	28	4	-56
1978	-1	33	1	-33
1979	0	35	0	0
1980	$+1$	41	1	41
1981	$+2$	45	4	90
1982	$+3$	47	9	141
1983	$+4$	52	16	208
	$\Sigma x = 0$	324	60	243
		Σy	Σd^2	Σdy

To find least squares regression line:

$$b = \frac{243}{60} = 4.05$$

$$a = \bar{y} = 36$$

Therefore the equation is:

$$y = 36 + 4.05x$$

(Alternatively, if $x =$ is given values 1, 2, ... 9, the equation is:

$$y = (15.75 + 4.05x)$$

Appendix C: Worked Answers

   (ii) Predicted value for 1985: $y = 36 + (4.05)6 = 60.3$
   The main assumption is that the past pattern of growth will continue over the two-year future period.

3. (a) The regression equation is a simple regression on sucessive quarterly data values, producing a straight line as a trend. The trend has a slope value of 10, i.e. furniture sales increase on average by £10 each quarter, with a starting (intercept) value of £150. The seasonal index factors, obtained from a multiplicative model, show that, for example, the forecast value for Quarter 1 will be only 0.8 of the estimated trend value, on a seasonally adjusted basis. Quarters 2, 3 and 4 will be 1.1, 1.4 and 0.7 respectively of the estimated trend values after seasonal adjustment.

   (b) 1985 and (c)

X	Y (est. values)	Seasonal Factor	Forecast
21	360	0.8	288
22	370	1.1	407
23	380	1.4	532
24	390	0.4	273

   (d) Any changes in the underlying assumptions affect the trend or seasonal factors. In addition, it should be remembered that the regression equation constants and the seasonal index factors are themselves 'average' relationships and hence actual values can normally be expected to differ from forecast values based on those relationships.

Exercises 12

1. (a)

Machine cost:	$4{,}000X + 12{,}000Y$	$\leqslant 120{,}000$
Operators:	$2X + 2Y$	$\geqslant 24$
Space:	$20X + 30Y$	$\leqslant 480$
Presses:	$X \geqslant 0,\ Y \geqslant 0$	

   (b) See graph (Figure A1).
   (c) Possible alternative combinations of X and Y are shown on the graph by the points A, B, C, D:

Alternative	(X, Y) combination	Production
A	(3, 9)	3,150
B	(18, 4)	3,900
C	(24, 0)	3,600
D	(12, 0)	1,800

(d) Maximum production is reached at alternative B, the acquisition of 18 X and 4 Y machines. This will produce 3,900 sheets per minute. The total cost of the presses in this case will be:

$$£18(4,000) + £4(12,000) = £120,000$$

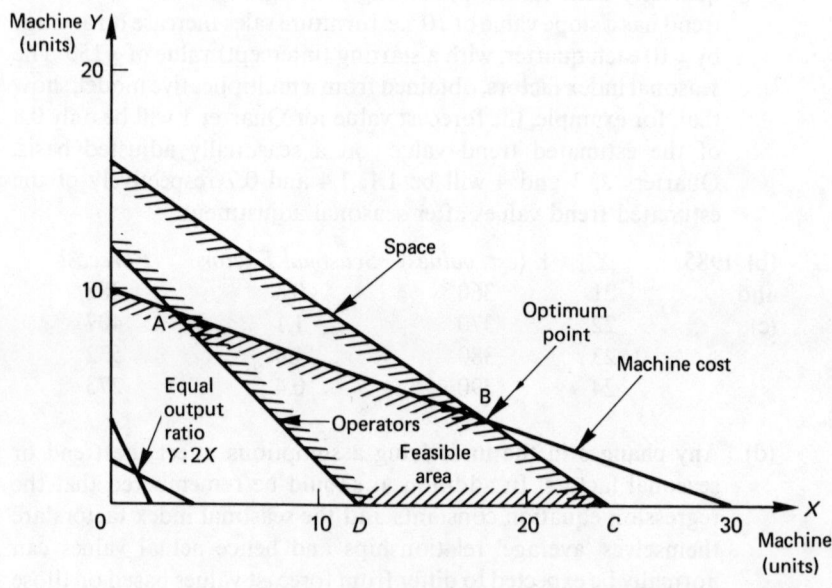

Fig. A.1

3. (a) The costs may be summarised thus:

	HT01	HT02	HT03
Price per unit	150	200	220
Direct labour at £4	100	120	132
Direct labour hours per unit	25	30	33
Direct materials at £20/kg	20	40	40
Material (kg.) per unit	1	2	2
Direct cost per unit	120	160	172
Variable overhead per hour	0.5	0.7	0.8
Variable overhead (£) per unit	12.5	21	26.4
Total variable cost per unit	132.5	181	198.4
Net contribution per unit	17.5	19	21.6

(b) In formulating the linear programming solution we take a, b and c to be the amounts of HT01, HT02 and HT03 respectively in the

Appendix C: Worked Answers

optimum plan. The problem becomes:

Maximise:
$$17.5a + 19b + 21.6c$$
subject to:

direct labour	$25a + 30b + 33c$	$\leq 257{,}600$
materials	$a + 2b + 2c$	$\leq 20{,}000$
demand	a	$\leq 16{,}000$
	b	$\leq 10{,}000$
	c	$\leq 6{,}000$

(c) (i) The optimum plan from the tableau is:

Produce only 10,304 units of HT01
Net contribution is £180,320
Shadow cost for HT02 is 2.0
 and for HT03 is 1.5 which indicates the losses of contribution per unit which would be incurred if these were to be produced instead of the optimum.

Unsatisfied demands are HT01 5,696
 HT02 10,000
 HT03 6,000

Unused material is 9,696 kg.

Shadow price for direct labour is £0.7 per hour so each extra hour of direct labour would increase contribution by £0.7 Examination of the penultimate tableau shows that a plan of 2,686 units of HT02 and 5,771 units of HT03 would give a contribution of £171,663. This is a substantially different product mix but a change of only about 5% in the contribution demonstrating a lack of sensitivity. This same result could have been predicted from the dual values in the optimum which showed that the loss of contribution from introducing the 5,771 units of HT03 would be:

$$5{,}771 \times 1.5 = 8{,}657$$

(ii) In the case of price increases in the materials we need to examine the contribution per unit of direct material:

	Product		
	HT01	HT02	HT03
Direct materials at £20/kg.	20	40	40
Material (kg.) per unit	1	2	2
Net contribution per unit	17.5	19	21.6
Net contribution per unit of material	17.5	9.5	10.8
Ranking	3rd	1st	2nd

So the order in which production of each product would cease to be worthwhile is:

$$HT02 \qquad HT03 \qquad HT01$$

		Products		
	A	B	C	D
4. Production (Tubs per machine per week)	14	4	3	6
Profit statement	£	£	£	£
Market price	390	390	450	570
Direct costs				
Process 1 Direct material	17	27	32	26
Direct labour	32	76	62	44
Process 2 Direct material	10	10	10	10
Direct labour	80	72	100	120
Transport	130	130	100	240
	269	315	304	440
Contribution	121	75	146	130
	× 14	× 4	× 3	× 6
Profit per week	1,694	300	438	780

At present: Available weeks A, B, C and D 10 × 48 = 480
B and C only 15 × 48 = 720

Total 1,200

Best mix currently:

         weeks
B (minimum) 500 tubs ÷ 4 = 125
C (minimum) 500 tubs ÷ 3 = 167 (501 tubs)
D (minimum) 500 tubs ÷ 6 = 84 (504 tubs)
A 7,000 ÷ 14   = 500
          876

A + D above = 584 but
maximum is 480
Therefore reduce A to 1,456 tubs (104)
and increase C to 312 tubs   104
then balance on C 972 tubs   324

         1,200

Appendix C: Worked Answers

(a)

Profit statement

Product		£
A	396 weeks × £1,694	670,824
B	125 weeks × £300	37,500
C	595 weeks × £438	260,610
D	84 weeks × £780	65,520
1,200	Total contribution	1,034,454
	Fixed costs	820,000
	Profit	£214,454

(b)

Maximum machine conversion

1 machine on Product A replacing Product C = 48 weeks per year

	£ per week
Contribution on Product A	1,694
Contribution on Product C	438
	1,256

or £60,288 per year

Is it worth investing £70,000 to modify machine?

Year	Cash flow £	DCF factor	Present value £
0	(70,000)	1.0	(70,000)
1	60,288	0.89	
2	60,288	0.80 } 2.40	144,691
3	60,288	0.71	
		NPV	74,691

This is clearly acceptable.
Lowest acceptable increased contribution per year

$$\frac{70,000}{2.40} = £29,167 \text{ per year for 3 years}$$

There are 428 surplus weeks on Product C and 104 extra weeks required on Product A

Therefore 2 machines should be modified × 48 weeks = 96 weeks
A third machine, if converted, could work 8 weeks on Product A plus 40 weeks on Product D
Is this acceptable?

$$\begin{aligned} \text{Product } A - C = £1{,}256 \times 8 &= £10{,}048 \\ D - C = £342 \times 40 &= 13{,}680 \\ \text{Additional contribution} &\ £23{,}728 \end{aligned}$$

As this is less than the £29,167 required as calculated above this is not acceptable.

Therefore only 2 machines should be converted.

Exercises 13

1. (a) The optimal production plan is   1,013.33 units of A
                                                    2,000   units of C

    Optimal contribution is   £42,080.00
    Unused resources are   2,460 machine hours
                                           740 despatch handling units

   (b) If Mega produces 1 unit of B there will be a net loss of contribution of £1.88. This will take units from A and C which will reduce their contribution by a total of £13.88 to offset the £12.00 contribution from B.

   (c) The shadow price for skilled labour hours is £8.00 and so each extra hour (up to the allowable limit of 123.33) increases contribution by £8.00. Each of these hours costs an overtime payment of an additional £1 and so the net gain is in fact £7.00 per hour.

   (d) The shadow price for storage is £7.52 per cubic foot and so each extra cubic foot (up to the allowable limit of 4,108.11) increases contribution by £7.52. The rental is an additional £5 per unit and so the net gain is in fact £2.52 per unit.
   There is a further factor, the internal costing of storage space which is not known but is taken to be a fixed cost.

   (e) The shadow price for storage is £7.52 per cubic foot and also operates down to a lower limit of 1,321.43 so a loss of 1,000 units decreases contribution by £7,520 with no change of product mix.

   (f) There is slack in the despatch handling resource and so there could be an infinite increase in this resource with no effect on the solution.

Appendix C: Worked Answers 327

(g) The despatch handling can be reduced by 740 units at which point it will become a binding constraint and any further reduction would change the optimal solution.

(h) A's contribution can change within the following limits without changing the product mix:

$$6-0.97 \quad \text{to} \quad 6+30.49, \text{ i.e.}$$
$$5.03 \quad \text{to} \quad 36.49$$

this gives a measure of the sensitivity of the solution to change in price of A. The total contribution will, however, change as the price changes.

(j) No, each change must be studied independently – we can study the changes to two variables only by evaluating a new solution with the changes to one variable and then studying the effects of changes in the other.

Further Reading

J. R. Ashford, *Statistics for Management*, 2nd edn (London: Institute of Personnel Management, 1984).

E. M. L. Beale, *Mathematical Programming in Practice* (London: Pitman, 1968).

J. A. Dallenbach, J. A. George and D. C. McNickle, *Introduction to Operations Research Techniques*, 2nd edn (New York: Prentice-Hall, 1983).

J. Dewhurst and P. Burns, *Small Business: Finance and Control* (London: Macmillan, 1983).

J. Johnstone, *Econometric Methods* (New York: McGraw-Hill, 1963).

S. Lipschutz, *Essential Computer Mathematics* (Schaums Outline Series) (New York: McGraw-Hill, 1982).

S. Vajda, *The Theory of Linear and Non-Linear Programming* (London: Longman, 1974).

H. A. Wagner, *Principles of Operations Research*, 2nd edn (New York: Prentice-Hall, 1975).

H. P. Williams, *Model Building in Mathematical Programming* (London: Wiley, 1978).

Index

abandonment values 194–5
absorbing states 206–8
abstraction in set theory 59
adjoint of cofactor matrix 82
algebra: elementary 5–6; Boolean 63; of sets 62; of matrices 76; in linear programming 251–3
algorithms 93–101; transportation algorithm 274, 279–90
Altman, Edward 242
American Standard Code for Information Interchange 23
annual percentage rate 37
annualised cost 46
annuities 40–3
arrow diagrams 68
artificial variable 257–8
asymptotic curve 134
augmented matrix 84–5
autocorrelation 234–5, 239
axes 122–3

back substitution 108–11
bankruptcy prediction 242
basic variables 251–2, 254–5
Bayes decision rule 193
Bernouilli distribution 219
big-M technique 258, 264
binary digits and calculations 14–17; binary decimal coding 22–3; binary relation 65–8; binary search 118
binomial theorem 212, 218–19
bits 14
Boolean algebra 63, 64

calculated value 233–4
calculus 147–78
canonical layout of matrix 208
Cartesian plane 65, 122
cash discount 8, 176
cash management 177

chain rule 152
change, rate of 147, 159, 165, 167
circles 128–9
Cobb–Douglas function 166–7
cofactors 80, 82
column vectors 74
combinations 208–12
combinatorial analysis 191, 208–9
complement of set 61, 187
confidence limits 227–8, 237
conic graph 128
constraints 170–2, 246–51, 274, 275
conversion, frequency of 36–7
co-ordinate diagram 67
co-ordinates 122–3
correlation matrix 235–6
costs: use of graphs 125–6, 133, 135, 136; use of calculus 147, 154–6, 159, 173–8; of stock out 191–2; relation to throughput 220–21, 225, 234–5; intermediate reduced costs 252; use of regression analysis 239–40
Cramer's rule 108
cubic equations 116
curves: in non-linear equations 127–8, 133; exponential 153; area under 162–3

decimals 3–4, 19, 22–3
decision trees 93, 194–5
decision variables 248, 251
degeneracy 255, 256
degrees of freedom 227, 232
De Morgan's laws 69
derivate 118
derivatives 150–1, 153, 156–60, 165–7, 173
determinants 79–82, 85
difference of sets 61
differentials 159, 162, 165–8
differentiation 147–51, 164–7, 230

329

discount factor 38, 45
discounted cash flow 45
discriminant analysis 240–2
duality 63–4, 274–9
dummy variables 238–9
Durban–Watson statistic 235

economic order quantity 173–8
efficiency frontier 201
elasticity of demand 172–3
ellipses 128, 143
empty set 59
equations 105–18, 122–32, 224
equiprobable spaces 188–90
events in probability analysis 187
expanded notation 2
expected value 193–5, 201–2
explicit function 164
exponential function 132–3, 153, 161
exponential smoothing 48–50
exponents 27–30
Extended Binary-Coded Decimal Interchange Code 23
extension in sets 58

factorial n 210
feasible region 249–52, 254, 256, 258
Fisher's separation theorem 141
floating point 30
flowcharts 93–7
four, power of, equations 117
function of a function rule 152–3

Gaussian elimination 108–11
geometric progression 34–5, 41
geometry in set theory 65
gradient *see* slope
graphs: 67, 105–6, 122–42; in calculus 147–50, 153–8, 162–5; in regression analysis 222; in linear programming 247–51
growth rate 147

hexadecimal system 17, 20–2
hyperbolas 128, 130–2

identity matrix 77
implicit functions 164
impossible event 187
independent event 190–1
indifference curve 139–41

induction, proof by 218
infinite geometric progression 35–6
inflexion, point of 158
input–output analysis 86–90
integer programming 290
integral, definite 162–3
integral, indefinite 160
integration 159–62
intercept 125
interest 10–11, 35–40, 140, 141
internal rate of return 45, 46
interpolation 38–40
intersection of sets 60, 61, 187
interval bisection method 118
inversion of matrices 78, 80–5, 111–12, 230
investment decisions 38, 41, 43–8, 105, 137–42, 198–203
iteration 98–9, 112–18, 255

joint probability 187, 197–8

lead time 173, 175–7
learning curve 135–7, 143, 237
least squares, line of 224
Legrange multipliers 170–2
Leontieff input–output analysis 87, 89
line of best fit 222–7
linear equations 105–8, 123–5, 221–2
linear programming 246–91
liquidity ratio 240–42
local maximum and minimum 158
logarithms 30–3, 153; in regression analysis 237
loops 97, 99, 100, 112

mantissa 29, 32
marginal analysis 173
marginal rate of return 139, 142
Markov chains 203–8
matrices: 68, 72–92; in regression analysis 228–9; inversion of 111, 230
maxima and minima 153–9, 168–71
minors 80
mixed integer programs 290
Monte Carlo technique 198
multicollinearity 235–6
multidimensional arrays 85–6
multiple determination, coefficient of 232–3
Myddleton, David 242

Index

n factorial 210
nanosecond 29
Napierian logarithms 31
'nature, states of' 201
net present value 142, 201
Northwest corner rule 284–5
null set 59

objective function 248, 252
observed values 233–4
octal system 17–20
operator precedence 6–7
opportunity cost 156
optimisation 170, 171
ordered pairs 64–8, 122
outcomes in probability analysis 186–7

parabolas 128, 143
partial derivatives 166–7, 173
partial differentiation 164–7, 230
Pascal's triangle 212, 218
payout ratio 240–42
percentages 7–9
perfect information 202
perodic function 133, 145
permutations 191, 208–11
pivot 110, 255, 256
power set 59
precision of computer 29
present value 11, 37–8, 43–6, 142, 201
probability 186–212
product rule 150–53
product sets 64
profitability index 46
progressions 33–6
pseudocodes 97–101

quintal system 20
quotient rule 152–3

ranging 264, 270–5
ratio test 254, 256
ratios 11–12
real arithmetic 30, 64–5
regression analysis 220–42
relation, binary 65–8
residuals 233–4
risk and uncertainty 192–203

rounding 4
row vectors 73–5

saddle point 168–9
sample space and sample point 186–8
scalar numbers 74, 77–8
scale, diseconomies of 126, 154
selection logic 98–9
sensitivity analysis 192–3, 202, 270–4
set theory 58–71, 186–7
shadow price 151, 153, 175, 179
simplex method 253–64, 274
slack variables 250, 251, 253, 278–9
slope 125, 127, 128, 133, 153–8, 162
square matrices 77–8, 81–2
standard deviation 200, 202
stationary value 158, 168–70
stock levels 173–8, 191–2
subscript notation 2, 72, 86
symbols 59, 62, 75–6, 93–7

t values 226–7, 231, 232
Taffler, Richard 242
tensors 86
three-dimensional functions 86, 164–5, 168–70
transformation matrices 82–5
transition probabilities 203–6
transportation algorithm 275, 279–90
transpose of matrix 82, 229
trigonometry 145–6
truncating 4–5, 30
two-phase technique 258, 264

uncertainty profile 198
union of sets 60–2, 187
unit matrix 77
universal set 59

variables: in equations 105–6; elimination of 111; dependent and independent 164, 221, 228–31, 251; dummy 238–9; decision 248, 251; artificial 257, 258
variation, coefficient of 202
vectors 73–4, 86
Venn diagrams 60, 62

Z score analysis 242
zero matrix 74